Product Research

N.R. Srinivasa Raghavan • John A. Cafeo
Editors

Product Research

The Art and Science Behind Successful Product Launches

Springer

Editors
N.R. Srinivasa Raghavan
General Motors India Pvt. Ltd
Whitefield Rd.
Bangalore-560066
Units 1-8, 3rd Floor, Creator Bldg.
India
srinivasa.raghavan@gm.com

John A. Cafeo
General Motors R & D Center
Manufacturing Systems
Research Lab.
30500 Mound Road
Warren MI 48090
USA
john.cafeo@gm.com

ISBN 978-90-481-2859-4 e-ISBN 978-90-481-2860-0
DOI 10.1007/978-90-481-2860-0
Springer Dordrecht Heidelberg London New York

Library of Congress Control Number: 2009929635

Cover design: eStudio Calamar S.L.

Printed on acid-free paper

Springer is part of Springer Science+Business Media (www.springer.com)

Editorial

1 Motivation for this Book

New product development is a highly creative exercise, often involving interdisciplinary decision making and execution. Some of the several functions in an organization responsible for chartering the course of this process including product design, R&D, engineering, manufacturing, marketing, procurement, planning, finance, and information systems. Not to ignore the role that strategic suppliers, customers and regulators can play in this process. The more complex the nature of the product, the more strategic is the impact of this process on the financial viability of a company. Given the complicated nature of product development, it is extremely important for an organization to invest in product research. Generally, product research concerns itself with processes, tools, and methods that aid in understanding customer needs (both expressed and latent), planning for the right product features to suit an appropriate portfolio of products, translating the needs into engineering specifications, and monitoring the consumer feedback so that better products can be sold to consumers in the future.

In this book, we have a compendium of research articles on various issues concerning product research, written by research scholars both from academia and industry. Our aim in this book is to highlight through these articles, the state-of-the-art in a number of areas within product research. We hope this book will be a useful reference material for both practitioners and researchers in academia.

2 Summary of Research Articles

This book has four clusters around which we thought it will be best to organize the papers we reviewed. The first cluster is on innovation and information sharing in product design. The second cluster is on decision making in engineering design. The third cluster is on customer driven product definition. The fourth cluster is on quantitative methods for product planning. We now summarize the contributions of authors in their respective clusters.

2.1 Innovation and Information Sharing in Product Design

New product development is all about innovation. Advanced decision making needs to be enabled by information as much as by intuition. Understanding the role of intuition, creativity, and appropriate customer and process information is critical for world class product design. In this cluster, authors address the above concerns in their papers.

For products where lead times for development can run into several years, it becomes important to make decisions that are 'intuitively' good. J. Hartley's paper deliberates on what exactly is intuition, how does one go about affecting it, and finally, how does one create a shared intuition among potentially divergent minds involved in product development. The second article in this cluster, by A. Chakrabarti, proposes a framework for improving chances of a successful product launch, by good creativity management. In the process, he answers questions such as: what do we mean by a successful product; how is a successful product created; how do we improve the chances of being successful, etc.

Designing customizable products for mass markets is indeed a challenge, especially for hi-tech products. In their work, J.Y. Park and G. Mandyam present a persona-based approach to wireless service design. The product design cycle is analyzed starting with the initial ethnographic research and ending with usability testing prior to a commercial launch. In the last paper in this cluster, J.A. Rockwell, S. Krishnamurty, I.R. Grosse and J. Wileden bring out the role of the design information and knowledge necessary for decision-based design, which may come from across multiple organizations, companies, and countries. Integrating distributed engineering information that allows decision makers to easily access and understand it is essential for making well informed decisions. The authors present a knowledge management approach for documenting and seamlessly integrating distributed design knowledge during the evaluation of design alternatives.

2.2 Decision Making in Engineering Design

Many products that customers use in their day to day lives like automobiles, are quite complex from an engineering perspective. It goes without saying that several design parameters co-evolve during the product development process. Uncertainty in design influencing factors needs to be given due treatment while optimizing the design variables. In this cluster, we have articles that address this issue.

G. Hazelrigg presents a treatise on the mathematics of prediction. He establishes the basic concepts of prediction in the context of engineering decision making. J. Donndelinger, J.A. Cafeo, and R.L. Nigel, present a case study using simulation in which three engineers were independently tasked with choosing a vehicle subsystem design concept from a set of fictitious alternatives. The authors acted as analysts and responded to the decision-makers' requests for information while also observing their information collection and decision-making processes. The authors then compare established theories of normative decision analysis, cognition, and psychological type.

S.S. Rao and K.K. Annamdas present a methodology for the analysis and design of uncertain engineering systems in the presence of multiple sources of evidence based on Dempster-Shafer Theory (DST) is presented. DST can be used when it is not possible to obtain a precise estimation of system response due to the presence of multiple uncertain input parameters. A new method, called Weighted Dempster Shafer Theory for Interval-valued data (WDSTI), is proposed for combining evidence when different credibilities are associated with different sources of evidence. The application of the methodology is illustrated by considering the safety analysis of a welded beam in the presence of multiple uncertain parameters.

Robustness can be defined as designing a product/service in such a way that its performance is similar across all customer usage conditions. R. Jugulum gives an overview of the principles of robust engineering or Taguchi Methods. The author describes the applicability of robust engineering principles in new product development with several case studies.

A. Deshmukh, T. Middelkoop, and C. Sundaram bring out the complex trade offs that are germane to distributed product development activities. They propose a negotiation protocol with an attendant optimization formulation that helps the distributed design team members to better explore globally optimal decisions. The authors conduct experimentations to verify the social optimality of the proposed protocol.

2.3 Customer Driven Product Definition

New products are seldom crafted without due diligence into understanding consumer behavior. Often frameworks like quality function deployment are adopted by product manufacturers, for a structured approach to prioritize product and technology features. Analytics is then carried out around the voice of customer that needs more focus. In this cluster, authors present their research centered on this theme.

Traditional automotive product development can be segmented into the advanced development and execution phases. In his paper, S. Rajagopalan focuses on three specific aspects of the Advanced Vehicle Development Process. The author highlights the different issues involved in understanding and incorporating the Voice of the Customer in the product development process. A catalog of questions is provided to help the product planners make informed decisions.

R.P. Suresh and A. Maddulapalli in their work formulate the problem of prioritizing voices of customer as a multiple criteria decision analysis problem and propose a statistical framework for obtaining a key input to such an analysis. They apply a popular multiple criteria decision analysis technique called Evidential Reasoning and also investigate a statistical approach for obtaining the weights of consumer surveys relevant to key voice analysis.

Primary and secondary market research usually deal with analysis of available data on existing products and customers' preferences for features in possible new products. S. Shivashankar, B. Ravindran and N.R. Srinivasa Raghavan present their work on how manufacturers need to pay immediate attention to the internet blogs as a valuable source of consumer tastes. The authors focus on application of web content analysis, a type of web mining, in business intelligence for product review.

M. Bhattacharyya and A. Chaudhuri, in their paper, observe that the product design team often needs some flexibility in improving the Technical Characteristics (TCs) based on minimum performance improvements in Customer Requirements (CRs) and the imposed budgetary constraints. The authors present a fuzzy integer programming (FIP) model to determine the appropriate TCs and hence the right attribute levels for a conjoint study. The proposed method is applied to a commercial vehicle design problem with hypothetical data.

2.4 *Quantitative Methods for Product Planning*

The role of advanced decision making using quantitative approaches has gained popularity over the last two decades. Product planners often need tools for scenario analysis for better new product risk management. In this cluster, authors present their research on innovative ways of addressing this issue.

J.B. Yang, D. Tang, D.L. Xu, K.S. Chinargue that New Product Development (NPD) is a crucial process to maintain the competitiveness of a company in an ever changing market. In the process of developing new products of a high level of innovation, there are various types of risks, which should be properly identified, systematically analyzed, modeled, evaluated and effectively controlled. The authors investigate the Bayesian Network (BN) method to assess risks involved in NPD processes. Their approach is discussed in a case study on a multinational flashlight manufacturing company.

Predicting the trajectory of customers' preferences for product attributes for long range planning, is a complex exercise. S. Bukkapatnam, H. Yang, and F. Madhavi aim to provide a mathematical model for this issue, based on Markov models. One and five step ahead prediction results of the proposed model based on simulated data indicates that the predictions are 5–25% more accurate than the models commonly used in practice.

Choosing products to launch from a set of platform based variants and determining their prices are some of the critical decisions involved in any NPD process. Commercial vehicles are products, whose sales are closely tied with the economic conditions. A. Chaudhuri and K.N. Singh present a mathematical model and a case study for profit maximization. The authors develop a two period model to choose the platform based variants, their prices and launch sequences within the two periods, spanning two economic conditions, boom time and recession.

Many thanks to Ms. Nathalie Jacobs and Ms. Anneke Pott of Springer for the great collaboration and assistance in preparation of this book.

Acknowledgements

Editing a book of this magnitude is unimaginable, without the solid and timely support and help from many people. It is our pleasant duty now to communicate our sincere thanks and appreciation to those concerned.

First and foremost, this book would not have been possible in the first place, without the express support sans fine print from Dr. B.G. Prakash, Director, GM R&D India Science Lab, Bangalore. When the first Editor brought this issue up, Dr. Prakash wholeheartedly offered his consent to get the necessary organizational acceptance and financial support for this endeavor. In fact, the idea of hosting a workshop on the book's theme, at Bangalore, was proposed by Dr. Prakash, when we were visiting the Eye of London (which you see on the cover page of this book!). The Eye of London is at once the expression of an artist, and an engineering marvel of recent times.

We would like to sincerely thank Alan I. Taub, Executive Director, GM R&D, Warren, for his unflinching support for this book. In spite of the troubled economic times that we have witnessed during the course of editing this book, Taub never ceased to prop us in our effort to advance our intellectual pursuits. Jan H. Aase, Director, Vehicle Development Research Lab at GM R&D, Warren, was a source of inspiration during the period.

Inviting outstanding research scholars to contribute to this book was indeed a pleasant and challenging job at once. We would like to thank, from the bottom of our hearts, all the contributing authors for their candid support and patience. The technical eminence of this book has indeed rested on the shoulders of world class researchers who have authored very interesting articles on product research.

Reviewing technical papers in quick time was made possible thanks to voluntary support from researchers working with the Customer Driven Advanced Vehicle Development Groups at GM R&D. We thank the following people for providing candid assessment of the papers:

Suresh P. Rajagopalan, Sai Sundarakrishna, Parameshwaran Iyer, Anil Maddulapalli, Shahid Abdulla, Sheela Siddappa, Krishna K. Ramachandran, Joe Donndelinger, Mark Beltramo, and R. Jean Ruth. The Editors would also like to thank their colleague group managers at GM R&D for the intellectual support. The office staff at GM R&D have been considerate in following up on paper work. Our sincere thanks to the team.

An effort of this magnitude needs support from family and friends. N.R. Srinivasa Raghavan would like to cherish the caring support he received from his wife N.S. Anusha during the course of editing this book. Anusha with her parents, N. Kannan and A.R. Lata, managed the two kids Sumedha and Hiranmayi, pretty well during this period. Raghavan also acknowledges his father, N.S.R. Tatacharya for being his role model to excel in intellectual pursuits. He also fondly remembers his late mother Jayalakshmi for all the kindness and blessings she bestowed on him. John A. Cafeo notes as with all of the technical activities that he is involved with, success would not be possible without the support and encouragement of his wife Karen and their children Anna, Johnny, Jacob and Maria. He avers that their forbearance is as necessary and important to these endeavors as anything that he does.

Contents

Part I
Innovation and Information Sharing in Product Design

Chapter 1
Improving Intuition in Product Development Decisions

Jeffrey Hartley

Abstract Market research has traditionally been aimed at collecting and delivering information to decision makers. A major problem has been that decision makers filter information according to its fit with their intuition. The present article maintains that market research must therefore directly target the intuition of decision makers. To do this, two problems must be solved. The first is to bring the intuition of the company mind into alignment with the thinking of the customer mind. Equally important is the need to bring the intuitions of the various functions within the company into alignment with each other. This research philosophy has led to two specific methods at General Motors: Inspirational Research and Iterative Design. Examples of these two approaches and evidence of their effectiveness are presented.

Keywords Intuition · Decision-making · Confirmation bias · Design · Iterative design · Data

> **"The human mind is not a container to be filled but rather a fire to be kindled."**
>
> **Dorothea Brande**

The prevailing metaphor for market research has been one of carrying information from customer to decision maker. This is evident in the ways market research is commonly defined:

> "Marketing research involves the specification, gathering, analyzing, and interpreting of information to help management understand its environment."
>
> Aaker, *Marketing Research*, 1986

> "Market research is the collection and analysis of information about consumers, market niches, and the effectiveness of marketing programs."
>
> investorwords.com, 2008

J. Hartley (✉)
Global Product Research, General Motors Corporation, 30200 Mound Road Warren, MI 48092-2025
e-mail: jeff.l.hartley@gm.com

N.R.S. Raghavan and J.A. Cafeo (eds.), *Product Research: The Art and Science Behind Successful Product Launches*, DOI 10.1007/978-90-481-2860-0_1,

> "Market research is the systematic and objective identification, collection, analysis, and dissemination of information for the purpose of improving decision making related to the identification and solution of problems and opportunities in marketing."
>
> www.nadbank.com, 2008

And it is evident in the fact that most of us treat market research as a corporate function more than a way of thinking. Most of the large firms employ a staff of market researchers, ready to venture out to the provinces and collect data on or from customers, then bring it back to the imperial decision makers in our companies. Information is the unit of exchange and the unspoken assumption is that it serves as some sort of nutrient which, if fed to the decision makers in abundant supply, will lead to wise decisions.

But this "information as food" metaphor breaks down under scrutiny. It breaks down for two reasons.

The first reason is most evident in design research (i.e., research meant to help designers develop successful innovative designs which appeal to customers) but I think that it holds for a much larger portion of market research. This problem stems from the fact that not all information is easily transported.

Some information is "stickier" than other information (von Hippel 1994) in that it is difficult or costly to move from one place or mind to another place or mind. Aesthetic information is a good example of this and that is why the "information as food" metaphor breaks down for design research.

Imagine that a customer says she wants a car interior to hold four people and she wants it to be "charming". The second piece of information is stickier than the first in that the recipient of the information understands what the first one means and can act upon it with a lot more certainty than they can for the second. Much of the customer information that a designer needs is sticky.

But there is a more fundamental problem which affects all market research. It is exemplified by an interesting finding that Rohit Deshpande, Sebastian S. Kresge Professor of Marketing at Harvard Business School, described at a Marketing Science Institute conference several years ago.

Dr. Deshpande interviewed senior managers at Fortune 500 companies. One question he asked was "Why do you do customer research?". Seventy percent said they did it to confirm their views. Interestingly, the more senior the person, the higher the percentage who said they did it to confirm their views.

When it was pointed out that their answer was not exactly the way a scientist formally uses data to test hypotheses, a representative answer was "that's why they pay me the big bucks. I can make the correct call, I just need data to support it afterward."

The executive's first reaction to countervailing data is typically to keep the hunch intact but explain away the dissonance by undervaluing the research. So when the results run counter to an executive hunch, then the research methods are often scrutinized.

> "Did we have the correct sample? Were the stimuli well done? Was the location representative? Were the procedures sound? Did we have the right competitors? Did the moderator lead the witness?"

Oftentimes these are stated not as questions but as declarations ("You must not have had the right sample!"). The perceived research quality pivots around the fit between the real and the desired results.

The inertia against changing a hunch is even stronger if that hunch has been publicly stated or has built up its own momentum. In our business, that momentum takes the form of a very expensive full size model of a future vehicle which has been fashioned and refashioned and adored in a design studio for several months. Meeting after meeting, various high ranking people venture in to see the artifact, and before long a reification of the hunch spreads to many people. It takes great personal strength to back down from one's hunch after proclaiming it publicly over a long period.

But the "information as food" metaphor is undermined even when the data *supports* an executive hunch. In such a situation, the market researcher gets either or both of two messages. First, the method is not questioned of course – there is no reason to question data in support of your view. But there may be a reason to then question why we "tested something this obvious."

Unfortunately, these executives are not unusual. In fact all humans behave this way. A large body of psychological literature documents this aspect of human thought, under the heading of "confirmation bias": the tendency to seek out things which confirm your preconceptions (i.e., intuitions) and to ignore things which contradict them. Humans are not passive in the intake of information. We are very active. We do not see, we look for. We do not hear, we listen for. We do not touch, we feel for. We form an expectation fairly quickly and then look for confirming evidence. This occurs at all levels of cognition from immediate perception to the more drawn out thought processes of conceptual thought and even scientific postulation.

> "It is the peculiar and perpetual error of the human understanding to be more moved and excited by affirmatives than by negatives."
>
> Francis Bacon

So all of us display the type of predisposed thought that characterizes our executives. We take in information differentially, allowing in things that support our intuitions and averting things that don't. We read things we agree with and avoid things we don't agree with. We watch or listen to media that agrees with our views. When "data" interferes with our world view, we tend to discount it.

Over time, we entrench in our views. But most of us do not even know this because the only data we accept supports us. In the political world, everyone is a moderate in their own minds, while others are radical. In the sports world, the referees are always tilted against our team.

So, given human behavior, the prevailing metaphor for market research leaves us with a quandary. When data runs counter to a hunch, the hunch usually wins. When it supports a hunch, it doesn't change thinking anyway. So one is inclined to argue, with some seriousness, why we do market research at all, since it is either superfluous or discounted. In either case, it adds little value.

One can see, then, that intuition acts as a filter to information. Information which "fits" the intuition passes through into the decision-maker's mind. Information which doesn't fit intuition is deflected. If there were a way to make the decision-maker's intuition somehow more likely to lead to successful product decisions, and if one could do this early on, before the intuition went off track and built up momentum, then wouldn't that pay dividends? For one, later research which would be more likely to lead to product success might be more likely to be accepted.

For this reason, I suggest that market research attack intuition itself and not see its role as simply the delivery of information to an ostensibly passive and even-handed mind. In fact the earlier-stated goals for market research might be replaced or augmented with this one:

1.1 The Goal of Market Research Is to Create Early and Accurate Intuition That Is Shared Across Functions

While my points are most directly aimed at design research, they also apply to market research in general to the extent that the information under consideration is sticky. I call this approach "nurtured intuition" as opposed to "information as food". And it begs 3 very important questions.

- First, what exactly is intuition? For market research to set as its goal the nurturing of intuition, a clear definition of intuition is critical. And what exactly is "accurate intuition"?
- Second, how does one go about affecting it? Researchers whom I have asked have acted as if data-delivery is the best way to affect intuition. I disagree, for the reasons already stated. But if data won't change one's intuition, what will?
- And finally, how does one create a *shared* intuition among potentially divergent minds? In the world of data delivery, the data is there for all to see and one at least has an objective foundation on which to build a shared conclusion. (My experience has been that data doesn't succeed as often as one might hope, probably due to the differing preconceptions.) But if we are serious about nurtured intuition leading to wise product decisions, we'll have to ensure that the cross-functional team shares the same intuition.

The remainder of this paper addresses these questions and presents two examples of how our research has shifted to align with this "nurtured intuition" approach.

1.1.1 What Is Intuition?

Historically, intuition has been defined as the faculty of knowing or sensing without the use of rational processes – an impression. It has been seen as immediate, somehow given to us rather than taken from facts. Oftentimes intuition has been treated as having a divine origin. Frequently it has been treated as a special skill that only psychics, mediums, and spiritual sages can have. It almost always has been treated as unlearnable. Pet psychics are born, not made.

The definition I am proposing is quite different.

1.2 Intuition Is the Abstract Knowledge That Comes Automatically from Guided Experiences – A Trainable Skill, Often "Beyond Words"

It is not mystical nor divine but rather a very common and prosaic activity. In fact I would say that one's intuition is constantly fed through experience and this happens thousands of times a day. A person's active participation is an absolute requirement for the development of their intuition – it is not achievable through book learning, charts, analyses or any of the sorts of things that make up most of the "information as food" approach.

Some examples will serve to make my point. Imagine that your child wants to learn how to ride her bike. You wouldn't ever think of reading her books or showing her charts about the physics of bike riding. Rather you encourage her to develop an intuition for what her body needs to do (you don't say it this way, but essentially that is what you are nurturing in her). She learns it through trial and error. Her vestibular, kinesthetic, and visual feedback helps her to learn what she needs her muscles to do to stay upright. And however this learning happens, it does not seem to be accessible through the verbal systems of the brain.

As another example, suppose you want to learn how to recognize the art of a new artist. If I used the "information as food" approach, I would give you statistics about the percentage of times the artist used various colors, subject matters, formats, media, sizes, and so on.

But if I wanted to improve your intuition, I would show you examples of this artist's works (and to be thorough, examples of other artists' works) and have you try to classify them. If told when you were right or wrong, you would gradually learn to recognize the style. The style might not be easily described, but your accuracy would get higher with more and more experience. When you saw a previously unseen painting, you would feel some sort of inkling, or hunch, or gut feel – an intuition – about who painted it.

The third example comes from a study by neuroscientists (Bechara et al. 1997). In this study, people played a gambling game with four decks, two of which were designed to create slight losses on average and two of which were designed to create slight wins on average, although it was not obvious on any hand if the deck were loaded for or against the player. At first, people would choose equally from the four decks, but over time they began shifting away from the decks which were stacked against them. They did this *before* they knew they were doing it or why they were doing it. The authors concluded that something in the brain was generating "intuitions" which guided behavior prior to conscious awareness or the ability to verbally articulate the basis for one's behavior.

So one way to define intuition is "the mental activity, based on experience, which may not be available to verbal articulation but nonetheless which influences decisions." This is the flavor of my definition.

What then is "accurate intuition"? In the business context, it is the ability to decide as your customer would. Of the thousands of decisions a company must make as it develops a new product, some will act to enhance the product's later appeal in the market and some will act to harm it. Most of these decisions are based on the decision maker's hunch and not on data, either because no data is available, the needed information is sticky, or as we have seen, the data is ignored if it disturbs the hunch. Those intuitive decisions which enhance the product's later appeal are, by definition, based on an accurate intuition of the market.

1.2.1 How Does One Nurture Intuition?

There are two paths by which knowledge can enter the mind. One is evolutionarily much older. Humans can take in knowledge through direct experiences, and the mind has natural mechanisms for representing the central tendencies and breadth of the experiences. This knowledge is connotative, contextual, and abstract, drenched with associations, and often difficult to reduce down to words. You might call this the *experiential* path to knowledge.

The other is evolutionarily more recent and follows from the development of language and symbols, primarily letters and numbers. Humans can take in knowledge through *symbols* – what others write or speak. This knowledge is denotative, usually lacks context, is non-experiential, and is very concrete (we call it "black and white" for a reason). It also suffers from the recipient wondering about the sender's motives, a problem which is not found with direct experience. You might call this the *symbolic* path to knowledge.

The former is similar to what psychologists call "episodic memory" (i.e., memory for episodes of experience) while the latter resembles "semantic memory" (i.e., memory for facts). Episodic memory is more compelling and memorable than semantic memory.

Intuition seems to accompany experiential knowledge much more than symbolic knowledge. An intuitive grasp for something comes with experience, not by reading or hearing about it from a secondary source.

This conceptualization of intuition parallels two topics in psychology. *Implicit learning* is the process through which we become aware of regular patterns in the environment without intending to do so, without being aware we are doing so, and in a way that is difficult to express (Cleermans 1993). It accompanies experience. *Category abstraction* is the process of learning to recognize members of an ill-defined category, such as an artist's style, by seeing examples of it (Hartley and Homa 1991).

My point in referencing these topic areas in psychology is to emphasize that intuition has been studied scientifically – it is not mystical. And all the research says that intuition improves with experience. But experience of what? That answer depends on what you want your decision maker to be intuitive about. It helps to state this as the *skill* you are trying to engender in the decision maker and then set up the experiences to grow this skill.

We want our designers to have an intuitive understanding of design tastes in our customers (which is very sticky information). But we also want them to understand what role design plays in customer decisions. We want our designers, while they are designing, to be able to envision what our customers would want if they were sitting next to the designer. In short we want our designers to develop the skills of thinking like a customer and of making design decisions that will ultimately delight the customer.

As one example, most designers think about design quite differently than our customers do. That shouldn't surprise anyone – designers have spent their lives immersed in design and so it is at the forefront of their product evaluations. For customers, it often is subjugated to function. When designers think of a car's interior, they typically ask themselves "Is it beautiful?". When customers think of a car's interior, they typically ask themselves "Is it functional?" and "Is it comfortable?" more than the designer would. This chasm between the thought worlds of designers and customers could lead to designs which sacrificed function for form and therefore displeased customers.

One way to reduce this chasm is to have designers go on a ride and drive with customers, who judge the interiors of our cars and those of our competitors. Somehow this direct experience – compelling, emotion-laden, and episodic – does more in one afternoon than all the data we may have provided about customer priorities.

1.2.2 How Does One Nurture Shared Intuition?

It is not just the chasm between the thought worlds of designers and customers that presents a problem. The chasms between the thought worlds of designers and engineers and marketers *within* the company also present problems.

Perhaps the best example of how to create shared intuitions among the functions is contained in the work of Dr. Edward McQuarrie, who won an MSI Best Paper Award for work he called *customer visits* (McQuarrie 1993). Although McQuarrie's work has been in business-to-business products, we have modified it for work with our products.

According to McQuarrie, customer visits are not "ad hoc visits to a few haphazardly selected customers". Rather they are characterized by three rules. The visits should be (1) made personally by the key decision makers (2) in cross-functional teams (3) according to a well articulated plan, with stated objectives, careful sampling, a focused discussion guide, and a plan for disseminating the results.

He highlights the benefits of cross-functional customer visits. The information derived from these visits is, in his words, credible, compelling, detailed, complex, novel, contextual, and nonverbal. This leads to higher motivation to respond, higher potential for change, deeper understanding, and more opportunity to generate innovations.

Because these visits are cross-functional, they decrease disputes and misunderstandings, and they increase consensus and integration between functions. In effect, each function improves not just their skill at thinking like a customer but also the skill of thinking like their colleague. They cross-pollinate their intuitions and optimize their actions at the group level and not the individual level.

We can now turn to the specific examples of how our work has changed to adopt the "nurtured intuition" approach.

Example 1: Inspirational Research

Our interpretation of customer visits is what we call Inspirational Research, which is defined more by its two goals than by specific methods. The first goal is to enkindle in the company's minds a deep empathy and understanding for the customer which should lead to products which connect with those customers.

We want to be able to go beyond our descriptive data and understand our customer in a *deep* manner. We want to understand their *dynamic* lives (as opposed to a static set of data). We want to know the customers *as individuals* (not as a collection of population averages). We seek to understand the customer using the product *in context* (not decoupled from that experience and placed in our contrived research setting). We want to know how their lifestyle leads to their product choices. In short, we want to know the personal side of our customers.

While sitting at their desks, the company minds may not recognize how today's customers think and make decisions. But they are certainly aware of state-of-the-art designs and engineering solutions and they probably are the in-house experts on design or engineering trends. Therefore a merging of these company and customer minds, especially when the dialogue is two way and lively, provides a fertile ground for ideas to spring forth in the creative minds of the company.

Second, we want to help the company minds converge. We want them to leave the research event with a focused vision of the product they want to develop and the work they will need to do to realize this vision.

We seek to have the team come away from the event with an agreed upon understanding of the customer and how the company's product and brand will fit into their customers' lives. But we also want them to come away with a better understanding of their colleagues and how their joint decisions can make the product better. By the end of the inspirational event, each functional specialty should clearly state its goals going forward and the other functions should understand them and agree with them. The package of goals should fit together nicely.

Activities Aimed at Seeing Through the Customers' Eyes. How do we nurture intuition? Despite my earlier assault on the "information as food" metaphor, we begin with a lay of the land: what we know about current customers, about our brand's strengths and weaknesses as well as those of our competitors. And we consider how the world will be changing over the course of the program. All of this is based on large sample, quantitative data. The team discusses what it thinks this data means and what questions it surfaces.

But this doesn't help us see through the customers' eyes. In fact this sort of analysis sanitizes all the personal aspects out of the data. And sometimes the most valuable, insightful information is impossible to capture and communicate.

So we make sure that we include activities where the company minds see what the lives of their customers are like. We see what their tastes are and how those tastes fit into their lifestyles. And we also see how our product fits into their lives.

Small cross-functional teams visit people's homes and examine or inquire about various aspects of the person's daily life, their key belongings, their tastes, and their favorite products. By studying what already connects with our customers, we are better able to develop products which connect (Homma and Ueltzhöffer 1990).

While at their homes, we also go on typical drives with them and observe how they use their vehicle, what emotions it conjures up and how it conjures them. We may also have them drive a set of vehicles to surface strengths and weaknesses of the competitors, while we watch. All of these activities help us see through their eyes.

We also have a more controlled setting to study their tastes through a line of work we call Value Cues Research. Every brand is meant to stand for some customer benefit or value, such as safety, elegance, or optimism. Designers and engineers want to know how to communicate these values to customers. We ask customers to take photographs of products they like which communicate each value and they discuss the results with our team. The results come, not in numbers but in images of real objects, the language that designers especially think in.

We encourage our team members to sketch solutions to problems while at the research event. Instead of market research data coming back in the form of numbers, we bring back an identification of a customer issue and the initial solution that a team member identified. In a sense, we are *stocking the shelves* with partially developed, customer-inspired ideas that we can later refine.

Perhaps the most common example of inspirational research has been to "take the studio to the customer". In this approach, we use the sketches and scale models in the studio as tools to generate discussion with customers who may other wise have difficulty articulating their "sticky information" but have no problem telling us which of several hundred sketches is appealing, or is most innovative, or conveys what we want our brand to convey.

As can be seen, the activities which attack this first goal are varied and growing, but everything is aimed at seeing through the customers' eyes.

Activities Aimed at Integrating the Company Minds. These activities are of two kinds. The first involves only the company minds. We meet in teams and surface our deeply held feelings and views about our brand and product as well as those of our competitors. For this we bring in all the information we have available from both quantitative and qualitative work. We have a "war room" and we place all the research questions we brought on the walls. After each meeting with customers we brainstorm what we are learning and in so doing, we evolve our group intuition.

We also rely at times on the use of metaphors (Clark and Fujimoto 1990; Zaltman and Zaltman 2008) for surfacing deeper thoughts and feelings held by team members. (This is a method we make heavy use of with customers too.) For example, we ask "If Cadillac were an animal which one would it be and why?" "If Mercedes were a party, who would attend? What would they wear? What would they do?, etc." "If our product concept were an animal, which one would it be and why?"

The reasons given reveal deeply held views and we seek to resolve differences among team members after we surface them. We also seek to clarify how our product or brand differs from its competitors. Out of repeated engagements with this sort of exercise, coupled with the customer-related activities, the team can agree on a core metaphor.

The second kind of activity involves in depth face-to-face meetings with customers as described in the last section. But because these are cross-functional, the *sticky information*is collected jointly which helps to integrate the conclusions drawn. And the engineer and designer, for example, reveal their own "questions of importance" and in so doing help each other understand how they think.

The culmination of all Inspirational Research is a set of "Key Insights" written jointly by the cross-functional team. The process of writing this is in and of itself a purifying act – it surfaces disagreements and forces resolutions. The final document is valuable too, but perhaps not as valuable as what endures in the minds of the team.

While it is difficult to quantify the success of this approach, one tell tale indication is pervasive. Everyone who takes part in Inspirational Research, from members of our team to the customers who we engage, has come away with a very positive appraisal of its benefits. Our teams always express a deeper understanding of who their customers "really are". Our teams are unified in the actions they take afterward. And customers are thrilled at the chance to engage directly with product teams.

There are of course concerns that this general approach raises. An obvious first concern is that the team can only meet with small samples of customers and the conclusions drawn may not be representative of the larger market.

Small Samples. We first seek to minimize this problem by coupling the deep understanding we get in face to face meetings with data drawn from larger samples. This occurs in the initial briefing and it also shows itself in the video results we bring back. On edited video records of our Lifestyle Research, we overlay large sample results to bring the individual's responses into perspective.

The concern about small samples begs the question of what "small" is for learning of this sort (i.e., direct experience with members of the category). The human mind is quite capable of abstracting out central tendencies from experience. For example, after seeing several paintings by Renoir, you are pretty good at recognizing others by him. How many examples of his paintings would you have to see to learn his style?

There is evidence in academic psychology that by encountering as few as 12 examples of a category, you can recognize later members very accurately (Hartley and Homa 1981). This finding is quite similar to results found by market researchers (Griffin and Hauser 1992; Zaltman and Higie 1993). Therefore, we make sure that every team member interacts with at least 12 customers.

Divergent Conclusions. Another problem with this type of work is that individuals within the company might come away with different conclusions. So we intersperse periodic cross checks to compare notes and summarize points of agreement among the company minds. For example, we might break into teams with each one visiting the homes of two different customers in the morning. We would convene for lunch and go over what we had been learning. Differences would be noted and hypotheses would be suggested.

Later interactions would be followed by other cross check meetings where we would continue to update our shared learnings. And of course we document our final conclusions in the Key Insights and distribute them to the team in the form of specific functional goals that everyone agrees to.

Selling the Results Back Home. It is one thing to be personally inspired in this activity. It is quite another to sell that inspiration to your colleagues who remained at work while you were away. We address this problem in three ways.

First, we ideally return with video records that can be used to bring our conclusions to life.

Second, because the team was made up of representatives from each function, the results are brought back by *disciples* who are distributed among the staffs. This personal envoy can also make the results come to life and assure the team that their viewpoint was represented in the discussions.

And finally, the "proof is in the pudding". Many of the results are in the form of sketches or ideas linked to customer needs and tastes. These ideas are thereafter subject to further testing with larger samples. So we let them speak for themselves.

Example 2: Iterative Design

A clear idea of the difference between the nurtured intuition philosophy and the "information as food" philosophy is seen in the evolution of theme research at General Motors. The traditional way of doing theme research (i.e., research aimed at understanding how customers react to designs) was for the market research department to show designs – or sometimes just one design – to customers, get quantitative ratings and qualitative discussions, and then write a report of "the official" interpretation of the results.

Because the market researchers saw their job as delivering information, and because they wanted this information to be as accurate, thorough, and unbiased as possible, they spent 2 weeks writing the report. Then the results were presented in a large meeting to interested parties from Design, Engineering, Marketing, and upper management.

Of course, in the intervening 2 weeks, many usually divergent conclusions had infiltrated the different functions. These conclusions were based only on what people had seen in the focus group discussions and so they were influenced greatly by confirmation bias. Typically, the design team drew their own conclusions, which "fit" their desired outcome, and took actions to evolve their design according to what they heard.

So when the meeting finally occurred and the official results were delivered, momentum at Design Staff had already mobilized behind their conclusions, which often went against the desires of other functions. This meeting then became the battleground for airing differences between the functional areas. The customer voice often was drowned out.

Traditionally, we would do two theme studies, separated by perhaps 6 months. Between theme studies, the design would undergo hundreds of modifications, with influences from dozens of sources, such as reactions from various layers of design management, marketing, engineering, upper management, and even from competitive activity. What we would not know however was how customers would react to these changes.

So when we returned to the next theme study, the differences between the theme tested at the first study and at the second would be numerous. Hence differences in the customer reactions could not be linked to any particular design change.

There were huge problems with this approach. Theme research had become a place for entrenchment and opposition between design and other functions. Customer feedback was wielded as a weapon between functions and not as a way to improve the design. Design decisions were made before the results of the study were unveiled and, once made, were resistant to change. Designers got essentially only two swings of the bat (and sat on the bench most of the time!). The linkage between a design action and the customer reaction was not learned.

So we set out to improve theme research. We felt that if designers could get the feedback instantly and directly, insulated from the battleground atmosphere that had sometimes characterized traditional theme research, they might use it more earnestly.

We set up a design studio at the study site itself and conducted what appeared to be a traditional theme study. But it differed in critical ways.

The research and design teams acted in unison, with design questions quickly fed into the research process and research results quickly fed into the design room. Negative customer reactions were probed to find out their exact source, leading to design hypotheses about what to change. Often we allowed the designer to ask questions directly to customers, which facilitated a more direct, intuitive form of learning.

The designers would react with new designs, based on these hypotheses, which would then be shown to customers during interviews. In short, designers took many "swings of the bat" with immediate feedback. This culminated in what was essentially a second theme study a few days later with both old and new designs shown so we could measure if the design hypotheses were right.

Because customers are much better at reacting to designs than telling us what to change and how to change it (because this information is sticky), the interjection of the designers' minds and hands into the research process itself was crucial. The result of all this activity was that designer intuition and their skills at connecting with customers through design were encouraged to ascend the learning curve.

The true output of this type of study is not a report, although we do issue one within 2 days. The real value is in the trajectories between the old and new designs and the customer influence that led to them. It is the gain in intuition that a designer gets by trying different design solutions to customer rejections. No designer in these iterative studies has ever done anything that runs counter to their own instincts. Rather they seem to find those designs which fit their own intuition and also fit the customers' desires.

To date, this approach has been extremely successful. The percentage of programs which are green at the final test (i.e., their appeal score is significantly higher than all designs tested) has gone from under 50%, when not using iterative design, to 92% when using iterative design. Similarly, the average percent of respondents who rated the design as "Very Appealing" has gone from 16% to 30%.

Perhaps most noteworthy though is that designers speak often about the knowledge gained from the experience of iterative design. We expect the intellectual capital of our designers to blossom, in terms of a growing empathy for customers, as we make this a standard practice.

1.2.3 How Will the Nurtured Intuition Philosophy Change Company Behavior?

Because the nurturing of intuition in decision-makers requires their active participation, they must change their own conceptualizations of what constitutes their "work". So market researchers will have to convince them that it is in their interest to take part in face-to-face dialogue with customers.

This is not an easy task, since few executives question their own intuition, as Deshpande showed, and their days are already full of other activities. Nonetheless, executives will have to rebalance their calendars to accommodate the added time needed to improve their intuition. The best market researchers in this "nurtured intuition" world must be good scientists, yes, but must also be capable of convincing executives to question their own intuition.

Market researchers must also come to judge their activities by how well they improve the decision-making *skills* of their colleagues, not by how much information they transmit nor even by how impeccable it is. Like any skill, the best way to learn it is through experience with direct feedback, and so researchers would need to arrange these.

A "nurtured intuition" researcher would first identify the skill they wanted to nurture (e.g., designing cars for women) and then set up guided experiences in which the colleague made a decision, got feedback, modified their decision, got additional feedback, and so on. In short, the researcher would help the decision maker take as many swings of the bat as he could, helping the swings get more productive and accurate over time.

The training of most researchers and executives matches the "information as food" model much better than the nurtured intuition model. The latter requires a much different set of "researcher" activities; in fact the term "researcher" is probably not broad enough. We now must facilitate exchanges between minds in ways that lead to better decisions.

All of which is to say that we have a lot of work ahead of us. In fact, there is evidence that husbands and wives are mediocre at intuiting the product choices of their mates (Davis, Hoch, and Ragsdale 1986). "How can we do better?", you may ask.

I truly believe that we must take on this apparently unassailable summit – intuition – directly. The first step is to replace the "information as food" metaphor with the nurtured intuition metaphor. Under this new conceptualization, information is still vital – note how we used it as feedback in iterative design – but it's value is closely tied to how well it can improve another's intuition and targeted skills.

References

Bechara, A., Damasio, H., Tranel, D., and Damasio, A. (1997). Deciding advantageously before knowing the advantageous strategy. *Science*, 275, 1293–1295.

Clark, K., and Fujimoto, T. (1990). The power of product integrity, *Harvard Business Review*, November 1990.

Cleermans, A. (1993). *Mechanisms of Implicit Learning*. MIT Press, Cambridge, MA.

Davis, D., Hoch, S., and Ragsdale, E. (1986). An anchoring and adjustment model of spousal predictions. *Journal of Consumer Research*, 13 (1), 25–37.

Griffin, A., and Hauser, J. (1992). The Voice of the Customer. Marketing Science Institute Working Paper 92–106.

Hartley, J., and Homa, D. (1981). Abstraction of stylistic concepts. *Journal of Experimental Psychology*, 7 (1), 33–46.

Hartley, J., and Homa, D. (1991). Abstraction of stylistic concepts. *Journal of Experimental Psychology: Human Learning and Memory*, 7, 33–46.

von Hippel, E. (1994). Sticky information and the locus of problem solving: implications for innovation. *Management Science*, 40 (4), 429–439.

Homma, N., and Ueltzhöffer, J. (1990). The internationalization of Every-Day-Life-Research: Markets and milieus. *Marketing and Research Today*, November, 197–207.

McQuarrie, E. (1993). *Customer Visits: Building a Better Market Focus*, Sage Publications, Newbury Park, CA.

Zaltman, G., and Higie, R. (1993). Seeing the Voice of the Customer: The Zaltman Metaphor Elicitation Technique. Marketing Science Institute Working Paper 93–114.

Zaltman, G., and Zaltman, L. (2008). *Marketing Metaphoria*, Harvard Business Press, Boston, Mass.

Chapter 2
Design Creativity Research

Amaresh Chakrabarti

Abstract We take *design as a plan* by which some undesired reality is envisaged to be changed into some desired reality. It is the plan for creation of an intervention, e.g., a product or a service, with which to bring about this change. *Designing*, or design process whereby the *plan is conceived and embodied*, starts with the perception of the need for a design. Products and the processes of their creation have undergone considerable changes over the last decades. Products have become more complex, and stronger customer awareness and stricter legislation resulted in shorter product life cycles and tighter requirements. Products have to be technically as well as commercially successful. In order to be able to cope with these changes and remain competitive, new approaches to improve effectiveness and efficiency of the product development processes are needed. The overall aim of design research is to support practice by developing knowledge, methods and tools that can improve the chances of producing a successful product. In this chapter, we provide an overview of the broad issues that are investigated in design research, introduce DRM - a design research methodology developed for systematic exploration of these issues, and provide an overview of research at IdeasLab, Indian Institute of Science (IISc) in the areas of design creativity. The following questions are addressed: What is creativity? How can it be measured? What are the major influences on creativity? How does exploration of design spaces relate to creativity? How well do designers currently explore design spaces? How can creativity be supported?

2.1 Design, Design Research and Its Methodology

We take *design as a plan* by which some undesired reality is envisaged to be changed into some desired reality. It is the plan for creation of an intervention, e.g., a product or a service, with which to bring about this change. *Designing*, or design process

---

A. Chakrabarti (✉)
Innovation, Design Study and Sustainability Laboratory (IdeasLab), Indian Institute of Science, Bangalore 560012, India
e-mail: ac123@cpdm.iisc.ernet.in

N.R.S. Raghavan and J.A. Cafeo (eds.), *Product Research: The Art and Science Behind Successful Product Launches*, DOI 10.1007/978-90-481-2860-0_2,

whereby the *plan is conceived and embodied*, starts with the perception of the need for a design. Products and the processes of their creation have undergone considerable changes over the last decades. Products have become more complex using new technological developments and integrating knowledge of various disciplines. Increasing competition, stronger customer awareness and stricter legislation resulted in shorter product life cycles and tighter requirements. Products have to be technically as well as commercially successful. As a consequence of product changes, the product development process has changed. Complexity, quality pressure and time pressure have increased. New approaches to improve effectiveness and efficiency of the product development processes are needed to be able to cope with these changes and remain competitive.

The overall aim of design research is to support practice by developing knowledge, methods and tools that can improve the chances of producing a successful product (Blessing et al. 1992, 1995, 1998; Blessing and Chakrabarti 2002, 2008). This aim raises questions such as

- What do we mean by a successful product?
- How is a successful product created?
- How do we improve the chances of being successful?

The first question leads to issues such as what criteria to be used to judge success, that is, what measures will determine whether our research has been successful. The second question leads to issues such as what the influences on success are, how these influences interact and how to assess them. Investigating these issues would increase our understanding of design which is needed to improve the design process. The third question gives rise to issues related to the translation of this understanding into design methods and tools and to the validation of these methods. Validation is needed to determine whether the application of these methods indeed leads to more successful products as determined by the criteria.

A pure product-focused research effort cannot resolve these issues. That this has been recognised is shown by the increasing number of studies of the way in which a design process actually takes place – to increase understanding on this process – both as a cognitive and a social process and in the organisation. Traditionally this is not the type of research conducted within engineering and it is not possible to transfer research methods directly from other disciplines – a new approach is required. To address these issues in an integrated and systematic way, a research methodology specific to studying and improving design as a phenomenon is needed.

Two characteristics of design research require the development of a specific research methodology. First, the selection of research areas is not straightforward due to the numerous influences and interconnectivity between them. Design involves, among others, people, products, tools and organisations. Each of these is the focus of a particular discipline with its own research methodology and methods, such as social science, engineering science, computer science and management science. Design research is therefore bound to be multidisciplinary. An additional complication is the uniqueness of every design project. This particularly affects repeatability in scientific research. The second characteristic of design research is that it not only

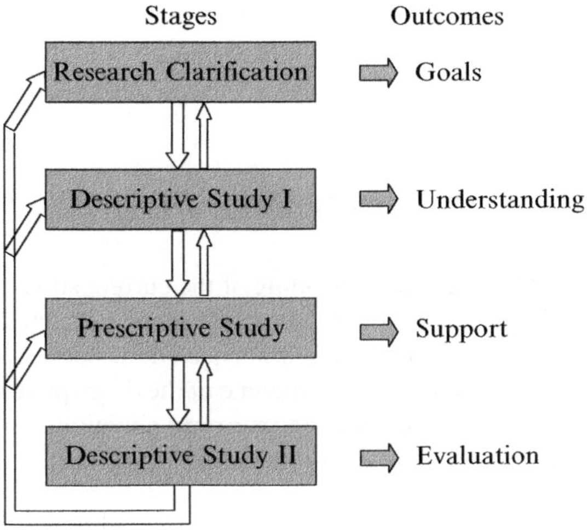

Fig. 2.1 DRM stages, links and outcomes

aims at understanding the phenomenon of design, but also at using this understanding in order to change the way design is carried out. The latter requires more than a theory of *what is*; it also requires a theory of *what would be desirable* and how the existing situation could be changed into the desired. Because this cannot be predicted, design research involves design and creation of methods and tools and their validation. Methods from a variety of disciplines are needed.

Figure 2.1 introduces DRM (Design Research Methodology) – arguably the most widely used methodology for design research. A simple example is used to clarify its main stages.

2.1.1 Research Clarification: Identifying Goals

The first stage is to clarify the aims and objectives of the research, with the resulting identification of the criteria for success of the research. For instance, in an example research a reduction in time-to-market may be identified as a criterion for success. This provides the focus for the next step and is the measure against which a design method or tool developed in the research would be judged.

2.1.2 Descriptive Study I: Understanding Current Situation

In this stage, observational studies are undertaken to understand what factors currently influence the criteria for success, and how. In the example case, a descriptive study involving observation and analysis may show that insufficient problem definition relates to high percentages of time spent on modifications, which is assumed to increase time-to-market. This description provides the understanding of the various factors that influence, directly or indirectly, the main criterion, in this case time-to-market.

2.1.3 Prescriptive Study: Developing Support

In this stage, understanding of the current situation from the last step is used to develop a support (methods, guidelines, tools, etc.) that would influence some of the factors to improve their influence on the success criteria. For instance, in the example case, based on the outcome of the descriptive study and introducing assumptions and experience about an improved situation, a tool is developed to encourage and support problem definition. Developing methods and tools is a design process in itself.

2.1.4 Descriptive Study II: Evaluating Support

In this stage the support developed is applied and a descriptive study is executed to validate the support. In the example case, this included two tests. The first test is whether problem definition is supported. The second test is whether less time was spent on modifications, and whether this, in turn reduced the time-to-market. There might be reasons as to why the second test fails, such as side-effects of the method.

Note that design research embraces both traditional, analytical research, and interventional, synthetic research. While its Descriptive Study stages involve understanding a given situation (with or without the support) as the primary motive, and therefore are primarily analytical in nature as in research in the natural sciences, its Prescriptive Study stage involves a synthesis activity, developing interventions to change the current situation. Unlike in the natural sciences, understanding a situation in design research is not per se the goal, but only a means to change the situation for better.

2.2 Objectives of This Paper

The rest of this chapter provides an overview of research at IdeasLab, Indian Institute of Science in the areas of design creativity. The following questions are explored:

- What is creativity? How can it be measured (Section 2.3)?
- What are the major influences on creativity (Section 2.4)?
- How does exploration of design spaces relate to creativity (Section 2.5)?
- How well do designers currently explore design spaces (Section 2.6)?
- How can creativity be supported (Section 2.7)?

2.3 Definition and Measures for Creativity

Creativity is essential in design. Definitions of creativity, however, are multiple and varied, and factors influencing creativity myriad and various. Moreover, the definition, the influences and their measures are not linked in a systematic way. Consequently, metrics for estimating creative potential of agents or methods are few and only as sound as the theories on which they are based. In this section, we explore what should be the:

- 'Common' definition of creativity
- 'Common' measures for assessing creativity

Unless stated otherwise, all references are cited from Davis (1999).

2.3.1 *What Is Meant by Creativity?*

There have been multiple attempts at qualifying and quantifying the main characteristics of a creative idea. Many see novelty as the sole essential characteristic of a creative idea, e.g., to Newell, Shaw and Simon, "creativity appears simply to be a special class of psychological activity characterized by novelty." For Rhodes, "Creativity ... is a noun naming the phenomenon in which a person communicates a new concept."

On the contrary, many others, like Davis, argue that an idea must have novelty as well as some sense of "appropriateness, value or social worth" for it to be considered creative. Perkins states that a "creative person by definition ..., more or less regularly produces outcomes in one or more fields that appear both original and appropriate." Hennessey and Amabile argue that "to be considered creative, a product or response must be novel ... as well as appropriate." In earlier papers, we defined (Chakrabarti 1998; Chakrabarti and Khadilkar 2003) creative outcomes as "new as well as interesting". However, these multitude of definitions of creativity leaves one wonder whether it is possible to arrive at an encompassing definition of creativity in a systematic way, rather than assuming allegiance to any particular definition. After all, for a research community that works on creativity to build on each other's work, such a 'common' definition and related 'operationalisable' measures are essential to have. This led us to undertake a more rigorous approach to understand what is meant by creativity, with the eventual aim of arriving at a 'common definition' and related measures. In the rest of this section, a summary of this work is given. For details, see (Sarkar 2007; Sarkar and Chakrabarti 2007a, 2008a).

2.3.2 *A 'Common' Definition*

Development of a 'common' definition requires that the research community is able to agree on what is meant by a 'common' definition, and is able to operationalise this meaning into a definition. In our view, a ‘common’ definition must embody what is common across existing definitions. Therefore, we collected a comprehensive list of creativity definitions (see Ideaslab 2007 for the list) from literature. Two possible meanings for ‘common’ definition were proposed. The first utilizes in the ‘common’ definition those concepts that are most frequently used across the current definitions, since the definition should reflect the views of the *majority* of the researchers in the domain. The second, alternative meaning is based on the possibility that the above – majority based – definition may not capture the rich, underlying *relationships* among the concepts used in the various definitions, and may not provide a ‘common’ definition that represents all the definitions. In this analysis, the features of the definitions are analyzed to identify the *relationships* between them and integrate the feature into hierarchies of *related features*. The overarching, high level features from the hierarchies that represent all the other features within the hierarchies are then integrated into a ‘common’ definition of creativity, thereby representing also those definitions that use these lower level features. The list of creativity definitions was analyzed using each of these approaches. The first approach is called *Majority* Analysis, and the second *Relationship* Analysis. The results from these two analyses were compared with each other in order to develop the proposed ‘common’ definition.

Using Majority Analysis, the ‘common’ definition of creativity was found to be the following: ‘*Creativity occurs through a process by which an agent uses its ability to generate ideas, products or solutions that are novel and valuable*’ (Definition 1).

Based on Relationship analysis, the proposed ‘common’ definition is: ‘*Creativity is an ability or process using which an agent generates ‘something’ that is ‘novel’ and ‘valuable’*. This ‘something’ can be a ‘problem’, ‘solution’, ‘work’, ‘product’, ‘statement’, ‘discovery’, ‘thought’, ‘idea’ or ‘judgment’ (i.e., evaluation). For design, ‘something’ is taken as ‘problem’, ‘solution’, ‘product’, ‘idea’ or ‘evaluation’ (Definition 2).

The difference between the two definitions lies in the meaning of ‘something’. In Majority Analysis, ‘something’ means ideas, solutions and products, while in Relationship Analysis it has a wider variety of meanings – in particular problems and evaluations. Since problem finding and evaluation are essential subtasks in any creative activity, we argue that, ‘generation of ideas, solutions or products’ already encompasses these subtasks and their outcomes. The definition of creativity from Relational Analysis is hence simplified as: ‘Creativity in design occurs through a process by which an agent uses its ability to generate ideas, products or solutions that are novel and valuable’. This is the same as Definition 1, from Majority Analysis, and is taken here as the general definition of creativity.

Note that the feature social ‘value’ in this definition can be more specific in the context of engineering, where it becomes utility value – or ‘usefulness’. Thus in the context of engineering design, the definition of creativity can be further specified as: ‘*Creativity is a process by which an agent uses its ability to generate ideas, solutions or products that are novel and useful*’ (Definition 4). We call this the definition for design creativity.

Together these two definitions (Definition 1 or 3 for creativity in general and Definition 4 for design creativity) provide an inclusive framework for creativity.

They also provide a justification for the various measures proposed by earlier authors for creativity, and how directly these relate to creativity, allowing most existing definitions to be subsumed and represented by the above two definitions, and at a greater degree of directness.

2.3.3 'Common' Measures

In order to operationalise the above common definition of engineering design creativity, we must be able to assess its two core components: 'novelty' and 'usefulness'. For this, the following information is needed:

- Candidate measures for novelty, usefulness and creativity, where creativity is a function of novelty and usefulness. Many measures are available in literature, but how they are related to one another is missing.
- Some way of independently assessing novelty, usefulness and creativity against which potential measures can be evaluated. This is also missing.

An ideal means for independent evaluation would be to use the collective knowledge of experienced designers from the domains to which the newly generated products belong. In a design house, creativity of new solutions is typically judged by experienced designers to decide whether to develop these solutions into products. In patent offices novelty and usefulness of products are judged by experts from related areas. We argue, like Amabile (1996) who suggest the use of experts to identify what is 'creative', that for any measure of novelty, usefulness or creativity to be valid, the results should reflect the collective notion of experienced designers. We use this as the benchmark for evaluating the potential measures.

2.3.3.1 Novelty

'New' is something that is recently created. 'Novel' is one that is socially new. 'Novelty' encompasses both new and original (Cambridge 2007). We need a direct measure of novelty. Developing a measure involves developing both a scale and a process of measurement. For detection of novelty of a new product, its characteristics need to be compared with those of other products aimed at fulfilling similar need (the process). The difference in these characteristics would indicate how novel the new product is. If no other product satisfied a similar need before, the new product should be considered to have the highest novelty (the maximum value in the scale). If the product is not different from previously known products, its novelty should be zero (the minimum value in the scale). Thus, to assess novelty of a product, one should know the time line of similar inventions and the characteristics of similar products. It must also be possible to determine the degree of novelty (resolutions in the scale). Existing literature on measuring novelty (Redelinghuys 2000;

Saunders 2002; Shah and Vargas-Hernandez 2003; Chakrabarti and Khadilkar 2003; Lopez-Mesa and Vidal 2006) deals mainly with identification of whether a product is novel or not. Patent offices often employ experts to determine 'novelty', 'usefulness' and other aspects of patent proposals. But, little work exists in identifying the degree of novelty in products.

Two major elements are missing in these current methods: history of ideas is not taken into account, and the scale without mention of its maximum possible value is potentially incomplete. Also, while all methods use some Function-Behaviour-Structure model (FBS) (Chandrasekaran 1994; Qian and Gero 1996; Goel 1997) of the artefact for determining novelty, we argue that FBS models alone are not sufficiently detailed to enable adequate assessment of degree of novelty. We use FBS as well as SAPPhIRE model (Chakrabarti et al. 2005) to achieve this.

2.3.3.2 Proposed Novelty Measure and Validation

To determine novelty of a new product with respect to available products, comparison of these products is carried out by comparing their features. FBS models are suitable for this. Since novel products are new and original, if the functions of a new product are different from those of available products, it must have the highest degree of novelty (we call this very highly novel). If the structure of the product is the same as that of any other product, it cannot be considered novel. If it is neither, the product has some novelty. To determine the degree of its novelty, a detailed model of causality – the SAPPhIRE (standing for State-Action-Part-Phenomenon-Input-oRgan-Effect) model (Chakrabarti et al. 2005) is used, see Fig. 2.2. It has seven constructs. *Action* is an abstract description or high level interpretation of a change of state, a changed state, or creation of an input. *State* refers to the attributes and their values that define the properties of a given system at a given instant of time during its operation. *Physical phenomena* are a set of potential changes associated with a given physical effect for a given organ and inputs. *Physical effects* are the laws of nature governing change. *Organs* are the structural contexts needed for activation of a physical effect. *Inputs* are energy, information or material requirements for a physical effect to be activated. *Parts* are the physical components and interfaces constituting the system and its environment of interaction. Parts are needed for creating organs, which with inputs activate physical effects, which are needed for creating physical phenomena and state change. State changes are interpreted as actions or inputs, and create or activate parts. *Activation*, *creation* and *interpretation* are the relationships between the constructs.

For detection of degree of novelty in products that are not 'very highly novel', state change and input constitute the next level of novelty ('high' novelty), physical phenomena and physical effect the following level ('medium' novelty), and organs and parts constitute the lowest level ('low' novelty) at which a product can be different from other products. Based on these, a method for novelty detection has been developed which employs FBS model first, and SAPPhIRE model thereafter to assess the degree of novelty of a product. The method was evaluated in terms of the

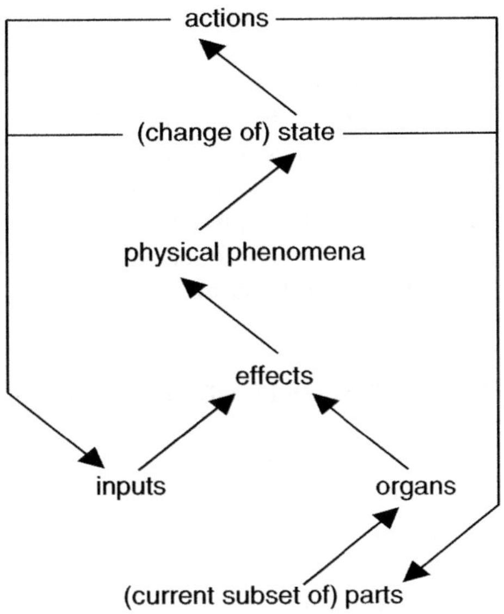

Fig. 2.2 The SAPPhIRE model of causality

degree to which its output (the degree of novelty of products as determined using the method) matched (it did with an average Spearman's rank correlation of 0.93, see (Sarkar and Chakrabarti 2007a, 2008a)) with the output of experienced designers (the degree of novelty of the same sets of products as perceived by these designers).

2.3.3.3 Usefulness

We argue that it is the actual use of a product that validates its usefulness. Thus, the usefulness of a product should be measured, whenever possible, by its actual use, and when this is not possible, value of its potential use should be used. Products could then be compared by assessing their degree of their usefulness – the second criterion for judging creativity.

Patent offices employ experts to determine both novelty and usefulness to ascertain validity and patentability of applications, but do not use explicit measures for these. Usability is the closest measure for usefulness available in literature. It denotes the ease with which people can employ a particular tool or other artefact in order to achieve a particular goal (Nielsen 1994; Green and Jordan 2002; Graham 2003). Various norms exist for its assessment such as ISO and SIS. The methods for evaluation of designs or products (Roozenburg and Eekels 1995) are the closest available for assessing usefulness of products. However, none of these are direct measures for usefulness. We therefore propose a new method for measuring usefulness, based on the following arguments:

- Usefulness should be measured in terms the degree of usage a product has in the society.

- The scale is provided by a combination of several elements to assess the degree of usage: the importance of the product's function, its number of users, and how long they use it for or benefit from it. Together these would give a measure of how extensive the usefulness of the product is to the society.
- Though usefulness should be ideally judged by taking feedback from a statistically representative collection of users of the product, this is best approximated by the collective opinion of experienced designers who are trained to understand users well. Hence, collective opinion of experienced designers is used as benchmark for corroborating results.

2.3.3.4 Proposed Usefulness Measure and Validation

As to how important the use of a product is depends on its impact on its users' lives. Some products are indispensable, and should have a higher value for their usefulness. We identified five *levels of importance* of products: extremely important (e.g., life saving drugs), very highly important (e.g., compulsory daily activities), highly important (e.g., shelter), medium importance (e.g., machines for daily needs), low importance (e.g., Entertainment systems). All other parameters being equal, the products that are used by a larger number of people – the *rate of its popularity* – should be considered more useful to the society. Finally, products that are used more *frequently* and have longer *duration* of benefit should be considered more useful to the society. Assuming that their 'level of importance' and 'rate of popularity' are the same, the 'rate of their usage' increases their usefulness. Together these parameters provide a measure for usefulness:

$$U = L\,(F\,D)\,R \tag{2.1}$$

U stands for usefulness; L stands for level of importance; F for frequency of usage (how often people use it); D for duration of benefit per usage; R for rate of popularity of use (how many people use it). Ranking of various product-sets using the proposed measure has been found to have consistently high correlation (Spearman's rank correlation average of 0.86) with that using experienced designers' collective opinion, showing that the proposed method captures well the designers' intuitive notion of usefulness.

2.3.3.5 Proposed Creativity Measure and Validation

With 'novelty' and 'usefulness' of products as the *only* two direct influences on creativity (as in the common definition), a measure for creativity must express creativity as a function of these two. For a list of creativity measures, see Sarkar (2007). We propose the relationship to be a product of the two influences, since absence of either should lead to perception of no creativity in the outcome (C: creativity, N: novelty, and U: usefulness):

$$C = N \times U \tag{2.2}$$

To assess the degree of creativity of products in a given set, the steps are to:

1. Assess novelty of each product (using method in Section 2.3.3.2) on the qualitative scale: 'Very high novelty', 'High novelty', 'Medium novelty' or 'Low novelty'.
2. Convert these qualitative values into quantitative values: Very high novelty = 4 points, High novelty = 3 points, Medium novelty = 2 points and Low novelty = 1 point.
3. Assess the usefulness of each product using the method described in Section 2.3.3.4.
4. Convert these qualitative values into relative grading using the following scale: if there are five products that are ranked 1–5, give them 1/5, 2/5, 3/5, 4/5, 5/5 points respectively.
5. Calculate creativity of each product as a product of its degree of novelty and usefulness using Eq. 2.2.

Once again, creativity ranks obtained using experienced designers' collective opinions are compared with those using the proposed method. The results (Spearman's rank correlation average of 0.85) show consistently high rank correlation between these, corroborating the proposed method. Further analysis shows no correlation between usefulness and novelty, indicating their independence, thus further corroborating our results.

2.4 Major Influences on Creativity

A wide variety of factors are cited in literature as influencing creativity. Rhodes (1961, see Davis 1999) group over fifty definitions of creativity into four Ps: product, people, process and press, the product factors being influenced by the factors of the other three Ps. Various factors related to each of these Ps have been identified, e.g., strong motivation (people), incubation (process), or relaxed work environment (press).

Several authors describe creativity as a special kind of information or knowledge processing (e.g., McKim 1980), and argue that information or knowledge must be a prime ingredient for creativity. For instance, Gluck (1985) sees as essential the "... possession of tremendous amount of raw information ...", as does Read (1955; cited in Davis 1999, p. 44) who describes this as "scraps of knowledge" in describing creative people who "... juggle scraps of knowledge until they fall into new and more useful patterns." Note the act of juggling in this description – one proposed to be described here with the generic name of 'flexibility'. Also note the mention of "new" and "valuable" patterns – the two aspects of creative outcomes. Various authors have also stressed the importance of flexibly processing knowledge. McKim (1980) speaks of flexibility in "levels, vehicles and operations", and argues that seamless use of and transfer between these are important in creative thinking. Gluck (1985) describes as essential in creativity the "ability to combine, order or connect" information. In C-K Theory (Hatchuel et al. 2004), the authors distinguish two different kinds of creative ideas: those that are dominated by knowledge requirement, and those that operate within existing knowledge but require imagination for conception. We interpret the first category as primarily requiring new *knowledge* while the second primarily requiring flexibility in thinking. In TRIZ (Terninko et al. 1998), children are described as capable of connecting all ideas to each other, while common adults connect only few – that too in the existing ways. In the light of flexibility and knowledge requirement for creativity, the act of children can be interpreted as having great flexibility in thinking with little knowledge of the constraints among them, while adults having far less flexibility with far more knowledge. In the four stage model of the creative process (see Wallas 1926, cited in Davis 1999, p. 44), the first stage – *preparation* is interpreted here as accumulation of knowledge – the "scraps" as described by Read. The second stage – *incubation* is one of transferring the task to the subconscious – a sign of flexibility (McKim 1980). The third stage – *illumination* – is when these two come together to create the idea. Note that 'mental blocks' (Adams 1993) are blocks against using knowledge in a flexible way.

We propose *knowledge, flexibility* and *motivation* (i.e., encompassing all motivational factors and indicators such as challenge, energy-level, single-mindedness and aggression) as the three factors essential for creative thinking, see Fig. 2.3. McKim has spoken of similar factors "for productive thinking" – information, flexibility and challenge. Perkins (1988, cited in Davis 1999, p. 45) describes creative people as "motivated", have creative "patterns of deployment" or "personal manoeuvres of thought" (both of which are interpreted here as flexibility) and have "raw ability in a discipline" (seen here as knowledge). Echoing somewhat similar notions, Torrance (1979; cited in Fox and Fox 2000, p. 15) argued that "prime factors" on creativity of people are their "abilities, skills and motivation".

The specific ideas proposed here in this regard are the following:

- Motivation, knowledge and flexibility are the broad, major factors influencing creativity.

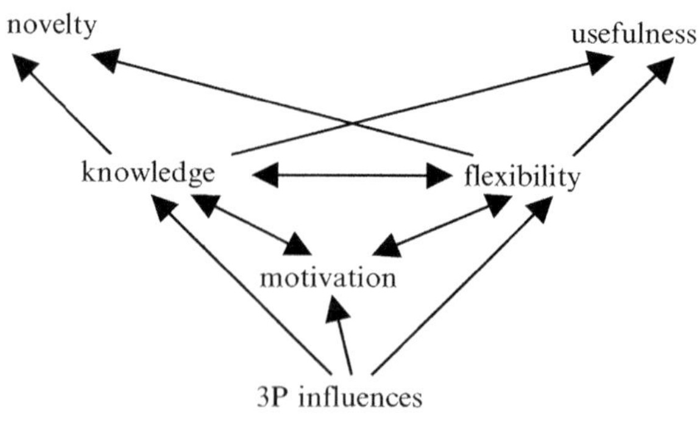

Fig. 2.3 Influences on creativity

- The factors are not independent of each other. Knowledge influences motivation, motivation may lead to acquiring of new knowledge; flexibility leads to development of new knowledge that may lead to more flexibility; motivation to utilise knowledge in a flexible way may lead to further flexibility leading to more motivation, etc. This idea of interdependence of factors is inspired by Lewis' model (1981) of influences on intelligence in children. Lewis sees intelligence as the ability to see and solve problems – at a broad level not very different from designing. In his model, motivation, self-image and attitude are all linked to a child's problem-handling skills, and vice-versa.
- Among these factors knowledge and flexibility directly affect the outcome of a creative problem solving process, while motivation assumes an indirect influence. Other factors, from the categories of people, process and press influence one of these factors, which in turn influence the novelty, purposefulness and resource-effectiveness of the product.

2.5 Effect of Search and Exploration on Creativity

The work reported in this section is primarily based on the work reported in Sarkar and Chakrabarti (2007b) and Srinivasan and Chakrabarti (2008). Design is seen by many as a phenomenon of exploration (de Silva Garza and Maher 1996) and search. Some see exploration or search as similar to idea finding since both are divergent processes, where many ideas need to be considered before selecting the best (Roozenburg and Eekels 1995). Exploration is an important part of design creativity (Gero and Kazakov 1996) since creative design is generation and exploration of new search spaces (Stal and George 1996). Exploration also improves a designer's problem understanding (de Silva Garza and Maher 1996). Thus, exploration and search are important influences on design creativity. We take 'Exploration' as a process by the space within which to search is determined. 'Search' is a process of finding improved designs in a given design space (Sarkar and Chakrabarti 2007b).

In order to understand how search and exploration takes place in design and how these influence creativity, we carried out and studied a series of design experiments, where various groups of designers solved various design problems in a laboratory setting (Sarkar and Chakrabarti 2007b). All utterances in the design experiments were video taped, transcribed and categorized into three phases: (i) problem understanding, (ii) idea generation and (iii) evaluation and selection. Each utterance was then classified into search or exploration. It was found that the number of utterances of the type 'exploration' was negligible (less than 1% in all protocols – see discussion later). Next, it was found that searches in the idea generation phase can be further classified into other sub-categories. We call these 'unknown search', 'global search', 'local search' and 'detail search'. These kinds of search are present not only in solution generation, but also in problem understanding and solution evaluation stages.

An 'unknown' or 'global search' represent search in a *global design space* that is less specific than that of the local and detailed spaces. A 'design space' consists of a set of ideas (which can be problems, solutions or evaluating criteria) that are similar to each other in some respect. Depending upon the relationship and level of abstraction used, a design space can overlap with, or subsume other design spaces. In a *global design space*, a solution is different from other solutions in terms of 'state change', 'input' 'Physical Effect', 'Physical Phenomenon', 'Organ' and 'Parts'. A local search space is within a global search space; the ideas are different in the 'Physical Effect', 'Physical Phenomenon' 'Organ' and 'Parts' used. Solutions in *detail search* are different only in the 'Organs' and 'Parts'.

The designers typically search first unknown or global, then local and ultimately detailed spaces, leading to the solutions becoming increasingly more detailed. While global, local and detailed search spaces are previously visited by designers while solving other similar problems, unknown spaces are not. Search at the higher level in the hierarchy (such as 'unknown' and 'global') include searches that are in the lower level of hierarchy (e.g., 'local' or 'detailed'). Each of these searches is either on finding potential problems, solutions or evaluation criteria. For instance, a 'global problem search' might contain many 'local problem searches' and 'detailed problem searches', leading to identification of several potential problems at various levels of detail. There can be many problem-, solution- and evaluation-searches possible for a given problem; their existence is established when the designers are found to identify a problem or generate a solution or evaluation criterion that belongs to a specific design space.

Analyses showed the following results:

- Each design process observed used all 12 variants of search –four types (unknown, global, local and detail) at three phases of problem solving (problem understanding, solution generation and solution evaluation).
- Higher levels of search had a strong influence on the lower levels of search, i.e., the number of unknown search influenced the number of global, local and detailed search, the number of global search influenced the number of both local and detailed search, and the number of local search influenced the number of detailed search.
- Assessment of creativity of the design groups using the creativity measures described in Section 2.3 of their final design outcome and its correlation with various searches carried out showed that the number of search at each phase variously correlate with creativity, with the highest correlation being with solution search (average correlation 0.85) and least with evaluation search (0.62), with problem search in between (0.67), see Sarkar and Chakrabarti (2007b) for more detail.

There is complementary evidence from the recent work of Srinivasan and Chakrabarti (2008) where video recordings of the design processes of various groups of designers were analysed using a model of designing called GEMS of SAPPhIRE as RS (Generation, Evaluation, Modification and Selection/Rejection of State-Action-Parts-Phenomena-Input-oRgan-Effects as Requirements-or-Solutions), and the number of ideas generated at the SAPPhIRE levels were correlated with the novelty of the solution spaces generated. The results showed that the number of ideas generated at higher levels of abstraction had a greater positive influence on the creativity of the solution space. A major conclusion from the above results is that carrying out search in greater depth at all design problem solving phases and at all levels of abstraction, in particular at the higher abstraction levels, substantially improve the creative quality of the solutions developed.

2.6 How Well Do Designers Currently Explore Design Spaces?

Sarkar and Chakrabarti (2007b) found that across all design cases studied, the design process follows a general pattern (with a correlation of 0.99), irrespective of whether it is in problem understanding, solution generation or solution evaluation. Observations indicate that unknown design search are generally less in number, followed by a larger number of global search but comparatively fewer local search, followed by a huge number of detailed search. This is contrary to the expectation that the number of searches should increase consistently as it gets more detailed.

There are two potential explanations for this anomaly. One is that the trend is due to progressive divergence and convergence in the number of searches performed, a commonly known means used by designers in order to control the amount of information handled as they go from less to more detailed phases of design (Liu et al. 2003). However, this does not explain why convergence has to be at the local search level only. The second possible explanation is that, once the required design functionality is established, designers work primarily at the device level. This is evidenced by the observation that designers frequently bring to fore past designs and try to mould them to do the current task. The sparse use of local searches likely to be due to *a lack of knowledge of phenomena and physical principles*, and due to the belief that working at the device level is likely to be more pragmatic in terms of creating realistic designs faster.

Further evidence of similar kind is found by Srinivasan and Chakrabarti (2008). In their design studies, they counted the number of ideas generated at each level of the SAPPhIRE, and found that for each team of designers, while the number of ideas at the action, state change and input levels steadily higher, the effects and organ level ideas are particularly low, before the number of ideas at the part level become high again. This is consistent with the findings of Sarkar and Chakrabarti (2007b). We argue that this indicates a serious deficiency in the uniformity and consistency with which search is carried out currently. This leaves substantial scope for bridging this gap and improving the creative quality of the solution space.

2.7 Supporting Creativity

Based on the findings discussed in the earlier sections of this chapter, we conclude the following. Since the two major direct influences on creativity are knowledge and flexibility, since creativity is enhanced if search is carried out uniformly in at all levels of abstraction of design at all phases of design problem solving, and since this is currently not followed, support is necessary to ensure that designers' knowledge and flexibility are enhanced to carry out search uniformly.

One way to support this is to provide stimuli as an inspiration for creativity. Inspiration is useful for exploration of new solution spaces (Murakami and Nakajima 1997). Literature provides evidence that presence of a stimulus can lead to generation of more ideas being during problem solving (Kletke et al. 2001), that stimulus-rich creativity techniques improve creativity (MacCrimmon and Wagner 1994), and that when stimulated with association lists people demonstrate more creative productivity than when not stimulated (Watson 1989). Both natural and artificial systems are seen as rich sources of inspiration for ideation. The importance of learning from nature is long recognized, and some attempts made (Vogel 1998; French 1998) to learn from nature for developing products. However, while artificial systems are routinely used for inspiration (e.g., in compendia, case based reasoning systems, etc.), natural systems are rarely used systematically for this purpose. Analogy is often proposed as a central approach to inspiring generation of novel ideas, and many methods and tools to support this are proposed (Gordon 1961; Bhatta et al. 1994; Qian and Gero 1996). Our objective is to support systematic use of biological and artificial systems as stimuli for aiding generation of creative designs.

2.7.1 Idea-Inspire

We developed a computational tool called 'Idea-Inspire' (Chakrabarti et al. 2005) for supporting designers to generate novel solutions for design problems by providing natural or artificial systems as analogically relevant stimuli to be used for inspiring ideation. It has two databases: a database of natural systems (e.g., insects, plants, etc.) exhibiting diverse movements, and a database of artificial systems (e.g., vacuum cleaners, clutches, etc.). The behaviour of these natural and artificial systems are described using the SAPPhIRE model of causality. Designers, with a problem to solve, are supported to describe their design problem using the constructs of SAPPhIRE – the software would search the databases for the entries in the databases that could analogically relevant for solving the problem.

The database of natural systems has over 300 entries from plants, animals and natural phenomena describing their motion behaviour. The motions analysed are varied in both the media in which they occur (air, water, land, desert, etc.), and the way in which they occur (leaping, jumping, walking, crawling, etc.). The description contains the function, behaviour and structure as well as a SAPPhIRE model

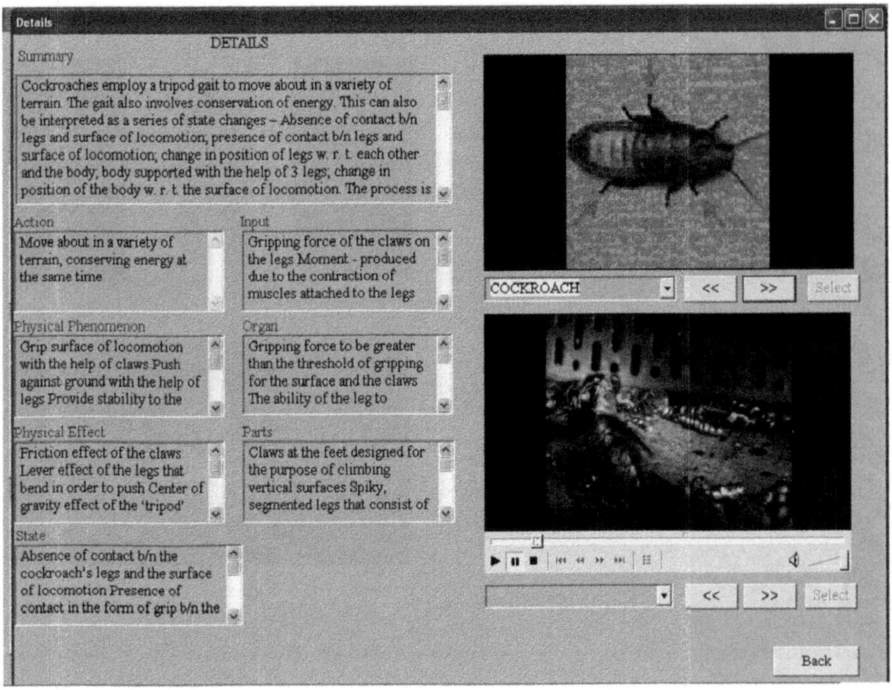

Fig. 2.4 A natural system as entry in idea-inspire

based description of each system, as well as their pictorial and video data about their behaviour. An example of an entry is given in Fig. 2.4. The database of artificial systems has over 400 entries and contains similar information as in the database of natural systems plus animation of the system behaviour for many mechanisms for which video is not available. An example of an entry is given in Fig. 2.5. Associated reasoning procedures are developed to help browse and search for entries that are analogically relevant for solving a design problem.

2.7.2 *Using Idea-Inspire*

Idea-Inspire can be used in two different modes:

- When a designer has a well-defined problem to solve. In this case, the designer defines the problem using the SAPPhIRE constructs, and uses reasoning procedures of the software for automated search for solutions. In some cases, the designer may try out different versions of the problem using the constructs until satisfactory solutions are obtained.
- When a designer does not have a well defined problem to solve. In this case, the designer can browse the databases and view related entries, then get interested in

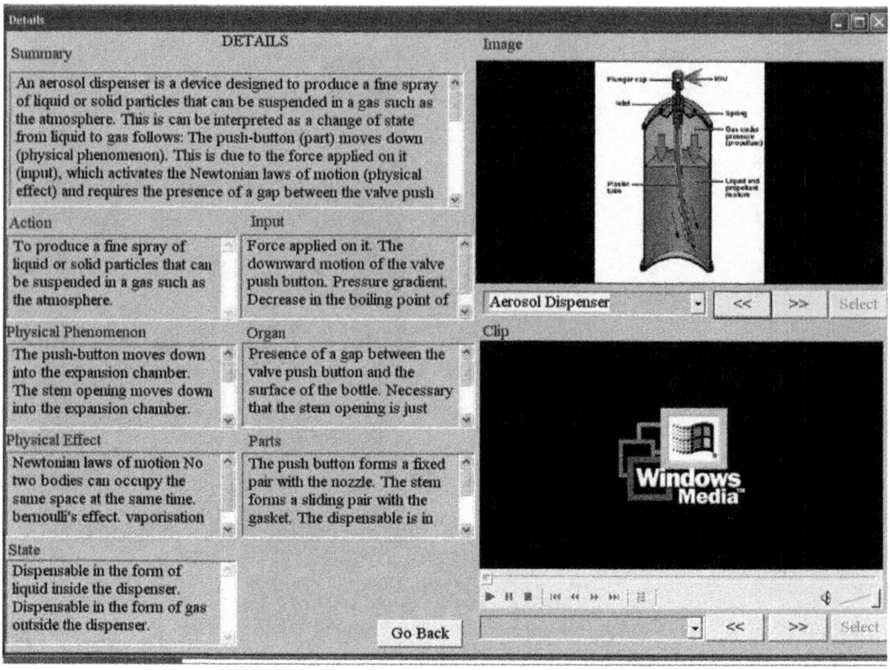

Fig. 2.5 An artificial system as entry in idea-inspire

some of these, and work on these in greater depth to solve the problem. Browsing may also help in understanding a problem better, as a designer gets exposed to a wider variety of related yet concrete solutions.

In each of these cases, the output from the software is a list of entries that match the constructs provided to the search engine as the problem. A design problem is often described using the *action* required to be fulfilled, and the search task is to retrieve all entries that have synonymous actions. An action is described using a verb-noun-adjective/adverb triplet. For instance, consider this design problem: *Design an aid that can enable people with disabled upper limbs to eat food.* A designer could describe the action required in many alternative ways- using different sets of verbs, nouns and adjectives. Some examples are

1. V = feed, N = solid, A = slow (put solid food in the mouth)
2. V = consume, N = solid, A = slow
3. V = take, N = solid, A = (nil)

   Alternatively, the problem can be decomposed into sub-problems and solutions for each can be searched. Some such combinations are
4. (V = hold, N = solid, A = quick) + (V = move, N = solid, A = slow) + (V = push N = solid, A = slow) (here, the device is intended to take the food in a container, move close to the mouth, and transfer to the mouth)
5. (V = get, N = solid, A = slow) + (V = swallow, N = solid, A = slow).

The entries retrieved for the first and last problem alternatives are

Case 1: List of some of the entries found by the software: Aardvark, barracuda, Duck, Clam Defence, and Pitcher plant, etc.

Case 5: List of some of the entries found by the software:

Sub-case 1: (V = hold, N = solid, A = quick) – Reciprocating lever gripper, Rack and pinion gripper, Hydraulic gripper.

Sub-case 2: (V = move, N = solid, A = slowly) – Camel moving, Millipede, Baboon, Crab walking, Transport mechanisms, Belt drives, etc.

Sub-case 3: (V = push N = solid, A = slowly): no entry was found for this.

Depending upon a designer's interest, various details of an entry could be explored. The problem may have to be redefined several times, using different VNA words, until satisfactory solutions are found.

2.7.3 Evaluation

In order to evaluate the effectiveness of the software in inspiring creative solutions in designers, a series of interventional case studies were undertaken. In the first study (Chakrabarti et al. 2005), two designers solved individually two design problems of their choice from a pool of problems given, first without using the Idea-Inspire software and then by using the software. The idea was to see if the intervention made any substantial difference in the number and kind of solutions generated. The number of ideas that were created as a result of being triggered by some entries (that are used as an inspiration) from the software, as well as ideas that were identical to some entries (that can be used directly as a solution), were both noted down. It was found that by using the software, each designer was able to create additional solutions for each problem after they completed created solutions without using the software. On average, the number of ideas created with the software constituted about 35% of all ideas created, that too with a database having a limited number of entries (about 200).

In a subsequent evaluation, three designers solved individually one engineering design problem, first without and then with the aid of Idea-Inspire (Sarkar et al. 2008). In each problem solving session, they first generated the ideas, and then selected those which they felt were worth developing further, and developed them into solutions. It was found that, despite some individual variations, the designers on average created 165% more ideas with the aid than without, that too after they felt (at the end of their session without aid) that they exhausted the ideas they could think of for solving the design problem. The size of the database used was about 500. It is also to be noted that about 40% of all ideas were chosen by the designers as worth developing further, indicating that the ideas generated with inspiration from Idea-Inspire were not only large in number but also similar in quality to those generated by designers on their own.

The software has been delivered to the Indian Space Research Organisation (ISRO) for aiding ideation of concepts to solve space-related design problems, and has been customised for use by IMI-Cornelius for aiding their designers both in individual and group ideation sessions. However, the work is far from over. While understanding has substantially increased over the years, it is still not well-understood what elements in the descriptions provided in the entries inspired ideation. Specific studies need to be undertaken to understand how best to support and encourage ideation for all types of search at all levels of SAPPhIRE.

While the above are the aspects of the content of the material for stimuli, another complementary but important aspect is the form in which a stimulus is provided. We provide the information about a stimulus (an entry in the database) in textual, graphical, animation/video and audio forms. The textual material is structured using FBS and SAPPhIRE models. How does the same information in different forms affect ideation differently?

To answer this question, a subset of Idea-Inspire entries were taken, and each entry was represented as several, alternative entries each of which was described using a different representation (textual only, graphical only, etc.). Different representations of these selected entries were placed in separate slides in a presentation form. The sequence of representations for each stimulus was randomized. Later, each slide was shown to six volunteer design engineers who solved the same, given problem, using each slide as a stimulus. The engineers were asked to capture each solution they generated in white sheets, along with the number of the slide that triggered the solution. The experiments were conducted in laboratory setting. Even though there was no strict time constraint, each slide was shown in the main experiment for about 5 min. The results (the stimuli and corresponding solutions created) provided the data required to answer the research question. It was found that in general non-verbal representations (graphical, followed by video) have a greater influence on the creative quality of the solutions generated than verbal means (Sarkar and Chakrabarti 2008b). However, each has its complementary, positive aspects. A video is inherently better in showing the dynamic aspects of the content while the verbal mode is better in terms of explaining the behaviour of a stimulus, and an image could be used for explaining its internal and the external structure. We argue that the 'non-verbal' representations of a stimulus should be shown first, followed by its 'verbal' representations, in order to draw attention first, and make all its aspects available for exploration.

2.8 Summary and Conclusions

The chapter strings together work from several research projects and PhD theses at IdeasLab in the last 7 years to provide an overview of the understanding and support developed in the area of design creativity. A 'common' definition for creativity has been developed after analysing a comprehensive list of definitions from literature. The two direct parameters for discerning creativity are found to be *novelty* and social *value*. The definition is further specified for engineering design creativity where value is taken as the utility value or *usefulness*. Both novelty and usefulness are operationalised into measures for creativity, and a relationship between these measures and creativity is established where creativity is assessed as a *product* of the value of these two measures. All of these are validated by comparing the outcomes of ranking product sets using the measures developed with that using the collective opinion of experienced designers. Using extensive analysis of work from literature, the three major factors influencing creativity were formulated to be *knowledge*, *flexibility* and *motivation*.

In order to understand how search and exploration affects creativity, design processes were recorded and analysed using protocol analysis methods. Four different types of search were identified, and all were found to be present in each of the three main phases of design problem solving. Searching design spaces well at all these levels were found to have a strong impact on creativity of the solution space. Ideas were found to be generated at all levels of abstraction modelled by the SAPPhIRE constructs, and search at all these levels, in particular at the higher levels was found to have a strong impact on creativity.

It was found that designers were consistently deficient in searching the effect and organ levels of abstraction – i.e., generate ideas in terms of the physical effects and properties of the products envisaged. This distinct gap, we felt must be bridged in order to enable a more uniform search of design spaces. Various forms of support are under development at IdeasLab; one of them – Idea-Inspire – has been described in some detail in this chapter.

For a given problem, Idea-Inspire searches its database of natural and artificial systems as entries to find relevant entries that can be used as stimulus for inspiring solutions to the problem. In the design cases studied, it consistently helped designers in ideation. The influence of both the form and content on ideation was studied. The work shows substantial potential, even though much is still to be researched.

References

Adams, J.L. Conceptual Blockbusting: A Guide to Better Ideas, 3rd Ed., Addison-Wesley, MA, 1993.

Amabile, T.M. Creativity in Context. Westview Press, Boulder, Colorado, 1996.

Bhatta, S., Goel A., and Prabhakar S. "Innovation in Analogical Design: A Model-Based Approach," Proc. AI in Design, Kluwer Academic Publishers, Dordrecht, The Netherlands, pp. 57–74, 1994.

Blessing, L., and Chakrabarti, A. DRM: A Design Research Methodology, in International Conference on The Science of Design – The Scientific Challenge for the 21st Century, INSA, Lyon, France, 15-16 March, 2002.

Blessing, L.T.M., and Chakrabarti, A. DRM: A design research methodology, Springer-Verlag, 2009.

Blessing L.T.M., Chakrabarti, A., and Wallace, K.M. Some issues in engineering design research, Proceedings EDC/SERC Design Methods Workshop, The Open University UK, 1992.

Blessing, L.T.M., Chakrabarti, A., and Wallace, K.M. A design research methodology, in 10th International Conference on Engineering Design (ICED'95), Prague, Czech Republic, 1, 50–55, 1995.

Blessing L.T.M., Chakrabarti A., and Wallace, K.M. An overview of descriptive studies in relation to a general design research methodology, in Designers – the Key to Successful Product Development, E. Frankenberger, et al. (eds.) Springer Verlag, pp 42–56, 1998.

Cambridge: 2007, Cambridge Advanced Learner's Dictionary, Cambridge University Press 2007. URL: http://dictionary.cambridge.org.

Chakrabarti A. A Measure of the Newness of a Solution Set Generated Using a Database of Building Blocks, and the Database Parameters which Control its Newness, CUED/C-EDC/TR64-April 1998, University of Cambridge, 1998.

Chakrabarti, A., and Khadilkar, P. A Measure for Assessing Product Novelty, Proc of the International Conf. on Engg. Design (ICED03), Stockholm, 2003.

Chakrabarti, A., Sarkar, P., Leelavathamma, and Nataraju B.S. A Functional Representation for Biomimetic and Artificial Inspiration of New Ideas, Artificial Intelligence in Engineering Design and Manufacturing, 19, 113–132, 2005.

Chandrasekaran, B. Functional representation and causal processes, in M Yovits (Ed.), Advances in Computers, pp. 73–143, 1994.

Davis, G.A. Creativity is Forever, 4th Ed., Kendall Hunt, Dubuque Iowa, 1999.

De Silva Garza, A.G., and Maher, M.L. Design by interactive exploration using memory based techniques. Knowledge based Systems, 9:151–161, 1996.

Fox, J.M., and Fox, R.L. Exploring the Nature of Creativity, Kendall Hunt, Dubuque Iowa, 2000.

French, M.J. "Invention and Evolution: Design in Nature and Engineering", Cambridge University Press, Cambridge, UK, 1998.

Gero, J.S., and Kazakov, V. An exploration based evolutionary model of a generative design process. Microcomputers in Civil Engineering, 11:209–216, 1996.

Gluck, F.W. "Big Bang" Management: Creative Innovation, 49–59, The McKinsley Quarterly, Spring, 1985.

Goel, K.A. Design, analogy and creativity, AI in Design, 62–70, 1997.

Gordon, W.J.J. "Synthesis: The Development of Creative Capacity", Harper & Row, New York, 1961.

Graham, I. A Pattern Language for Web Usability. Pearson Education, 2003.

Green, S.W., and Jordan, W.P. Pleasure with Products: Beyond Usability (Eds.), CRC, 2002.

Hatchuel, A., Le Mason, P., and Weil, B. C-K Theory in Practice: Lessons from Industrial Applications, International Design Conference, Dubrovnik, May 18-21, 2004.

Ideaslab (2007) URL: http://www.cpdm.iisc.ernet.in/ideaslab/research/creativity.html

Kletke, M.G., Mackay, J.M. Barr, S.H., and Jones, B. Creativity in the organization: the role of individual creative problem solving and computer support, Int. J. Human-Computer Studies, 55, pp. 217–237, 2001.

Lewis, D. You Can Teach Your Child Intelligence, Souvenir Press, London, 1981.

Liu, Y. C., Bligh, T., and Chakrabarti, A. Towards an'Ideal' Approach for Concept Generation, Design Studies, 24, (2):341–355, 2003.

Lopez-Mesa, B., and Vidal, R. Novelty metrics in engineering design experiments in International Design Conference – Design 2006, Dubronik, Croatia, May 15-18, 2006.

MacCrimmon, K.R., and Wagner, C. Stimulating Ideas through Creative Softwares, Management Science, 40, 1514–1532, 1994.

McKim, R. Thinking Visually, Dale Seymour Publications, 1980.

Murakami, T., and Nakajima, N. "Mechanism Concept retrieval Using Configuration Space", Research in engineering design, 9:9–11, 1997.

Nielsen, J. Usability Engineering, Morgan Kaufmann, 1994.

Qian, L., and Gero, J.S. Function-Behavior-Structure paths and their role in analogy- based design. Artificial Intelligence for Engineering Design, Analysis and Manufacturing, 10(4), pp. 289–312, 1996.

Redelinghuys, C. 2000, Proposed criteria for the detection of invention in engineering design, Journal of Engineering Design, 11(3), pp. 265–282.

Roozenburg, N.F.M., and Eekels, J. Product Design: Fundamentals and methods, John Wiley and sons, 1995.

Sarkar, P. Development of a Support For Effective Concept Exploration to Enhance Creativity of Engineering Designers. Ph.D. thesis, Indian Institute of Science, Bangalore, India, 2007.

Sarkar, P., and Chakrabarti, A. Development of a Method for Assessing Design Creativity, International Conference on Engineering Design (ICED07), Paris, France, August 2007 (2007a).
Sarkar, P., and Chakrabarti, A. Understanding Search in Design, International Conference on Engineering Design (ICED07), Paris, France, August 2007 (2007b).
Sarkar, P., and Chakrabarti, A. Studying Engineering Design Creativity – Developing a Common Definition and Associated Measures, in Studying Design Creativity, J Gero (Ed.), Springer Verlag, 2008a (In Press)
Sarkar, P., and Chakrabarti, A. The effect of representations of triggers on design outcomes, Special Issue on Multi-Modal Design, A.K. Goel, R. Davis and J. Gero (Eds.), Artificial Intelligence in Engineering Design, Analysis and Manufacturing (AI EDAM), 22(2), 101–116, March 2008 (2008b).
Sarkar, P., Phaneendra, S., and Chakrabarti, A. Developing Engineering Products using Inspiration From Nature, ASME Journal of Computers in Information Science and Engineering (JCISE), 8(3), 2008 (2008).
Saunders, R. Curious design agents and artificial creativity a synthetic approach to the study of creative behaviour. Ph.D. thesis, Department of Architectural and Design Science, University of Sydney, 2002.
Shah, J., and Vargas-Hernandez, N. Metrics for measuring ideation effectiveness, Design Studies 24, 111–143, 2003.
Srinivasan, V., and Chakrabarti, A. Design for Novelty – A Framework?, International Design Conference (Design2008), Dubrovnik, Croatia, 2008.
Stal, D.M., and George, T.M. Skeleton based techniques for the creative synthesis of structural shapes, In Artificial Intelligence in Design (AID'96), John S. Gero (Eds.), Kluwer Academic Publishers, Dordrecht, The Netherlands, 3761–780, 1996.
Terninko, J., Zusman, A., and Zlotin, B. Systematic Innovation: An Introduction to TRIZ, CRC Press, 1998.
Torrance E.P. Unique needs of the creative child and adult, in AH Passow (ed), The gifted and talented: Their education and development, 78th NSSE Yearbook, 352–371, Chicago: National Society for the Study of Education, 1979.
Vogel, S. Cat's Paws and Catapults: Mechanical Worlds of Nature and People, WW Norton, NY, 1998.
Watson, D.L. Enhancing Creative Productivity with the Fisher Associated Lists, the Journal of Creative Behavior, 23, pp. 51–58, 1989.

Chapter 3
User Experience-Driven Wireless Services Development

Jee Y. Park and Giridhar D. Mandyam

Abstract Cellular systems have undergone many technological advances since the beginning of the decade, and this development coupled with the arrival of afforable feature-rich smartphones into the mobile market has resulted in a renewed interest in development of wireless data services and applications beyond the limited categories that most end users experience today. However, many wireless services have not attained desired uptake despite the fact that devices and access technologies are not the impediments to user acceptance that they once were. A critical reason for this is the failure to define a target user experience when developing a new service.

In this paper, a product development cycle is discussed that seeks to address user experience during initial stages of development. Based on an initial definition of target demographics along with associated sample user personas, case studies can be developed that can result in concrete product requirements. As an example, this development cycle is applied to the area of mobile social communities. With the target user experience driving product requirements and technology development, wireless services can be developed that can result in improved market penetration.

Keywords Mobile · Wireless · Data services · Cellular · Personas · User-centered design · Storyboard

3.1 Introduction

Data services that target wireless devices have been available for nearly two decades. These services have only increased in adoption over this period as improvements in cellular access technology as well as devices have come to market. Like all software services, the need for user-centric design is critical in determining adoption. Unfortunately, much development in the area of wireless services tends to focus on simply

J.Y. Park (✉) and G.D. Mandyam
Qualcomm Incorporated, 5775 Morehouse Drive, San Diego, CA 92121, USA
e-mail: jee@qualcomm.com

N.R.S. Raghavan and J.A. Cafeo (eds.), *Product Research: The Art and Science Behind Successful Product Launches*, DOI 10.1007/978-90-481-2860-0_3,

extending existing desktop services into the mobile space instead of carefully considering the challenges and benefits of mobility in crafting a compelling service. As a result, the average user has difficulty using a service that was not originally designed for a handheld, mobile device and therefore is quickly alienated from using the handset for anything more than voice or messaging.

An example of the pitfalls of ignoring the end user in mobile services is shown from the experiences gathered in the late 1990s with the introduction of the Wireless Applications Protocol (WAP) as a means of transcoding web content for mobile devices. WAP started with what seemed to be a reasonable premise that the end user would want to access the same web content on a handheld device as a desktop. However, arbitrary approaches to limiting or altering content and interaction with content led to poor usability of most WAP sites.

A user study by the Nielsen Norman group released in 2000 (Hafner 2000) provided a negative review of existing WAP services. Although some of the problems that users encountered were related to data speeds of the time, it was also concluded that much of the problem arose from the misapplication of web design principles to mobile content and context. In addition, none of the users were unhappy with the actual devices being used (implying that the users could distinguish when a poor experience was the result of limitations of their devices as opposed to shortcomings of the service). Ironically, the report is still available to this day because in part (as the authors put it) "the user attitudes to mobile services and their frustrations with difficult services that are documented in the report continue to be relevant for current designs" (Nielsen Norman Group December 2000).

The authors' statement suggests that defining target users and understanding their needs may be the way to create a compelling, useful, and usable service. However, defining the target end user segments is not trivial. It becomes particularly difficult if the segments are diverse, as designing a single product that meets all of the collective needs could result in a product that actually pleases no-one. As a result, Alan Cooper in the 1990s proposed the use of target *personas* for software design (Cooper 1999). A persona is an archetype representing the needs, behaviors, and goals of a particular group of users. Personas are based on real people but not an actual real person.

The description of the persona is precise, with much more associated information than a market segment. The persona makes for an attractive design target because of this precision, as there is little room for debate as to what the persona requires from the product. As a consequence, the persona cannot be as *accurate* as typically required for interaction design since it is a representative personality. The persona must be *goal-oriented*, with a related set of tasks that it must accomplish to meet those goals. Once the personas are created, they provide a common aim for all members of the service design and development team: to solve a problem for these personas. The personas are helpful during discussions about features in that the feature under discussion can be measured for relevance with respect to the personas. As a result, personas provide an objective way to make decisions about requirements.

In this work, a persona-based approach to wireless service design is discussed. The product design cycle is provided, starting with the initial ethnographic research and ending with usability testing with actual users prior to a commercial launch.

The application of this process to the area of mobile social community networks is provided as an example. Finally, a discussion of the benefits and limitations of persona-based mobile service development will serve to better provide an understanding of where and when this kind of methodology can be applied.

3.2 Persona-Based Mobile Service Design

Ethnographic research, often in the form of *in situ* observation, is the ideal starting point in the persona-driven product design cycle. Observing the end user in his/her own environment is key to creating relevant personas for a few reasons. First, users cannot always identify or articulate their own behavior. A user may not recognize that he/she has an unmet need since he/she has created a workaround for the problem. For instance, a colleague observed a user solving the problem of finding a place for an iPod® in her automobile by putting it in her cooling vent. The slots of the vent conveniently fit the iPod®, and the vent is positioned just under eye-level so that the user can access the iPod® while driving, without danger (see Fig. 3.1). The user creatively solved her stowage problem by making use of an available tool in her car, the air vent.

Second, the context in which a user employs a tool is important to know while designing the tool. For example, when designing a tool such as a web-based email application, the designer needs to understand that the user may have interruptions while performing a task such as composing an email. The user may consider these

Fig. 3.1 User-generated workaround for a problem

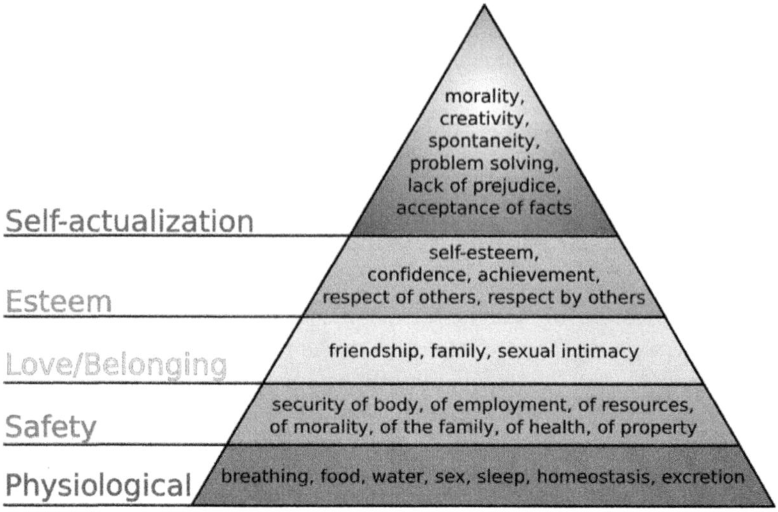

Fig. 3.2 Maslow's hierarchy of needs (Finkelstein 2006)

interruptions routine and may not mention them during an interview in a controlled environment such as a lab. Knowing that the user has frequent interruptions that last several minutes at a time suggests that the email application should automatically save drafts of the email before the user sends it. By observing the user in context, the designer becomes aware of the interruptions, which may or may not have been articulated by the user.

Third, understanding user goals, values, and needs is aided by the artifacts with which the user surrounds himself/herself. Service designers and developers should also consider these goals, values, and needs as they relate to Maslow's hierarchy of needs (Maslow 1943) (see Fig. 3.2), since the service should ultimately address one or more of those. It requires a fairly self-aware user to identify or admit that belonging to a group (i.e., social need) is an important value. However, by observing the user in context, clues such as the way he/she dresses and how it compares to others, photographs of family and friends at the workplace, and items indicating affiliations such as college paraphernalia, provide information about the user's values that would have otherwise gone unnoted.

3.3 Stakeholders

In any wireless service deployment, there are multiple stakeholders. When considered as individuals, the number of stakeholders can be very large. One way to prioritize the stakeholders is by reviewing market intelligence data. Market segmentation is helpful in identifying personas to target, as it supplies aggregate information on usually a large sample set of users. Once key segments are identified, representative personas are created for each. This can significantly simplify the problem of designing to these stakeholders.

Stakeholders are persons who use and therefore interact with the product. The stakeholders of most types of traditional software service design can be grouped into two categories: the end users (or consumers) of the service, and the service administrators (i.e., service deployers). As a result, two sets of personas can be created that would describe these two stakeholder categories. These personas will be denoted as *end users* and *trade customers*, respectively.

Mobile services differ in that many times the services are not directly deployed to the end user, and as a result an intermediary stakeholder is necessary to ensure the service extends over cellular access. In particular, cellular operator stakeholders often enter the picture. Therefore, a third set of personas is necessary to capture these stakeholders. These personas are designated as *operator users*. In spite of the fact that there are three distinct sets of personas, the same process of in situ observation is employed for creating all groups.

3.4 End Users

Arguably, the best consumer products and services have emphasized the end users during design and development. This is the same for mobile consumer products and services. A number of companies, including Qualcomm Incorporated (Qualcomm), employ a user-centered design and development approach, which integrates the end user throughout the process. Understanding the end user is the first step in this process, and this can best be done through observation and in context interviews. The aforementioned are the ideal methods in which personas are created, but even at Qualcomm, resources and schedules sometimes challenge our ability to conduct observational research. We are at times faced with making compromises in our persona creation process. One of the ways we have addressed these challenges is by interviewing people who are familiar with a particular group of users. For example, one of our recent projects required insight into a particular Indian market, yet we were unable to do any first hand observational research to create personas for this market. In this case, the best alternative was to interview colleagues who were familiar with the group of interest. Those interviews yielded enough information to create a believable end consumer persona (see Figs. 3.3–3.5).

3.5 Trade Customers

Many of the services provided by Qualcomm are often used by trade customers. Such an example is a service called the BREW® Delivery System (BDS) that enables trade customers (e.g., BREW® developers) to submit and manage applications that are offered to the operator users (e.g., wireless carriers). Since in this case, the

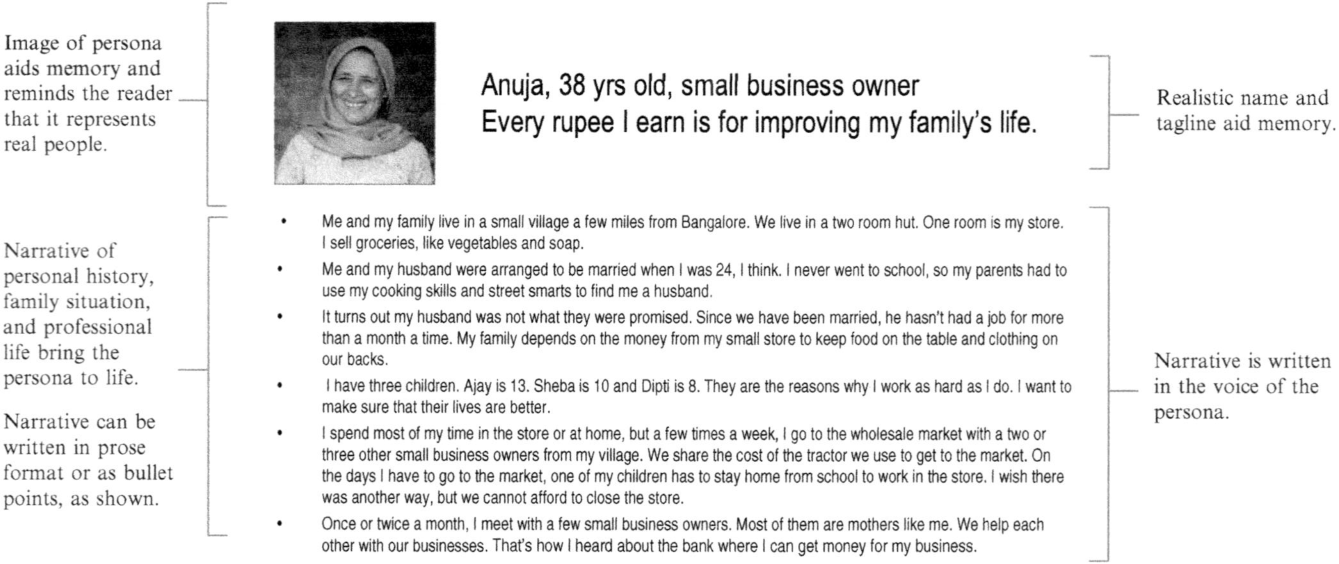

Fig. 3.3 Sample end user persona for India – Personal narrative; Photo courtesy of istockphoto.com

Anuja, 38 yrs old, small business owner
Every rupee I earn is for improving my family's life.

A "day in the life" provides a snapshot of the persona's daily rituals and schedule.

On most days I wake up around 6 to help my children get ready for school. After they're gone, I get ready to open my store, which is connected to my home.

My store is small, but it is the only steady income for my family. I work there all day, usually without a break until my children come home.

Around 4pm, my children come home from school. I take a quick break to go to the bathroom and then make sure they do their homework. I know education is the key to a better life for them.

In the early evening, when it starts to cool down, I have Ajay, my oldest child, work at the store and watch his siblings while I make dinner.

We eat dinner with my husband. I keep the store open while we eat and watch for customers.

Sometimes, Ajay watches the store at night for a little while while I go to the self-help group for small business owners like me.

Fig. 3.4 Sample end user persona for India – A Day in the Life; Photo courtesy of istockphoto.com

BREW® developer interacts with the BDS, it is critical to understand their behaviors, workflow, and goals. Through secondary research on trade customer persona creation, it was found that the relevant information for this group was different from that of the end users. For example, it may not be relevant to know socio-economic information since it is unlikely to affect business related objectives. Hence, the information contained in non-consumer personas varies slightly from that of the consumer persona. An example of a trade customer persona is shown below in Figs. 3.6 and 3.7.

Depending on the service, trade customer may or may not directly interact with the service. As a Qualcomm example, a BREW® developer may interact directly with the BDS to submit and manage an application or a carrier may do these tasks on behalf of the developer. Hence, the trade customer persona may or may not be considered during the design and development of a service.

3.6 Operator Users

Another set of primary users of the BDS is the operator users. The operator has a variety of users who interact with the system and therefore require a persona for each high priority user. One of the key users of the BDS is the operator catalogue manager; as a result, a persona was developed for this person (see Figs. 3.8 and 3.9).

In the case of the BDS, the operator user personas are particularly important to understand since the BDS must interact with existing operator infrastructure. Hence, the BDS must be designed with consideration for the operator user personas' business goals and objectives in order to be successful.

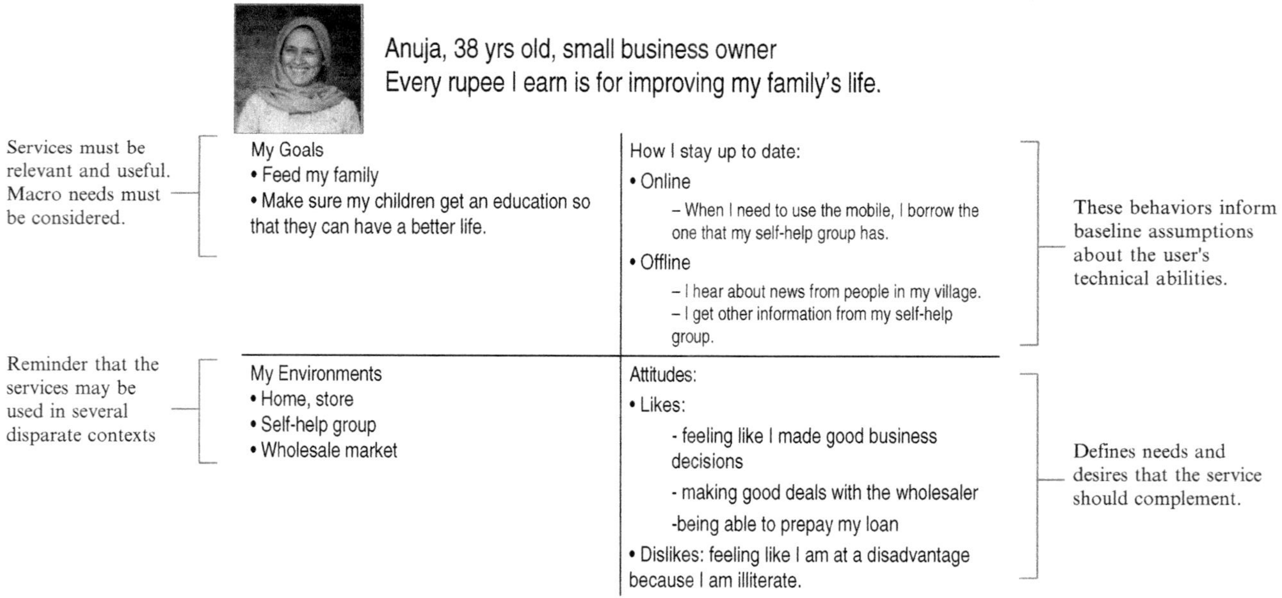

Fig. 3.5 Sample end user persona – Other information; Photo courtesy of istockphoto.com

Persona Description

Ajay, Software Developer, Age 28

Ajay is a software developer for a media company. He is concerned with delivering quality software and on time. Sometimes he has to perform other duties outside of development but his passion is software. He enjoys software development because it requires him to analyze problems from varied perspectives.

Narrative summary provides a short, memorable story of the persona.

Potential Organizations: Brand, Developer.

Education: College educated with an post-grad degree.

Technical skills/level: Expert

Experience:
- Knowledge of various coding languages : C, C++, CSS, HTML, etc.
- Knows BREW® (if app developer).
- Knows the products that are already deployed. Knows ins and outs of current workflow.

Attitude:
- Self-teaching.
- Intolerant of hype/marketing resistant.
- Ambitious-works long hours/weekends

Knowing attitudes helps guide the service's tone and content.

Business Objectives (roles):
- Understand system architecture.
- Focused on product APIs/SDK.
- Coding and bug fixing
- Provide technical support to non-engineering folks to help with business decisions
- May have a supervisor role (leads)
- Create test signatures and class IDs (app developer)
- Collect app development tools via BDS.

Fig. 3.6 Trade customer persona example – Introduction; Photo courtesy of istockphoto.com

Ajay

Job Priorities (goals):
- Timely delivery of products and features.
- Produce efficient and bug free code.
- Create apps that work across multiple platforms (app developer).
- Minimize validation costs (app developer)
- Configure recommendations so that demographic information is integrated appropriately.
- Integrate with existing systems to ensure recommendations are configured correctly.

Service should align with persona's goals, as they are measures of success for the persona.

Likes:
- Quick access to Qualcomm® support to resolve issues.
- Easy to use Qualcomm® tools.
- Ability to easily interact with Qualcomm® interfaces

Dislikes/Problem Areas:
- Sometimes gets frustrated with the tight project schedules and disagrees with some business priorities.

Service should minimize dislikes and problem area

Insights & Product Considerations:
- Ajay likes private time to work on the code and think through the scenarios and doesn't want to be distracted. He enjoys the software challenges arise and sees them as personal battles.
- Ajay gains satisfaction knowing that his work will eventually be in the hands of many users.

Fig. 3.7 Trade customer persona example – Additional information; Photo courtesy of istockphoto.com

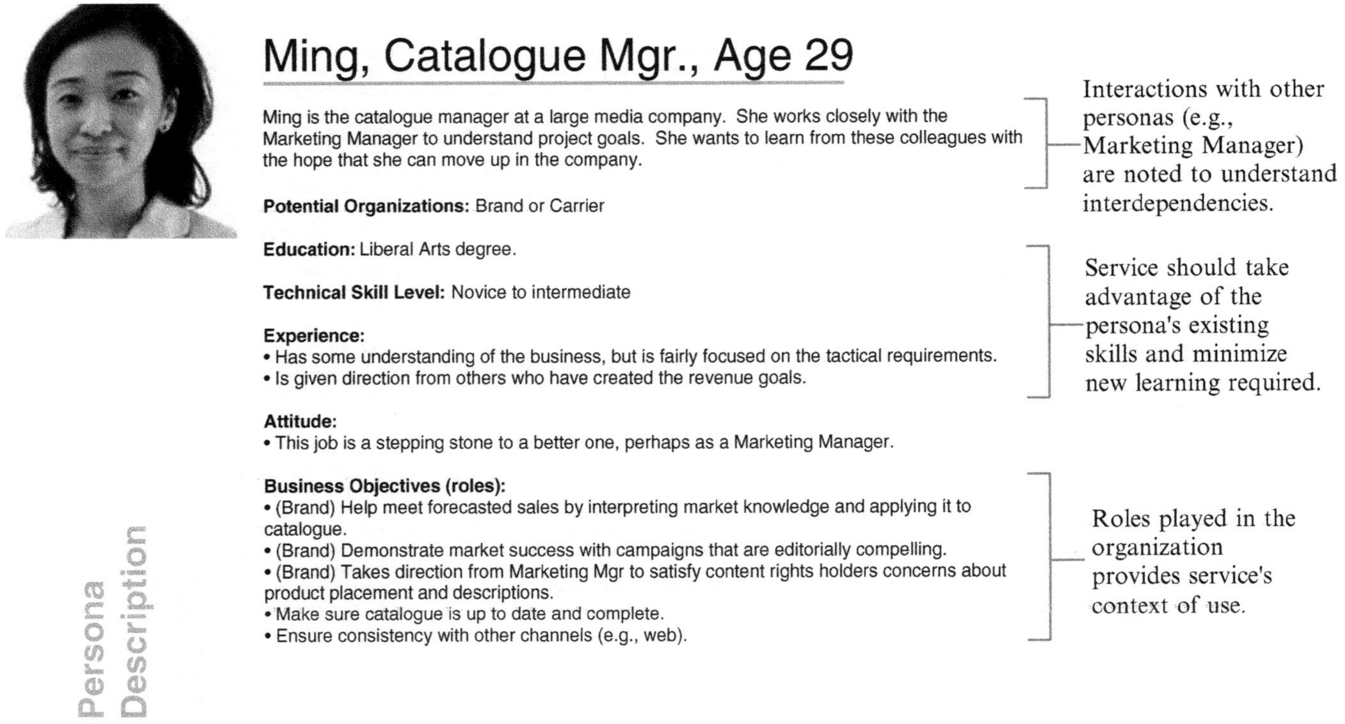

Fig. 3.8 Operator user persona example – Introduction; Photo courtesy of istockphoto.com

Persona Description

Ming

Job Priorities (goals):
- Make content go where content should be.
- Ingest and publish new content.
- Position item in catalogue.
- Publish catalogue.
- Manage and maintain price points and packaging (e.g., create consumer pricing)
- Maintain and share content inventory
- Retire old content
- Work with content adaptor (e.g., submit metadata to content).
- (Brand) Run transactions reports and analytics reports.
- (Brand) Maintain list of variants
- Enable, configure, place, track recommendations

Likes:
- Being able to log in remotely to the system.
- Being able to make changes to the catalogue easily .
- Being able to preview a catalogue and recommendations before publishing it.
- Easy to use and responsive system.
- (Brand) Being able to see what the content adaptor has inputted into the system and to communicate with the content adaptor through the system.

Service should address likes, since they are conscious opinions of the persona.

Dislikes/Problem Areas:
- Getting calls in the middle of the night.
- Too many touch points to go through workflow.
- Having to republish the entire catalogue after making a change.
- Difficult to automatically upgrade applications.
- Inefficient management of a catalogues or multiple catalogues.

Insights & Product Considerations:
- Ming is on call for catalogue issues, so she wants to be able to log in remotely. She would like to get alerted about issues on her mobile, as she is sometimes away from the computer.
- Ming wants to be able to preview catalogues before publishing.
- Ming wants to be able to run a script that flags any obvious problems in the catalogue before she publishes.

Service should align with the persona's insights and product considerations, as they may influence purchase and usage of the service.

Fig. 3.9 Operator user persona example – Additional information; Photo courtesy of istockphoto.com

3.7 Mobile Social Community Example

An area of rapid growth in mobile services is social communities. There are several reasons for this phenomenon. One is that handsets are being accepted by end users in greater numbers as a means of maintaining connectivity to their online communities. Another is that several existing online communities have opened their content and made them available to 3rd-party applications developers (some of whom target the mobile market). In addition, the end users for online communities have become more diverse in recent years, expanding beyond the original 20 and under crowd that originally popularized these communities.

In the case of mobile communities, target end user personas are chosen that capture a majority of online users. Three target end user personas are shown in the ensuing pages (see Figs. 3.10–3.14).

Once the target personas have been identified and studied by the design and development team, the team starts the process of creating an overall vision for the service. Part of the visioning process is to create user scenarios (user stories in prose format) and storyboards to illustrate common uses for the service. As shown in the samples below, storyboards come in different styles and lengths (see Figs. 3.15 and 3.16). The style of the storyboard often matches the attributes of the service depicted.

The purpose of the storyboard is to build consensus about one part of the vision for the service. The storyboard always incorporates a target persona and is kept at a high level. That is, details about button presses on the phone and similar granular information are avoided at this stage, as they distract the reader from understanding the overall message. Moreover, the images used in the storyboard are intentionally simple and are rendered in grey-scale instead of color in order to reinforce the point that the storyboard is a work in progress. It is intended to be discussed and altered until agreed upon by the team.

A storyboard must be easily consumable by the reader; this is the rationale for keeping it to one page. Moreover, the scenario's author should be mindful of the number of service elements that are included in one story, since the memorability and overall messaging may be compromised. Hence, a collection of storyboards is required to create the entire vision. Once the design and development team reaches a consensus on the collection of storyboards, the result is an illustrated vision of the product that contains, at a macro-level, the key features, attributes, and functions of the service.

From a design perspective, storyboards serve a few distinct purposes. For the visual designer, storyboards provide guidance about the important attributes to convey in the service. Using the storyboard in Fig. 3.15 as an example, attributes such as "fun", "dynamic", and "trust worthy" are relayed. For the interaction designer, storyboards are an illustrated narrative of the end user goals. Referencing the storyboard in Fig. 3.15, these user goals include sharing personal content (e.g., annotated digital photos) with close friends.

From the development perspective, storyboards help to define requirements and priorities for the product management and engineering teams. Each frame in the

Fig. 3.10 Mobile Community end consumer target persona 1 – Title page; Photos courtesy of istockphoto.com

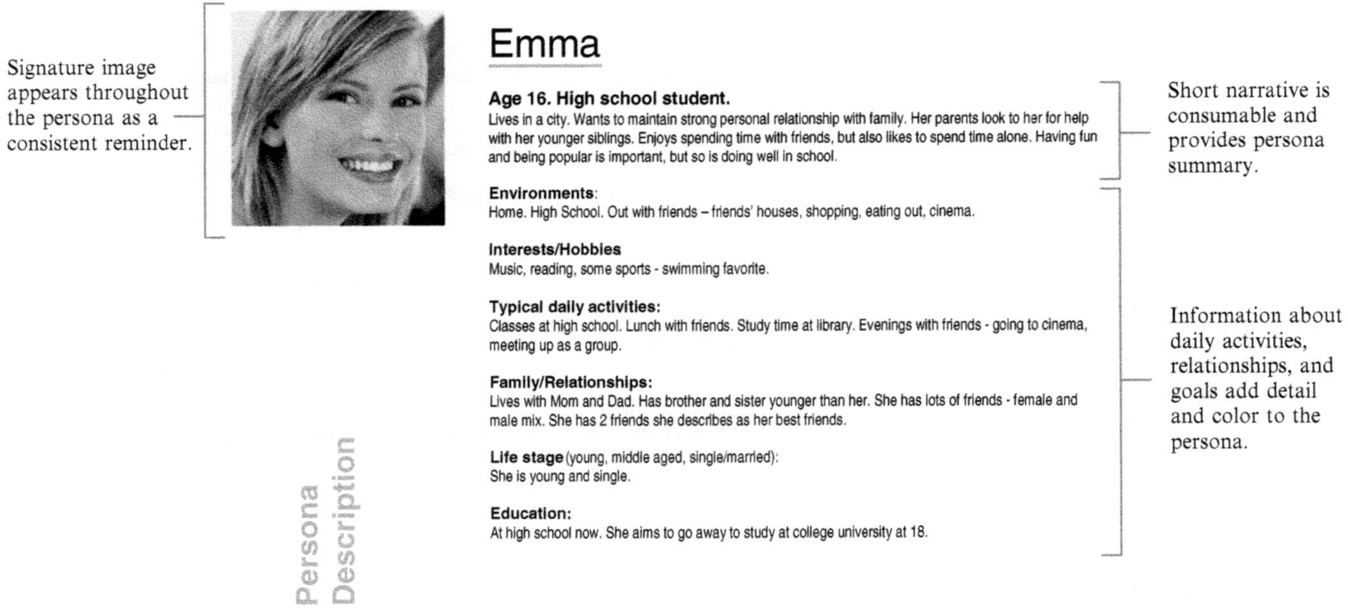

Fig. 3.11 Mobile Community end consumer target persona 1 – Introduction; Photo courtesy of istockphoto.com

Media/Products Usage:

Media consumption: (newspaper, online, magazines, TV, etc)
MySpace to keep up with friends and bands and as personal blog.

Why using mobile phone?:
Communication with friends and family. Make a statement through personalizing with ringtones, wallpapers, etc. Music. Taking and sharing photos.

3 core goals?:
Calls. Messaging – chat and texts. Listening to music.

Other devices:
PC at home and uses PC at high school. iPod® mini.

Acceptance of innovation:
Early mainstream.

Specific attitudes and goals relating to the phone and technology provide clues to when a persona will adopt the service and the features that are most important to the persona.

Fig. 3.12 Mobile Community end consumer target persona 1 – Service specific information; Photo courtesy of istockphoto.com

Fig. 3.13 Mobile Community end consumer target persona 2; Photos courtesy of istockphoto.com

Fig. 3.14 Mobile Community end consumer target persona 3; Photos courtesy of istockphoto.com

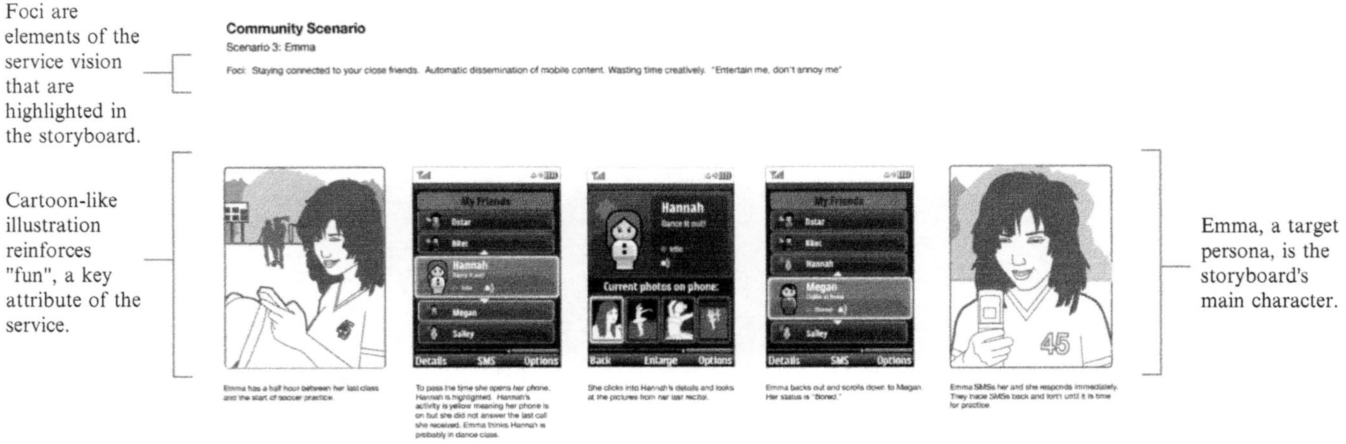

Fig. 3.15 Storyboard sample – Illustrative

Fig. 3.16 Storyboard sample – Photorealistic; Photos courtesy of punchstock.com and istockphoto.com

storyboard can be translated into requirements necessary to complete the end user goals. A table as shown in Table 3.1 may be used to facilitate the creation of requirements. The table contains features that are directly connected to the end user needs. For example, the end user goal of communicating with available close friends that is illustrated in the Fig. 3.15 storyboard is related to the section of the table called "Communication".

With the personas in mind, the design and development team assign priority values of 1, 2, or 3 (1 is the highest priority) for each feature. These values are used in two calculations. The first is an average for each persona (average of the values in a column). The persona average indicates the persona that is most likely to use the service with its current set of features. In Table 3.1, this persona is Emma, since her average of 1.167 is closest to 1 among the three personas. Product management may use this information to determine if the service is best designed with this persona in mind or if the feature set should be modified to address other personas. The second calculation, which is shown in the column labeled "Weight (Average)", aids in determining the priority of the feature of the service in that particular row.

Examining the end user goal of staying connected to close friends and family provides an example of how the trade customers and operator user personas come into play in the service design and development process. The end user goal of staying connected to close friends requires that trade customers and/or operator users provide an infrastructure to maintain a unique list of contacts per subscriber (i.e., end user). In order to create the infrastructure, several other requirements are necessary, such as mechanisms for approving, disapproving, and blocking contacts on the list. Furthermore, a relationship between this unique list of contacts and the existing address book, another unique list of contacts on the same mobile device, must be determined. That is, the service deployer (e.g., trade customer and/or operator user) must determine how these lists will coexist, be combined, or otherwise relate to each other. Keeping these two persona types in mind allow for the service design to accommodate any decision.

3.8 Caveats in the Use of Personas for Mobile Service Design

There exist certain pitfalls that one must strive to avoid in using personas for any kind of product or service design. It is critical that the personas are defined with as much precision as possible and that these personas are consulted in every stage of the design process. In other words, the persona is not meant to be discarded after product requirements are in place and development has started. Changes or tradeoffs will undoubtedly surface during the development process and personas provide guidance in making decisions around them. Although this is understood among interaction designers, often these principles are not implemented.

In "Best and Worst of Personas, 2007" Dorsey (2007), an evaluation of the use of personas in several web design companies was conducted in which the personas that each company used were subject to the following questions:

Table 3.1 Sample requirements 0; Photos courtesy of istockphoto.com

Component	Sub-Component	Feature	 Naomi – 24 *Marketing*	 Emma – 16 *Student*	 Rajesh – 32 *Finance'de*	Weight (Average)
...	...	...	...	...	...	...
Communication						
	Address Book					
		Add/del contact	Required	Required	Required	Required
		Modify contact	Required	Required	Required	Required
		Accept friend	1	1	1	1.0
		Accept new #s	1	1	1	1.0
		Accept presence	1	1	1	1.0
		Set # priority	2	2	1	1.7
		Set speed dial	2	1	2	1.7
	Presence					
		Manual status	2	1	2	1.7
Average			1.5	1.166666667	1.333333333	
Number of default features			2	2	2	

I. Does the persona sound like a real person?
II. Is the persona's narrative compelling?
III. Does the persona call out key attributes and high-level goals?
IV. Is the persona focused on enabling design decisions?
V. Is the persona usable?
VI. Does the persona have appropriate production values?

The results were not positive, with very few companies receiving an acceptable grade based on subjective scoring of the answers to the above questions. Some of the common characteristics of problematic personas were:

A. Persona description has contradictions.
B. Persona narrative was insufficient and sometimes omitted.
C. Persona has defined details specific to the product, and as a result were not believable.

Personas that came close to the original ideals set aside by Cooper (Cooper 1999) were found to be almost indistinguishable from real people, had engaging stories, and highlighted key details.

Another pitfall is the use of stale personas. Personas should be revisited on a regular basis to ensure their validity as well as their usability. With careful examination of the personas shown in this paper, some inconsistencies, such as the use of first and third person in the narratives, may be observed. These variants were included intentionally to illustrate the evolution of personas at Qualcomm. Related to this point is that the consumers of personas are a diverse group including engineers, product managers, and designers. Each group has a particular point of view, which must be considered when creating the content and determining the language and formatting used in the personas. Fortunately, personas help to provide a common point of view, that of the users.

When it comes to mobile services, an additional pitfall is incomplete inclusion of all categories of end users. From the beginning, personas that represent all categories of end users must be included. In particular, the trade customer or operator user, if not considered at the beginning of the design process, can result in the product being dramatically altered in later stages of development when entities such as mobile operators are consulted about product features. For instance, in (Ronkko et al. 2004) the authors describe a case in which a company that develops UI-centric platforms for the Symbian™ mobile operating system, UIQ Technology AB, ran into issues in application of personas in the design process because the enterprise user (denoted as the "client") was not considered from the very beginning of the product cycle. However, each new client had a large influence on the feature set of the product. As a result, the final product did not look like it had much correspondence to the personas originally identified for the product.

Furthermore, the use of personas works best when the personas have well-defined goals or are task-oriented. However, certain products are targeted towards audiences who have needs, but not necessarily goals. These kinds of situations require personas to be developed with needs and desired experiences, not only possible tasks. Moreover, this could require that personas be used in conjunction with actual live subjects. For instance, in (Antle 2006), the author describes a persona-driven web service developed for and targeted to children (CBC4Kids.ca, by the Canadian Broadcasting Corporation). Although development using child personas was possible, given the subjective nature of developing these personas from needs rather than tasks it became necessary to include live subjects to augment the development process.

Finally, it is vital to mention that the use of personas is just one tool in applying a user-centered approach to design and development of services. Ideally, other tools such as participatory design, a technique where live users work with designers and developers to define and design products and services iteratively, and usability testing, a process in which live users provide feedback on designs at various stages of completion in a controlled lab environment, should complement the use of personas. Regardless of the user-centered design tools used, they all focus on keeping the user's needs, desires, and goals at the core of the design, ensuring that the service designed is not only a product, but a solution to a legitimate user need.

3.9 Conclusions

Mobile service development is certainly possible using personas, but like all other persona-driven product development certain critical practices must be followed:

1. The personas must be realistic and sufficiently well-defined.
2. The personas must be goal-oriented.
3. The personas must be consulted in every phase of the development process, from beginning to end.

In addition, for mobile services it is essential to define personas from the start that capture the enterprise users (oftentimes the mobile operators). If this does not happen, then the design process may end up taking a sometimes significant detour to capture these users' associated requirements and could end up neglecting the critical requirements of end user personas.

Nevertheless, design and development with personas has the potential to provide a new form of wireless application that endears users in ways that have never been seen with mobile data services. As illustrated in the example of mobile social communities, the potential to interact with one's own community is expanded beyond the limited experiences of desktop world when taking into account not only the constant accessibility of the handset, but also the unique capabilities (e.g., location, presence) that are not always readily available in desktop environments. This kind of design methodology is also sufficiently versatile to be applied to wireless services beyond those targeted to handheld devices (e.g., telematics).

References

Antle, A.N. Child-Personas: Fact or Fiction? *ACM Conference on Designing Interactive Systems*. June 2006. pp. 22–30, University Park, PA, USA.

Cooper, A. The Inmates are Running the Asylum. *Why High-Tech Products Drive us Crazy and How to Restore the Sanity*. Indianapolis, IN: SAMS Publishing, 1999.

Dorsey, M. Best and Worst of Personas, 2007. *Forrester Research*. July 19, 2007.

Finkelstein, J. Maslow's hierarchy of needs. http://en.wikipedia.org/wiki/Image:Maslow%27s_hierarchy_of_needs.svg 2006.

Hafner, K.A. A Thumbs Down for Web Phones. *The New York Times*. November 30, 2000.

Nielsen Norman Group. *WAP Usability Report*. http://www.nngroup.com/reports/wap/ 2000.

Maslow, A.H. A Theory of Human Motivation. *Psychological Review*, 50, 370–396, 1943.

Ronkko, K. et al. Personas is not Applicable: Local Remedies Interpreted in a Wider Context. *ACM Participatory Design Conference*. 2004. pp. 112–120, Toronto, Ontario, Canada.

Chapter 4
Integrating Distributed Design Information in Decision-Based Design

Justin A. Rockwell, Sundar Krishnamurty, Ian R. Grosse, and Jack Wileden

Abstract The design information and knowledge necessary for decision-based design may come from across multiple organizations, companies, and countries. Integrating distributed engineering information that allows decision makers to easily access and understand it is essential for making well informed decisions. In this work we present a knowledge management approach for documenting and seamlessly integrating distributed design knowledge during the evaluation of design alternatives. Our approach takes advantage of emerging Semantic Web technologies to improve collaboration during engineering design through increased understanding of content and representation of knowledge in a manner that is easily shareable and distributable. This Web-based, collaborative approach for decision-based design is demonstrated through an example involving the re-design of a transfer plate for an aerospace circuit breaker. The geometrical forms, structural analyses, and optimization techniques for several proposed re-designed alternatives are captured and combined with enterprise information to create a knowledge repository. A modified conjoint-analysis based decision matrix is then utilized to evaluate, compare, and realize the most optimal design from amongst the alternatives. Salient features of this approach include an easy to access knowledge repository that improves consistency of information and facilitates collective comparison of the design alternatives based on specified evaluation criteria.

Keywords Decision support · Distributed design · Ontology · Semantic web

4.1 Introduction

The decision-based design research community addresses the notion of design as primarily a decision making process (Howard 1971, 1986). Engineering designers are continually making decisions based on system information and evaluating

---

J.A. Rockwell, S. Krishnamurty (✉), I.R. Grosse, and J. Wileden
Department of Mechanical and Industrial Engineering, University of Massachusetts, Amherst, Amherst, MA 01003, USA
e-mail: skrishna@ecs.umass.edu

N.R.S. Raghavan and J.A. Cafeo (eds.), *Product Research: The Art and Science Behind Successful Product Launches*, DOI 10.1007/978-90-481-2860-0_4,

the outcomes. This is consistent with *the normative decision analysis* (Keeney and Raiffa 1976; Decision-based Design Open Workshop 2008; Lewis et al. 2006). Recent works towards the development of decision based design methods can be found in the works by (Thurston 1991; Antonsson and Otto 1995; Ullman and D'Ambrosio 1995; Hazelrigg 1996; Hazelrigg 1998; Tang and Krishnamurty 2000; Chen et al. 2000; Jie and Krishnamurty 2001; Wassenaar and Chen 2001).

Decision methods provide a rational and systematic procedure for applying critical thinking to information, data, and experience in order to make a balanced decision under conditions of uncertainty (Baker et al. 2002). At the highest level, the two most essential aspects of decision making are: (1) available information as it relates to the specific decisions, and (2) a decision method to process the information and facilitate the identification of the most optimal alternative from an overall system perspective. It is, therefore, apparent that the management and communication of design information are critical to the decision making process.

Traditionally, engineers have captured and communicated design information through such means as text documents, data worksheets, slide show presentations, and CAD models. The widespread use of the Internet (*c*. 1998) has allowed distributed engineers to collaborate by allowing engineers to easily exchange these documents. However, the Internet does not inherently improve the management of information contained within the documents. That is, it is neither straight-forward nor automatic that the information within these documents can be easily retrieved and used. This can lead to the following shortcomings: (1) the structure/layout from one document to the next may vary, thus making it cumbersome to quickly locate information; (2) the quality of information documented may be incomplete or inconsistent – moreover, an engineer may have to sift through multiple documents to get the full understanding of a concept/process; and (3) the use of such documentation is limited as it requires engineers to manually search, decipher, and interpret information (Caldwell et al. 2000). For these reasons engineers in industry sectors are forced to spend 20% to 30% of their time retrieving and communicating information (Court et al. 1998). Often times these types of documents and the information stored within them become strictly historic records. As engineering design continues to become a distributed process, the management of design information in a manner that supports the decision making process becomes an immediate issue that must be addressed. Improved management of design information could greatly reduce the time spent finding information, provide decision makers with better access to distributed information, and improve the consistency of documented design information.

To improve the management and integration of design information there has been an increased focus on the methods for representing and storing design information (Szykman et al. 1998). Documented information needs to be stored in a manner that is easy to share and understand. Academia and industry have spent significant time and effort developing improved means for the documentation, sharing, and retrieval of the design information gained during the design process. The National Institute of Standards and Technology (NIST) has developed a design repository (Szykman et al. 1998, 2000) with these goals in mind. The NIST design repository uses an object-oriented representation language and accommodates standardized representations such as ISO 10303 (commonly known as STEP) and XML to provide improved exchange of information. Commercially available, there are also product data management (PDM) and requirement management (RM) systems that offer approaches for the management of design information. Although these systems clearly provide improved means for the integration of design information, problems often due to predefined vocabularies or proprietary internal representation limit the usefulness of such systems. These problems are well recognized and NIST has taken efforts to resolve some XML integration issues in (Morris et al. 2008). An alternative approach to managing engineering design knowledge, which is gaining momentum, is the use of ontologies (Moon et al. 2005; Grosse et al. 2005; Kanuri et al. 2005; Ahmed 2005; Kim et al. 2006; Kitamura 2006; Nanda et al. 2006; Witherell et al. 2007; Fernandes et al. 2007; Li et al. 2007; Lee and Suh 2007; Crowder et al. 2008; Chang and Terpenny 2008; Rockwell et al. 2008; Witherell et al. 2008; Fiorentini et al. 2008) for representing engineering design information.

Our research has been at the forefront of developing ontologies for engineering design (Grosse et al. 2005; Kanuri et al. 2005; Witherell et al. 2007; Fernandes et al. 2007; Rockwell et al. 2008; Witherell et al. 2008). Here an ontological framework for integrating design information and implementing preferred decision methods is presented. The framework benefits from the use of domain knowledge and Semantic Web technologies for representing design information. This enables (1) computers to "understand" information so that new information can be logically inferred, (2) provide information retrieval methods based on meaning and context; (3) information to be easily integrated through established Web protocols to improve accessibility of information; (4) information to be represented in an easily extendable manner so that it can evolve with the design process; and (5) information to be structured by domain concepts so that it is easily understood by decision makers (Rockwell et al. 2008). The premise being that by improving a decision maker's ability to access, retrieve, and understand design information the decision maker is able to make well-informed decisions.

Section 4.2 discusses relevant emerging technologies followed by the presentation of our ontological approach. Section 4.3 details the decision support ontology and how information about the decision making process is documented. Section 4.4 presents a case study to illustrate how the ontological framework facilitates and documents the decision making process. Section 4.5 provides a summary of the research.

4.2 Integrating Distributed Design Information

This research utilizes ontologies for representing design information. An ontology is an explicit specification of a conceptualization where "conceptualization" refers to the entities that may exist in a domain and the relationships among those entities (Gruber 1992; Farquhar et al. 1996). A common ontology language is the Web Ontology Language (OWL) (World Wide Web Consortium 2004c). OWL is a

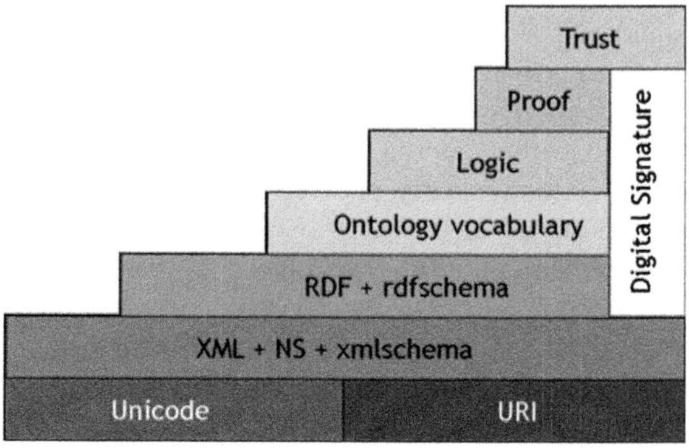

Fig. 4.1 Semantic web layer cake (used with permission from World Wide Web Consortium 2005)

developing information technology of the Semantic Web, which is an extension of the World Wide Web (herein referred to as Web) that aims at explicitly representing information in a computable manner so that information can be automatically processed and integrated. Tim Berners-Lee, the inventor of the Web, identifies three basic components of the Semantic Web (Berners-Lee et al. 2001): (1) the eXtensible Markup Language (XML); (2) Resource Description Framework (RDF); and (3) Ontologies. Figure 4.1 is an illustration of the Semantic Web Layer cake that shows how these technologies build upon each other. To understand how ontologies will help engineers integrate distributed information the capabilities of the information technologies being used must be understood. Here we briefly discuss some of the more important aspects of these technologies.

4.2.1 Emerging and Existing Information Technologies

4.2.1.1 Unicode and URI

The first (bottom) layer includes Uniform Resource Identifier (URI) and Unicode technologies. A URI is a string which identifies the location of a specific resource on the Web. URIs are unique and they enable the use of hyper-linked text. Unicode is a standard for representing and manipulating text expressed in most of the world's writing systems. These two technologies are the foundation for the Web.

4.2.1.2 XML

The Web enables users to easily exchange and understand information. XML was developed to represent data in a manner that is interpretable by a computer rather than a human. This is done by giving data an arbitrary structure through the use of tags. The tags define the information that exists between the tags. XML is composed of syntax and schema. The syntax defines a set of principles and rules that govern how a statement should be structured. The schema defines what the tags mean. XML is extensible as it allows users to define any custom schema, or vocabulary (World Wide Web Consortium 2006). To exchange data using XML schema all collaborators must agree upon and accept the vocabulary.

Agreement upon a predefined schema for representing data is considered a limitation of XML. Sharing information between those that agree upon the schema is facilitated but, integrating information beyond that group of users is not easily achievable. This is due to the fact that others outside of the group define their vocabulary differently. "Stated more succinctly, XML standardizes syntax; it was never designed to even capture, much less standardize, semantics" (Ray and Jones 2006). Furthermore, making changes to a schema to adapt to an evolving knowledge structure or changing needs can pose major difficulties (Morris et al. 2008). Although there are limitations to XML schema, XML is the current state of the art in sharing distributed data and its use has been widely adopted (Szykman et al. 1998). Further, XML syntax is a generic framework for storing any information whose structure can be represented as a tree and this syntax can be adopted by other languages that aim to integrate distributed information.

4.2.1.3 RDF

The Resource Description Framework (RDF) is the second basic component of the Semantic Web (World Wide Web Consortium 2004a). RDF takes advantage of XML-based syntax, but unlike prescriptive XML schema, RDF schema is descriptive. That is, instead of defining what data can be captured RDF further enriches the description of the data (World Wide Web Consortium 2004b). Relationships between information are expressed through sets of RDF triples, each consisting of a *subject*, a *predicate* and an *object*. These triples can be conceptually thought of as similar to the subject, verb and object of a sentence. The assertion of a RDF triple says that a particular relationship, as indicated by the predicate, holds between the things denoted by the subject and object of the triple (World Wide Web Consortium 2004a). This can be represented graphically as in Fig. 4.2(a). In Fig. 4.2(b), a specific example, which identifies that a "table" (the subject) "has material" (the predicate) "wood" (the object) is illustrated.

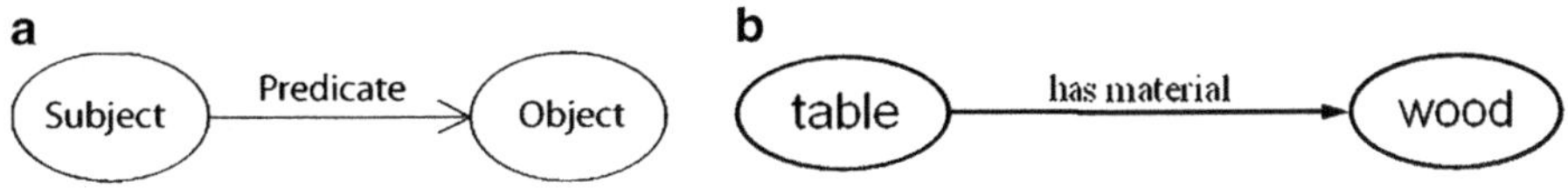

Fig. 4.2 RDF triple (**a**). Generic RDF triple. (**b**) Specific example of RDF triple

Thus, RDF makes assertions that particular things (subjects) have properties (predicate) with certain values (objects). Each part of the triple (subject, predicate, and object) is identified by a URI. URIs enable concepts to be identified by defining a URI for where that concept exists on the Web. In this manner, ambiguous meanings of content can be avoided through the specification of a unique URI.

4.2.1.4 Ontology

Layered on top of RDF is the third basic component of the Semantic Web, ontologies. Ontologies are domain specific knowledge structures that improve the users understanding of information and enable computers to logically reason upon and infer a useful understanding of captured information. Formally defined classes represent concepts and are the backbone of an ontology. Relationships are established for each concept by assigning properties to classes. Properties may have values (i.e., instances) that are data-type (e.g., integer or string) or object-type (i.e., another instance). Therefore, object-type properties are used to create relationships between concepts. Sub-classes are required to inherit properties from their respective super-class(es). OWL class axioms allow further description of classes by stating necessary and/or sufficient characteristics of a class. An ontological knowledge structure captures and stores information in the context of a specific domain, thereby improving a user's ability to understand and use the documented information. Class axioms explicitly represent domain specific knowledge in a computer interpretable way. Mechanisms (i.e., reasoners) may then be used to reveal implicit characteristics of a concept based on defined axioms; in so doing, allowing computers to "understand" the captured information. (World Wide Web Consortium 2004c).

For example, referring back to Fig. 4.2(b), suppose that the definition of the tag "table" is unclear. Using an ontology, it then becomes possible for a computer to deduce what "table" means (a piece of furniture or a set of data arranged in columns and rows). Since the "table" has a relationship "has material" and it has a value of "wood" then through appropriate class axioms it could be logically reasoned that "table" refers to a piece of furniture. This is possible despite the fact that upfront the meaning of the tag "table" was unknown. Inferring this type of distinguishing information is not possible in languages, such as XML, that do not have embedded domain logic.

Furthermore, logical axioms can be used to integrate ontologies that describe the same concepts but use different vocabularies. This is achieved by using axioms to formally specify a meta-information model. Through a meta-information model, or mapping ontology, it is possible to explicitly define relationships such as *equivalent classes* and *equivalent properties*. A mapping ontology can be created separate from the ontologies that it maps together and in the same OWL representation. This allows a mapping ontology to be easily reused, changed, and shared. (Dimitrov et al. 2006)

4.2.1.5 Information Technology Summary

Currently XML is a widely accepted and adopted technology for the exchange of distributed data. OWL builds upon XML and RDF to facilitate the integration of databases and knowledge-bases. OWL is easily shareable across the Web as it uses XML syntax. Integration of information is improved through OWL by increasing the extent to which a computer can understand captured information. This is achieved by using classes to represent concepts, class axioms to explicitly define domain knowledge, and relationships to further enrich the information.

4.2.2 An Ontological Approach to Integrating Design Information

Recently we have developed an approach for representing and integrating design knowledge gained during the design process that utilizes modular ontologies (Rockwell et al. 2008). In the past a single ontology has been used to represent a single concept within engineering design (Grosse et al. 2005; Kanuri et al. 2005; Kim et al. 2006; Nanda et al. 2006; Witherell et al. 2007; Fernandes et al. 2007; Chang and Terpenny 2008). To create an integrated knowledge-base that models a larger range of the engineering design process we have proposed an approach that links multiple domain ontologies via the Web. Such an approach enables the linking of an ontological knowledge-base to decentralized information resources.

4.2.2.1 Engineering Design Ontologies

Figure 4.3 summarizes the ontologies developed to capture and integrate engineering design information. The ten ontologies in Fig. 4.3 were developed using Protégé (Noy et al. 2001) and the representation chosen was OWL – specifically OWL DL (Description Logic). Description logic provides the expressiveness to explicitly represent domain knowledge to develop a consistent formal model (Fiorentini et al. 2008).

The ten ontologies developed can be broadly grouped into three categories: *Enterprise*, *Engineering* and *Standard* (Fig. 4.3). Here, ontologies in the enterprise grouping are considered to be largely dependent upon the actual company/organization that is using the ontology. For example, it would be expected that a *Product* ontology for a company that sells electronic sensors would be greatly different from a company that sells aircraft engines. Similarly, ontologies in the engineering category are specific to engineering design. They will include the common divisions of *function, form*, and *behavior* (Szykman et al. 1998), as well as *optimization* and *decision support* ontologies (Fig. 4.3). Although not an exhaustive set of all possible engineering concepts, we have developed this semantic framework such that the consistency of information is maintained and easily shared. As such, additional concepts can be added to extend a knowledge-base beyond. Lastly,

Engineering Design

Enterprise Ontology

Organization - Includes such concepts as *people, projects, tasks,* and *requirements.* Captures information about a company's resources.

Component - Documentation of in-house and off-the-shelf components. Includes assembly information. Identifies component models,materials, weight, size, etc.

Product - Classifies a company's products. Identifies related components, assemblies and models. Also documents materials, weight, price, purchasers, etc.

Standard Ontology

Units - Identifies measurement units. Published by (NASA Jet Propulsion Laboratory 2008)

Materials - Classifies materials. Mechanical, physical, thermal, and electrical properties of materials are identified.

Engineering Ontology

Function - Functional model information about components, assemblies and products. Input and output energy, materials, and signals. Used to decompose overall product function into sub-functions. Ontology is based on functional basis developed in (Hirtz et al. 2002).

Form - Information about the geometry of an artifact. Type of information varies from text description and images to links of CAD files. Form models can be 2D or 3D models. Any related models are also identified.

Behavior - Detailed information about the behavior of an artifact. The different analysis techniques are classified and the details of the model are captured (i.e. for a FEM: element type, applied force, model symmetries, etc. are captured). Model inputs and model outputs (results) are recorded. Modeling assumptions and idealizations are also captured. Based on the work from (Grosse et al. 2005)

Optimization - Classifies optimization technique(s) used for an artifact. Inputs and outputs of the optimization are captured. Other related models are identified. Based on the work in (Witherell et al. 2007)

Decision Support - Decision making methods are classified creating a library of different techniques for the decision maker to use. Information captured in other ontologies is integrated together to facilitate a decision made based on the collective knowledge. Design alternatives, evaluation criteria and decision rationale are all documented.

Fig. 4.3 Engineering design ontologies

standard ontologies represent concepts that are independent of application. For example, measurement unit is a concept that is clearly not confined to engineering design alone.

Each modular ontology facilitates the documentation of information through a domain specific knowledge structure. Information contained within any of these ontologies can be integrated to create a design knowledge-base. With an integrated knowledge-base created the aggregated design information can be easily accessed and used to provide decision makers with information of high value.

The number of classes and properties in each ontology varies depending on the completeness of each ontology and complexity of the concept. The more complete and complex ontologies, such as the Optimization Ontology, can have over fifty classes and just as many properties. Alternatively, some of the simpler ontologies can have five to ten classes and ten to twenty properties. Figure 4.4 is a part of the class hierarchy for the Optimization Ontology and illustrates one of the more complex class structures.

4.2.2.2 Linking Distributed Information

OWL enables multiple ontologies to be linked together via Web URIs to combine distributed information. The ontologies presented in Fig. 4.3 were made accessible via the Web using Apache Tomcat (The Apache Software Foundation 2008), an open-source Web-based Java application server. The modular ontologies were then linked together by specifying each URI. By linking ontologies together the class structures, properties, and any instantiated knowledge are incorporated. With multiple ontologies connected, relationships were specified between distributed information, and logic rules were established so the aggregated design knowledge could be queried and reasoned upon.

Figure 4.5 illustrates how the ontologies are linked together to create an integrated knowledge base. From Fig. 4.5 it is seen that for this research we have used the units ontology developed at the NASA Jet Propulsion Laboratory by linking to their ontology (NASA Jet Propulsion Laboratory 2008) while the rest of the ontologies we have developed internally. The units ontology only has a few classes but includes around one-hundred instances. Additional classes, properties, and knowledge instances can also be added to the knowledge-base. New classes can be added at the root level or as a sub-class of an existing class. New properties can be added to any class to create relationships between the different domain ontologies. This enables users to easily extend the modular ontologies to include additional concepts.

In summary, an approach for capturing and integrating distributed design information has been presented. This approach takes advantage of ontologies and the web ontology language to capture and easily integrate design information. Next we would like to support decision makers in using this information to implement a decision method.

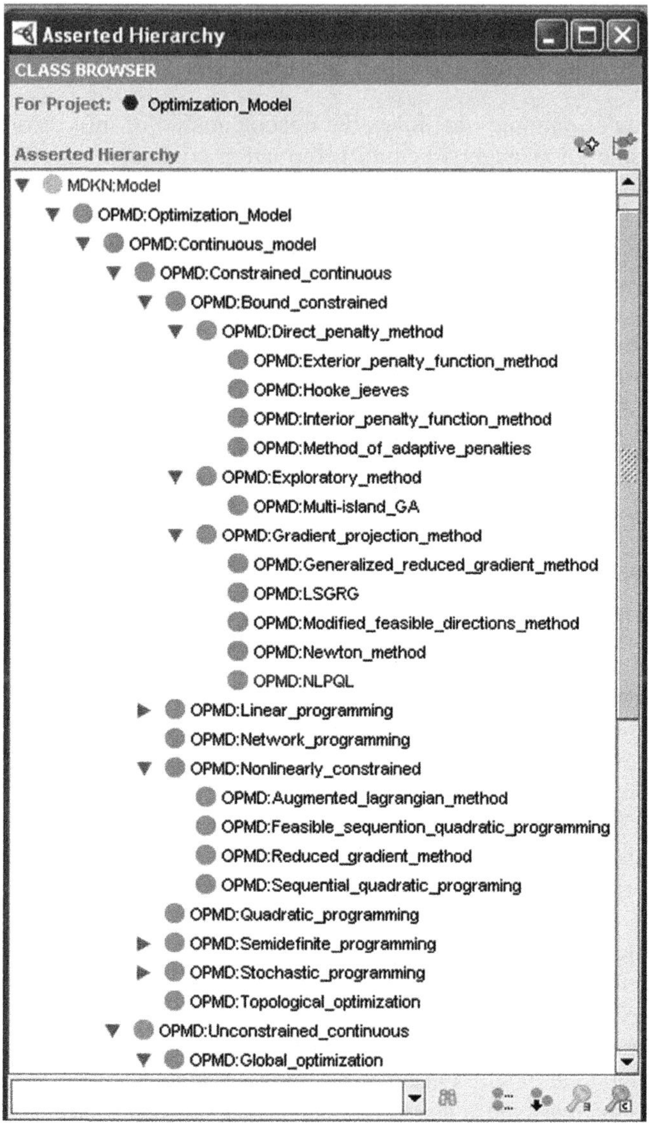

Fig. 4.4 Class structure of optimization ontology

4.3 Modeling Decisions in a Distributed Environment

Through the use of modular ontologies, we have presented an approach that facilitates the integration of design information. The next step is to design, develop, and implement a design method that can readily use the information contained in

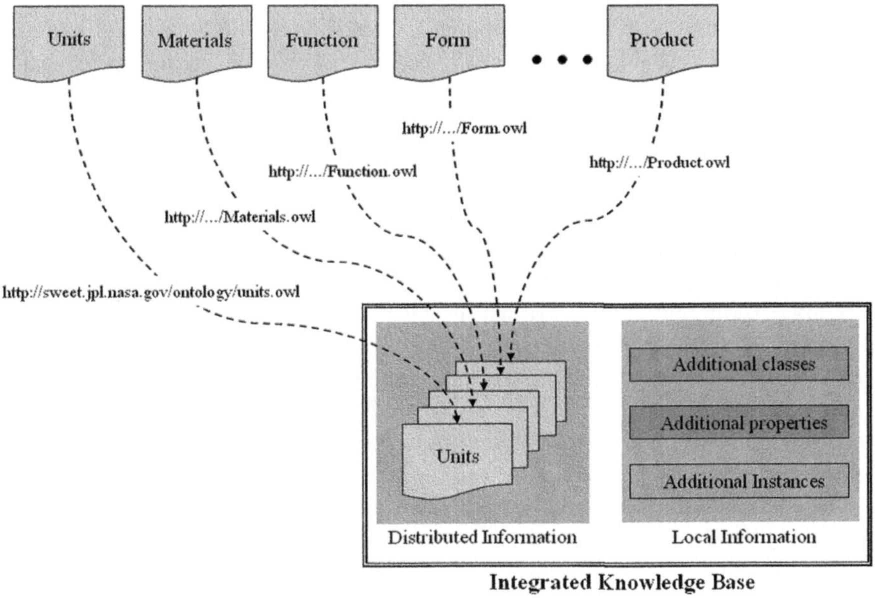

Fig. 4.5 Integration of distributed design information

the ontological framework in making decisions. The authors recognize the lack of consensus within the DBD community in the choice of a decision method or in how DBD should be implemented (Wassenaar and Chen 2001). In truth, any number of different selection and evaluation methods can be needed during the design process to go from customer requirements to a set of manufacturing specifications. Determining the most suitable method for a particular task can be a difficult decision and is outside the scope of this research. Instead, the development of our framework is such that it is independent of a specific decision method and its implementation.

A flexible decision support platform for modeling decisions within a distributed design environment is developed to ensure that: (1) the decision method and the decision process are documented; and (2) related design information within an ontological knowledge base is seamlessly accessible. For the purposes of illustration, the Conjoint House of Quality (HoQ) method presented in (Olewnik and Lewis 2007) was implemented. It is, however, important to recognize that the approach is independent of a particular decision method as it aims to capture the generic information relevant to decision making, irrespective of the decision method used. The DSO has been developed to allow documentation of this generic information while also being flexible enough to accommodate the documentation of method-specific information. For example, there must be at least two alternatives in a decision making process. Similarly, all decision methods must aim to identify the most preferred alternative based on a set of criteria, or, the overall utility based on the set of criteria. Therefore, the generic information pertaining to design alternatives, design attributes, and the designer's preferences is common to all decision methods. The DSO has been developed to capture such information and is based on the findings of (Ullman and D'Ambrosio 1995; Pahl and Beitz 1997; Baker et al. 2002). The following section details the development of the decision support ontology (DSO).

The DSO documents design alternatives in a manner that clearly explains how the alternative uniquely solves the defined problem. This is achieved through a combination of a written description, functions performed, and geometrical form models. Since it is likely that the alternatives may be performing the same functions, documentation of the physical phenomena, or working solution (Pahl and Beitz 1997), explicitly distinguishes how each alternative achieves functionality. Additionally, the abstraction level of each alternative (Ullman and D'Ambrosio 1995) is introduced to specify the design information as quantitative, qualitative, or mixed (both quantitative and qualitative). Here, the determinism is used to specify the information as deterministic (point-valued) or stochastic. Table 4.1 details some of the more important properties that are used to model each alternative.

Once the alternatives have been defined, a set of evaluation criteria must be created. Ultimately, the evaluation criteria greatly influence the outcome of any decision method. Consequently, it is important to understand *why* each criterion is present and *how* each criterion is measured. Each criterion should be based on a design goal. The DSO defines a relationship between each criterion and a corresponding design goal, thus making it transparent *why* a criterion is included. Furthermore, it becomes simple to check that each goal is represented by at least one criterion. To understand how each criterion is measured, an *objective parameter*

Table 4.1 Generic information captured for all alternatives

Property	Type	Description
Description	Data	Text description of alternative
Image	Data	Important images of the alternative
Has abstraction level	Object	Information about the alternative is either quantitative, qualitative or mixed
Has determinism	Object	Specify if information is being considered as deterministic (point-valued) or distributed
Is alternative for	Object	Specify the decision model that will evaluate this alternative
Has design parameter	Object	Specify important design parameters to consider when evaluating alternative
Has design summary	Object	Specify the design summary for this alternative. The design summary provides links to any functional, form, behavior, and optimization models of the alternative
Has working solution	Object	Identifies the physical phenomena that the solution is based upon. This information is directly instantiated from the functional model

Table 4.2 Generic information captured for all criteria

Property	Type	Description
Description	Data	Text description of alternative
Has objective parameter	Object	A design parameter that can be used to characterize a specific goal
Has importance weight	Data	Relative measure of importance. This value may change during the design process as new information is obtained
Is criterion for	Object	Identify the decision model that uses this criterion
Has units	Object	Specify the unit of measure that will be used to evaluate the objective parameter
Has goal	Object	Relates the criteria to the established. goals of the design

Table 4.3 Generic information captured for all preference models

Property	Type	Description
Consistency	Data	Specifies if the preference model represents a single view point (consistent) or multiple view points (inconsistent)
Comparison_basis	Data	Specifies if the evaluation of alternatives is absolute or relative
Has_objective_function	Object	Specifies the objective function being used for evaluation

and a *unit of measure* are identified for all criteria. An objective parameter characterizes a criterion or goal. For example, an objective parameter may be "maximize fuel efficiency" with a unit of measure of "miles per gallon". Lastly, associated with each criterion is an importance scaling factor (weight) that provides a relative measure of importance of each criterion. Table 4.2 details some of the more important properties that are used to model each criterion.

Finally, the evaluation of how well each alternative satisfies the criteria must also be documented. This evaluation is typically quantified through an objective, utility or cost function, based on designer's preferences, and it provides a measure of preference of one alternative relative to others. Evaluation may be done quantitatively based on a desire to maximize, minimize, or achieve an internal optimum value for a given objective parameter. Alternatively, it can also be done qualitatively based on experience and judgment. If the decision is made by a group, then the resulting preference model may be consistent, representing a single viewpoint, or inconsistent, representing conflicting viewpoints. Similarly, the evaluation of each criterion may be compared absolutely to a target value or relatively to other alternatives. The DSO captures all these aspects of a preference model. Table 4.3 provides an overview of these properties used to capture this information. (Ullman and D'Ambrosio 1995).

In Tables 4.1–4.3, the properties in *italic* indicate that the range of that property refers to a class outside of the decision support ontology (DSO). For example, the *has_design_summary* property has a range of 'Task summary' so a relationship between the class 'Alternative' and the class 'Task_summary' is created. The task summary of a design provides detailed information about an alternative's function, form, behavior, and optimization models. This relationship allows a decision maker to easily access detailed information about alternatives developed by others, thus facilitating the integration of design information in a distributed design environment. Here, consistency is maintained by linking to distributed information, thus avoiding the creation of duplicate information. Similarly, by creating the property *has_goal* for the class 'Criteria,' an explicit relationship between the criteria and the design objectives is established. Maintaining the association of the objectives and criteria is important for two reasons: (1) It is important to remember why and what a criterion is supposed to be measuring; and (2) The association helps clarify if all the essential criteria were identified, and if additional criteria are needed (Baker et al. 2002).

With the generic decision information represented, we must also consider how to model specific decision methods. Specific decision methods are represented as subclasses of the class 'Decision Model'. As subclasses of 'Decision Model' all decision models inherit relationships to the concepts of alternatives, criteria, and preference model as outlined above. Additional properties for each subclass of 'Decision Model' will vary for each method. These models of different methods need to be carefully developed, consistent with both the uses and limitation of specific methods.

The DSO thus provides a platform to capture generic information pertinent to all decision problems and it is easily extendable to capture method-specific decision information. Additionally, the DSO facilitates the integration of distributed information. Section 4.4 introduces an industry provided case study to demonstrate the utility of this approach.

4.4 Case Study

The case study of the re-design of a transfer plate is used to demonstrate the integration of design information with the DSO. First an introduction to the re-design problem is presented and then a brief description of the decision method implemented is presented.

4.4.1 Problem Setup

The case study is an industry inspired problem that focused on the re-design of a transfer plate for a circuit breaker intended for aerospace applications. The transfer plate, illustrated in Fig. 4.6, is a simple lever that transfers the motion of the current sensing element to trip the aircraft circuit breaker. Information about the re-designs (e.g., function, form, behavior, and optimization) was documented within a

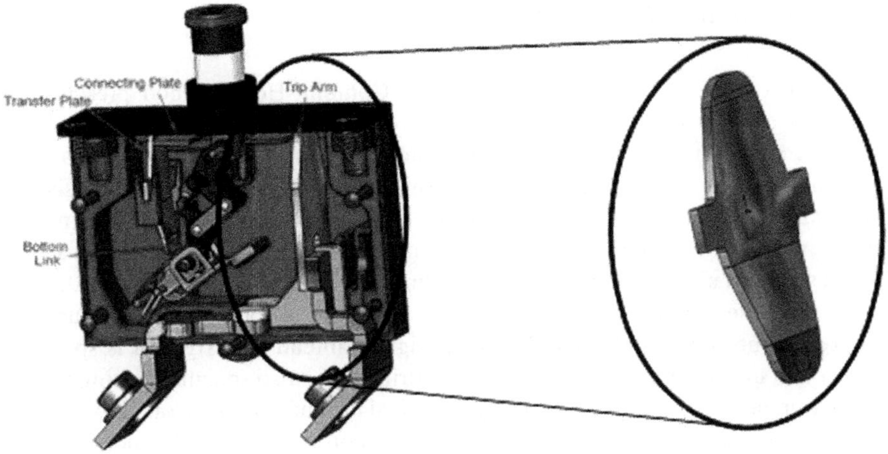

Fig. 4.6 Circuit breaker and transfer plate

Fig. 4.7 Knowledge base of linked ontologies

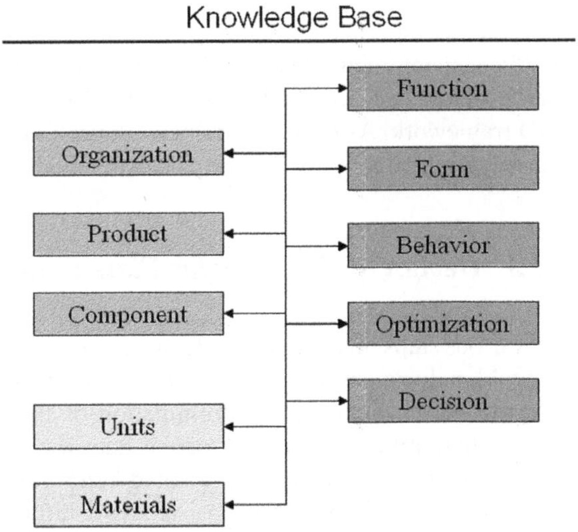

knowledge-base created by linking together the ontologies presented in Section 4.2. Figure 4.7 shows the ontologies used in creating the knowledge-base. The arrows in Fig. 4.7 represent relationships between the different ontologies. With the multiple re-designs documented in the knowledge-base the task was to select a re-design to be used.

4.4.2 Conjoint-HoQ Method

The decision method implemented was the Conjoint-HoQ method presented in (Olewnik and Lewis 2007). The Conjoint-HoQ method by Lewis and his research group is a consumer-centric approach to decision-based design. Through a combination of HoQ (House of Quality) (Terninko 1997) and conjoint analysis, the importance weighting for each criterion are determined and then used within a decision matrix to evaluate the overall performance of each alternative. Using a HoQ enables systematic identification of the technical attributes that will satisfy the customer attributes. In this approach, for each technical attribute, a target value level that represents a nominal value that the designers intend to design for is specified. Lower and upper limits for each technical attribute are also specified. Using the target, lower, and upper limits, a set of hypothetical products are generated. A conjoint analysis approach is then used to present each hypothetical product to potential customers. The potential customers rate each hypothetical product. This information forms the basis in the generation of scaling factors or weights for the different attributes.

This is achieved by setting the problem in a decision matrix representation with a set of simultaneous linear equations and using the least squares linear regression to solve for the importance weighting factor for each technical attribute. Further details on this decision method can be found in (Olewnik and Lewis 2007). Note that the conjoint-HoQ decision method is used to illustrate the application of the proposed DSO framework. As such, the framework itself is independent of decision methods and will work equally well for any decision method.

4.4.3 Design of the Transfer Plate Using DSO Framework

The various steps involved in the decision-based design of the transfer plate includes the problem formulation or the decision model development, the identification of the design criteria, mathematical formulation of the design objective, generation of design alternatives, application of the decision method, and identification of the most optimal design solution. As stated before, information plays a pivotal role in decision making. Improving consistency of information and facilitating collective comparison of the design alternatives based on specified evaluation criteria can enhance the decision making process. Accordingly, we employ the DSO approach that includes an easy to access knowledge repository to capture the geometrical forms, structural analyses, and optimization techniques, as well as enterprise information associated with the design problems.

Step 1: *Problem Statement*

The problem statement for this re-design is straight forward: *Select a transfer plate for a circuit breaker intended for aerospace application.* The problem statement is captured as a data-type property of the class 'Decision Model'.

Step 2: *Determine Requirements*
The engineering requirements for the transfer plate are documented within the Organization ontology. The property *has–requirements* of the class 'Decision Model' identifies specific requirements for each decision. For the transfer plate re-design the requirements are: (1) the manufacturing process must be stamping; and (2) the mounting means and force location are fixed and can not be changed.

Step 3: *Determine Goals*
The goals for the transfer plate re-design have been identified as: (1) light weight and (2) high reliability. Similar to the engineering requirements these goals are captured in the Organization ontology.

Step 4: *Identify Alternatives*
Alternative designs were developed and preliminary optimization and analysis models were created for each re-design. The details of all re-designs have been documented within the appropriate ontology. So, for example, the analysis models are documented in the Behavior ontology. The three re-designs being considered are illustrated in Fig. 4.8.

Within the DSO the design alternative information presented in Section 4.3 was captured for each alternative. A design summary for each alternative is also specified. The design summary links to an instance from the Organization ontology that contains links to information about the details of the alternative. The decision maker is able to use these links to find detailed information about each alternative. Figure 4.9 illustrates captured information, within Protégé, for the design alternative 'Circular Cutouts' and how distributed information is linked together in a way that is easy for decision makers to access the information. In Fig. 4.9 the arrows represent the URI links between concepts in different ontologies. The documentation of the Circular Cutout alternative refers to a task

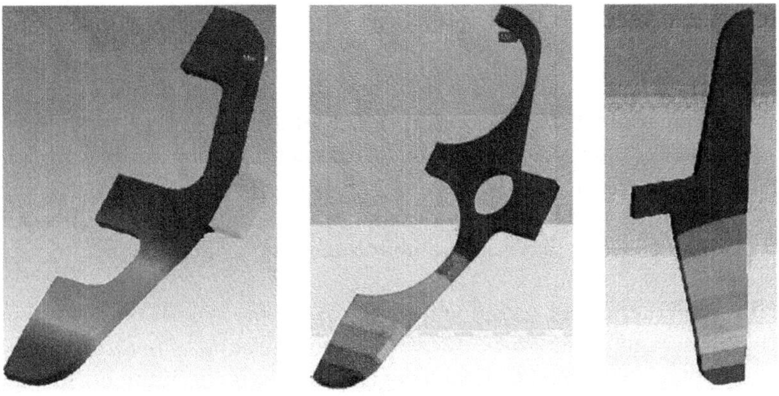

Rectangle Cutouts (RC) Circle Cutouts (CC) Slim Down (SD)

Fig. 4.8 Transfer plate design alternatives: rectangle cutouts (RC) circle cutouts (CC) slim down (SD)

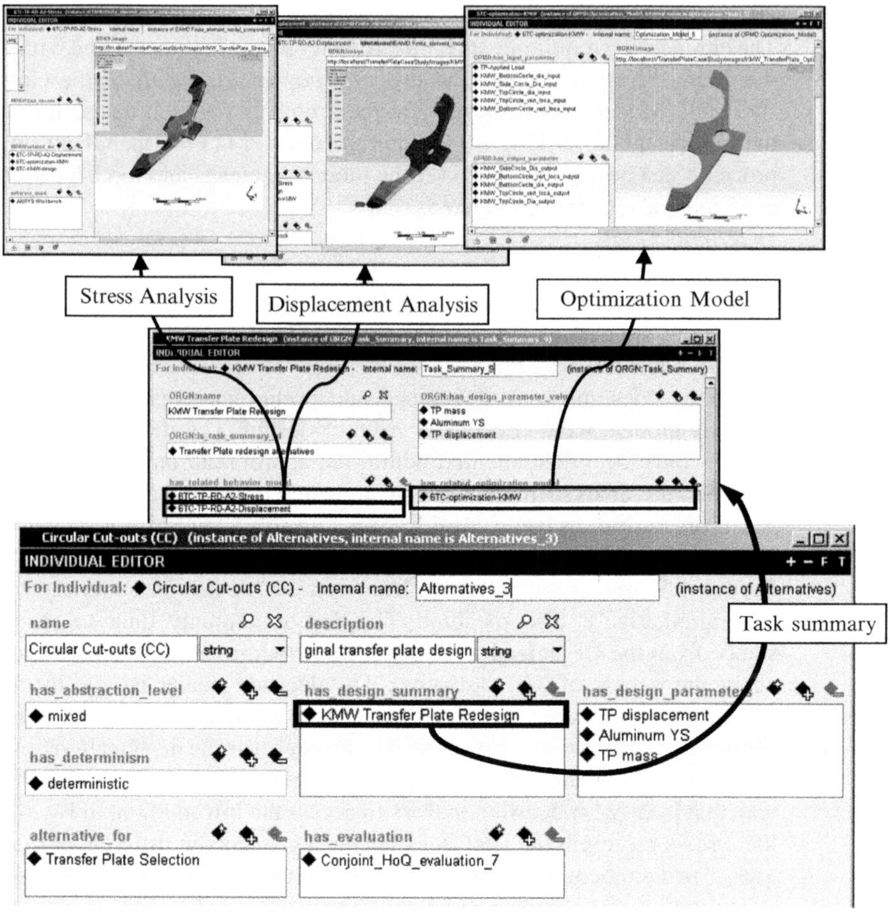

Fig. 4.9 Documentation of design alternative and connection to related information

summary that then links to detailed information about the optimization model, stress analysis, and displacement analysis for that design. Also documented for the Circular Cutouts alternative are design parameters important to that re-design. This linked information allows the retrieval of information that is relevant for that specific alternative.

Step 5: *Identify Criteria*

The criteria for evaluation were to (1) minimize deflection, (2) minimize mass, and (3) minimize stress. The minimization of mass criterion corresponds to the goal of light weight and the criteria of minimize deflection and minimize stress correspond to the goal of high reliability. These criteria were measured in units of inch, pound, and pound per square inch. Using the Conjoint HoQ method, importance weights for each criterion

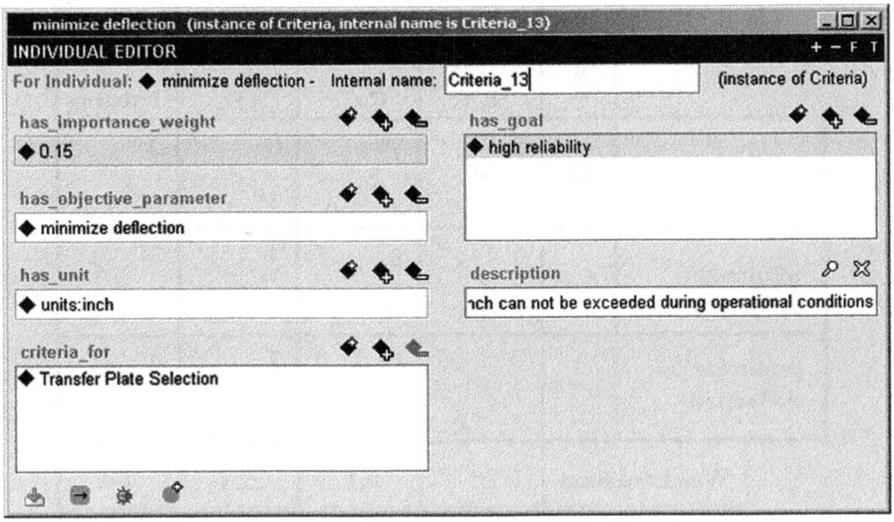

Fig. 4.10 Documentation of decision criterion

were calculated. The deflection criteria had a weight of 0.15, the mass criteria 0.70, and the stress criteria 0.15. Figure 4.10 is an example of how criteria are captured in the decision support ontology.

Step 6: *Select Decision Method*

For this problem we have chosen to use the Conjoint HoQ with a decision matrix to evaluate the alternatives. Criteria were evaluated across all alternatives before proceeding to the next criteria. The evaluation rating scale used was 1–10. The alternative that satisfied the criterion best was given a 10. The other alternatives were given a rating based on comparison to that alternative. Weighted sum scores were calculated by multiplying the importance weighting for a criterion by each alternative rating for that criterion and then summing scores by alternative.

Step 7: *Evaluate Alternatives*

The decision matrix used with the completed evaluations is presented in Fig. 4.11. For all criteria each alternative has a rating in the upper left and the weighted score in the lower right. The weighted sum for each alternative is at the bottom of each column. Also the existing transfer plate design has been included in the evaluations to provide a relative measure of improvement. Through the decision matrix the Circular Cutout re-design has been determined to be the preferred design. Although currently we do not have a user interface that allows the decision maker to directly evaluate alternatives in a table as shown in Fig. 4.11 and directly store that information in the DSO, the DSO is capable of capturing and storing such information. This is a simple implementation issue and not a limitation of the approach. In Fig. 4.12 documentation of this information within the DSO (using Protégé) is illustrated. For each alternative-criterion pair

	weight	R.C.	C.C.	S.D.	Existing
minimize mass	0.7	8 / 5.6	10 / 7.0	6 / 4.2	2 / 1.4
minimize stress	0.15	6 / 0.9	8 / 1.2	4 / 0.6	10 / 1.5
minimize deflection	0.15	7 / 1.0	6 / 0.9	7 / 1.0	10 / 1.5
Weighted Sum		7.5	9.1	5.8	4.4

Fig. 4.11 Evaluation of design alternatives

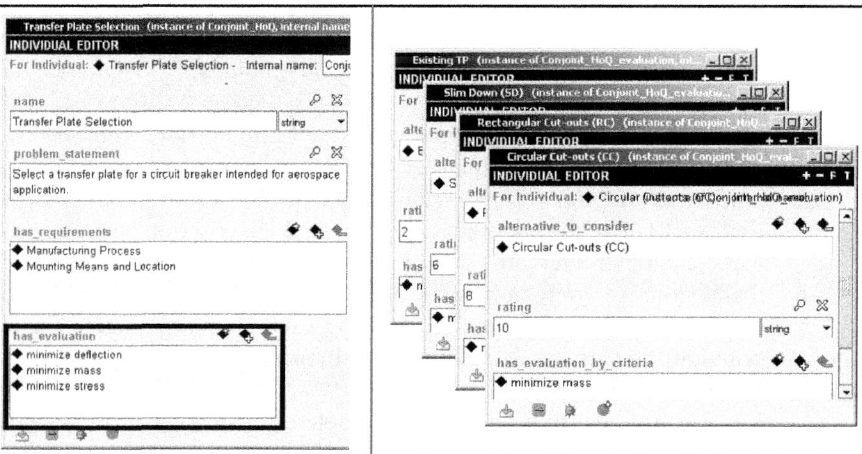

Evaluation documented based on criteria and then alternatives

Documentation of the evaluation of the four designs for the *minimize mass* criteria

Fig. 4.12 Documentation of evaluations within the DSO

there is an instance that documents the evaluation rating for the alternative based upon the criterion. It is important to note that the illustrated representation is an internal representation of information to demonstrate how information is stored in an ontology. This representation allows information to be documented and stored in a structured manner that increases computer interpretability. It is expected, as with any data or knowledgebase (e.g., XML), that the end user does not actually interact with the information in its internal representation but that user interfaces filter and render information in a display that reduces the cognitive demands of the decision maker so better decisions can be made. Studies have shown the extent to which features of displays can influence decisions (Wickens and Hollands 2000). This is a consideration that must be addressed before an end-user application can be realized.

Step 8: *Validate Solution*
The weighted scores show that the Circular Cutout design best meets the stated goals. Checking the Circular Cutout design back against the problem statement and requirements we see that the design satisfies all the stated conditions and is the preferred alternative.

4.4.4 Case Study Summary

This case study demonstrates how information captured in domain ontologies can be integrated together to facilitate decision making by providing easy access to distribute information resources.

A knowledge-base was created by linking a set of modular ontologies together and a decision method was modeled by documenting both generic decision information and method-specific information. The case study revealed how improved representation of design information can facilitate the seamless access and use of design information within the decision making process.

4.5 Summary

Based on the premise that the two underpinnings of engineering design are information and a decision method, an ontological approach was presented to support decision making in engineering design in a distributed environment. The framework presented uses ontologies to capture and store design information in a domain-specific structure with underlying description logic axioms to define concepts within the domain. Thereby improving the user's understanding of documented design information and storing information in a computer interpretable representation. The framework facilitates improved consistency of information by enabling multiple ontologies to be mapped together using logical axioms. Information stored in a knowledge-base consisting of multiple ontologies can be seamlessly accessed and searched based on the meaning of the content to improve the retrieval and reuse of documented information. A decision support ontology (DSO) that captures generic information and provides decision makers access to detailed information about alternatives and criteria was developed. The DSO is also extendable to include method-specific decision information. The usefulness of the approach was demonstrated through the decision-based design of an industrial transfer plate.

Using an ontological knowledge-base the geometrical forms, structural analyses, and optimization techniques for several proposed alternatives, as well as the related enterprise information, was documented. A modified conjoint-analysis based decision matrix was utilized to evaluate, compare, and realize the preferred design from amongst the alternatives. Ongoing research includes further development of the DSO system and the accommodation of company "best-practices" or "lessons-learned" into the decision process, as well as development of a user-friendly interface.

Acknowledgements This material is based on work supported by the National Science Foundation under Grant No. 0332508 and by industry members of the NSF Center for e-Design.

References

Ahmed, S., Kim, S., Wallace, K. (2005). *A Methodology for Creating Ontologies for Engineering Design*, Proceedings of ASME 2005 IDETC/DTM, Sept., DETC2005-84729.

Antonsson E.K., Otto, K.N. (1995). Imprecision in Engineering Design, ASME Journal of Mechanical Design, Special Combined Issue of the Transactions of the ASME commemorating the 50th anniversary of the Design Engineering Division of the ASME, v.17(B), 25–32.

Baker, D., Bridges, D., Hunter, R., Johnson, G., Krupa, J., Murphy, J. and Sorenson, K. (2002). *Guidebook to Decision-Making Methods*, WSRC-IM-2002-00002, Department of Energy, USA. http://emi-web.inel.gov/Nissmg/Guidebook_2002.pdf.

Berners-Lee, T., Hendler, J., Lassila, O. (2001). *The Semantic Web*, Scientific American Magazine, May 17, 2001.

Caldwell, N.H.M., Clarkson, P.J., Rodgers, P.A., Huxor, A.P. (2000). Web-Based Knowledge Management for Distributed Design, Intelligent Systems and Their Applications, IEEE, v15, i3, pp. 40–47, May/Jun 2000.

Chang, X., Terpenny, J., (2008). *Framework for Ontology-Based Heterogeneous Data Integration for Cost Management in Product Family Design*, ASME 2008 International DETC and CIE Conference DETC2008-49803, Brooklyn, New York, USA.

Chen, W., Lewis, K.E., and Schmidt, L. (2000). *Decision-Based Design: An Emerging Design Perspective*, Engineering Valuation & Cost Analysis, special edition on "Decision-Based Design: Status & Promise.

Court, A.W., Ullman, D.G., and Culley, S.J. (1998). *A Comparison between the Provision of Information to Engineering Designers in the UK and the USA*, Int. J. Information Management, 18 (6), 409–425.

Crowder, M.R., Wong, S., Shadbolt, N., Wills, G. (2008). *Knowledge-Based Repository to Support Engineering Design*, ASME 2008 International DETC and CIE Conference DETC2008-49155, Brooklyn, New York, USA.

Decision Making in Engineering Design, Editors: Lewis, K., Chen, W. and Schmidt, L. C., ASME Press, New York, 2006.

Decision-based Design Open Workshop, Last viewed October 2008, http://dbd.eng.buffalo.edu/.

Dimitrov, D.A., Heflin, J., Qasem, A., Wang, N. (2006). *Information Integration Via an End-to-End Distributed Semantic Web System*, 5th International Semantic Web Conference, Athens, GA, USA, November 5–9, LNCS 4273.

Farquhar, A., Fikes, R., Rice, J. (1996). *The Ontolingua Server: A Tool for Collaborative Ontology Construction*, Knowledge Systems, AI Laboratory.

Fernandes, R., Grosse, I., Krishnamurty, S., Wileden, J. (2007). *Design and Innovative Methodologies in a Semantic Framework*, In Proceeding of ASME IDETC/CIE, DETC2007-35446.

Fiorentini, X., Rachuri, S., Mahesh, M., Fenves, S., Sriram, R.D. (2008). *Description Logic for Product Information Models*, ASME 2008 International DETC and CIE Conference DETC2008-49348, Brooklyn, New York, USA.

Grosse, I. R., Milton-Benoit, J.M., Wileden, J.C. (2005). Ontologies for Supporting Engineering Analysis Models, AIEDAM, 19, 1–18.

Gruber, T.R. (1992). Toward Principles for the Design of Ontologies Used for Knowledge Sharing, In Int. J. of Human-Computer Studies, 43, 1992, 907–928.

Hazelrigg, G. A. (1996). *System Engineering: An Approach to Information-Based Design*, Prentice Hall, Upper Saddle River, NJ, 1996.

Hazelrigg, G.A. (1998). A Framework for Decision-Based Engineering Design, ASME Journal of Mechanical Design, Vol. 120, 653–658.

Hirtz, J., Stone, R.B., McAdams, D., Szykman, S., Wood, K. (2002). A functional basis for engineering design: reconciling and evolving previous efforts, Research in Engineering Design 13(2), 65–82.

Howard, R. (1971). *Decision Analysis in System Engineering*, System Concepts – Lectures on Contemporary Approaches to Systems, Systems Engineering and Analysis Series, pp. 51–86, 1971.

Howard, R.A. (1986). *The Foundations of Decision Analysis*, IEEE Transactions on System Science and Cybernetics, September 1968.

Jie W. and Krishnamurty, S. (2001). Learning-based Preference Modeling in Engineering Design Decision-Making, ASME Journal of Mechanical Design, 191–198, June 2001.

Kanuri, N., Grosse, I., Wileden J. (2005). *Ontologies and Fine Grained Control over Sharing of Engineering Modeling Knowledge in a Web Based Engineering Environment*, ASME International Mechanical Engineering Congress and Exposition, November 2005, Florida.

Keeney R. L. and Raiffa H. (1976). *Decisions with Multiple Objectives: Preferences and Value Tradeoffs*, Wiley and Sons, New York.

Kim, K.-Y., Yang, H., Manley, D. (2006). Assembly Design Ontology for Service-Oriented Design Collaboration, Computer-Aided Design & Applications, v3, no. 5, 2006, 603–613.

Kitamura, Y. (2006). *Roles of Ontologies of Engineering Artifacts for Design Knowledge Modeling*, Proc. of 5th International Seminar and Workshop Engineering Design in Integrated Product Development, Sept. 2006, Gronow, Poland, pp. 59–69, 2006.

Lee, J., Suh, H. (2007). *OWL-Based Ontology Architecture and Representation for Sharing Product Knowledge on a Web*, In Proc. of ASME IDETC/CIE, DETC2007-35312, Las Vegas, Nevada, September 4–7.

Li, Z., Raskin, V., Ramani, K. (2007). *Developing Ontologies for Engineering Information Retrieval*, In Proc. of ASME IDETC/CIE, DETC2007-34530, Las Vegas, Nevada, September 4–7.

Moon, S.K., Kumara, S.R.T., Simpson, T.W. (2005). *Knowledge Representation for Product Design Using Techspecs Concept Ontology*, Proc. of the IEEE International Conference on Information Reuse and Integration,, Las Vegas, NV, August 15–17.

Morris, K.C., Frechette, S., Goyal, P. (2008). *Development Life Cycle and Tools for Data Exchange Specification*, ASME 2008 International DETC and CIE Conference DETC2008-49807, Brooklyn, New York, USA.

Nanda, J., Simpson, T.W., Kumara, S.R.T., Shooter, S.B. (2006). A Methodology for Product Family Ontology Development Using Formal Concept Analysis and Web Ontology Language, Journal of Computing and Information Science in Engineering, v6, i2, pp. 103–113, June.

NASA Jet Propulsion Laboratory, Last viewed October 2008. Semantic Web for Earth and Environmental Terminology (SWEET), http://sweet.jpl.nasa.gov/ontology/.

Noy, N.F., Sintek, M., Decker, S., Crubezy, M., Fergerson, R.W., Musen M.A. (2001). Creating Semantic Web Contents with Protege-2000, IEEE Intelligent Systems 16(2): pp. 60–71.

Olewnik, A.T., Lewis, K.E. (2007). *Conjoint-HoQ: A Quantitative Methodology for Consumer-Driven Design*, ASME IDETC/CIE2007-35568, Las Vegas, Nevada, USA.

Pahl, G., Beitz, W. (1997). *Engineering Design: A Systematic Approach, Springer-Verlag*, Berlin.

Ray, S.R., Jones, A.T. (2006) Manufacturing Interoperability, J Intell Manuf (2006) 17:681–688.

Rockwell, J.A., Witherell, P., Fernandes, R., Grosse, I., Krishnamurty, S., Wileden, J. (2008). *Web-Based Environment for Documentation and Sharing of Engineering Design Knowledge*, ASME 2008 International DETC and CIE Conference DETC2008-50086, Brooklyn, New York, USA.

Szykman, S., Sriram, R.D., Bochenek, C., Racz, J. (1998). *The NIST Design Repository Project*, Advances in Soft Computing- Engineering Design and Manufacturing, Springer-Verlag, London.

Szykman, S., Sriram, R.D., Bochenek, C., Racz, J.W., Senfaute, J. (2000). Design repositories: engineering design's new knowledge base, Intelligent Systems and Their Applications, IEEE Volume 15, Issue 3, May-June 2000 Page(s):48–55.

Tang X., Krishnamurty, S. (2000). On Decision Model Development in Engineering Design, Special issue on "Decision-Based Design: Status and Promise," Engineering Valuation and Cost Analysis, September 2000, Vol. 3, pp. 131–149.

Terninko, J., 1997. *Step-by-Step QFD Customer-Driven Product Design*, 2nd Edition.

The Apache Software Foundation, Apache Tomcat, Last viewed October 2008, http://tomcat.apache.org/.

Thurston, D.L. (1991). A Formal Method for Subjective Design Evaluation with Multiple Attributes, Research in Engineering Design, Vol. 3, pp. 105–122, 1991.

Ullman, D.G., D'Ambrosio, B. (1995). A Taxonomy for Classifying Engineering Decision Problems and Support Systems, Artificial Intelligence for Engineering Design, Analysis and Manufacturing, 9, 427–438.

Wassenaar, H.J., Chen, W. (2001). *An approach to decision-based design*, ASME 2001 DETC and CIE Conference DETC2001/DTM-21683, Pittsburgh, Pennsylvania.

Wickens, C.D., Hollands, J.G. (2000). *Engineering Psychology and Human Performance*, 3rd Edition.

Witherell, P., Krishnamurty, S., Grosse, I.R. (2007). Ontologies for Supporting Engineering Design Optimization, Journal of Computing and Information Science in Engineering, June 2007, v7, i2, 141–150.

Witherell, P., Grosse, I., Krishnamurty, S., Wileden (2008). *FIDOE: A Framework for Intelligent Distributed Ontologies in Engineering*, ASME 2008 International DETC and CIE Conference, Brooklyn, New York, USA.

World Wide Web Consortium, (2004a). *Resource Description Framework (RDF) Concepts and Abstract Syntax*, http://www.w3.org/TR/2004/REC-rdf-concepts-20040210/#section-Concepts.

World Wide Web Consortium, (2004b). *RDF Primer*, http://www.w3.org/TR/REC-rdf-syntax/.

World Wide Web Consortium, (2004c). *OWL Web Ontology Language Overview*, http://www.w3.org/TR/2004/REC-owl-features-20040210/#s1.1.

World Wide Web Consortium, (2005). *An Introduction to RDF*, http://research.talis.com/2005/rdf-intro/.

World Wide Web Consortium, (2006). *Extensible Markup Language*, http://www.w3.org/TR/REC-xml/.

Part II
Decision Making in Engineering Design

Chapter 5
The Mathematics of Prediction

George A. Hazelrigg[1]

Abstract Prediction is the basis for all decision making, and it is particularly important for engineering and especially engineering design. However, the mathematics of prediction is neither trivial nor obvious. No prediction is both precise and certain, and a significant question is how to incorporate data and beliefs from disparate sources to formulate a prediction that is consistent with these inputs. The purpose of this paper is to establish the basic concepts of prediction in the context of engineering decision making and to make these concepts accessible to engineers.

Keywords Decision making · Bayes · Dutch book · Probability · Engineering design · Belief · Kolmogorov · Prediction

5.1 Introduction

The practice of engineering is largely centered around decision making. Engineering design involves a great deal of decision making: the layout or conceptual design of a product or service, what materials to use for a product, dimensions, tolerances, the whole manufacturing process, and so on. Indeed, decision making is the main thing that sets engineering apart from the sciences. It is why professional engineers are licensed, but not scientists. Decisions impact society directly. They affect health and safety, and societal welfare overall.

A decision is a commitment to an action, taken in the present, to affect a more desired outcome at some time in the future. Thus, all decisions demand some ability to forecast or predict the future: if Alternative A_1 is chosen, Outcome O_1 results; if Alternative A_2 is chosen, Outcome O_2 results; and so on. The preferred alternative is that alternative whose outcome is most preferred. If the outcomes all could be

G.A. Hazelrigg (✉)
National Science Foundation Arlington, VA, USA
e-mail: ghazelri@nsf.gov

[1] The views expressed in this paper are strictly those of the author and do not necessarily reflect the views either of the National Science Foundation or the Federal Government.

N.R.S. Raghavan and J.A. Cafeo (eds.), *Product Research: The Art and Science Behind Successful Product Launches*, DOI 10.1007/978-90-481-2860-0_5,

known with precision and certainty at the time of the decision, decision making would be easy. But outcomes can never be known with precision and certainty at the time of any decision – no one is ever certain of the future. This is what makes decision making mathematically interesting. The principal goal of prediction is to enable good decision making.

The goal of any mathematic is self-consistency, and this certainly is the case in prediction. A decision maker always begins with some notion of what the future holds and how it will be influenced by a particular decision. But then, evidence, perhaps in the form of observations, experimental data or computational results, may be introduced. Evidence influences the decision maker's initial notions. A goal of the mathematics of prediction is to assure that evidence is accounted for such that its influence on the decision maker's initial notions is entirely consistent with the evidence itself. A formal mathematics of prediction is needed because consistency is neither easy to achieve nor does it yield results that are intuitively obvious. Indeed, enforcement of consistency often leads to conclusions that can be quite counterintuitive.

5.2 Basic Concepts

Basic concept 1: A decision is an individual act, taken only by an individual. Decisions are not made by groups. Groups have an emergent behavior that is the result of the decisions of the members of the group *and* the rules (formal, tacit, or whatever) of the interaction between the members of the group.

This concept is very important to the correct use of the mathematics of decision theory. Many people, for example, Thurston (1999), think that groups make decisions, and they try to apply the mathematics of decision theory to group "decisions." They then complain that decision theory doesn't work in many instances. The fact is, however, that the mathematics of decision theory works for all decisions, recognizing that only individuals make decisions. One branch of mathematics that deals with group behavior is called game theory.

Basic concept 2: Everyone holds a different view of the future, but the view that is important in any given decision making context is only the view of the decision maker.

It may seem strange to insist that everyone holds a different view of the future. But, in the stochastic sense, the future is quite complex (it contains many elements). For example, perhaps you want to cross the street. If you go now, what will happen? It's not so simple as, "I'll get to the other side." You could be injured in any number of ways. You could be struck by a meteorite in the middle of the street. Your shoe could fall off. You could be subject to harassment by a malicious bird en route. In making your decision, you assign probabilities (maybe subconsciously) to all these outcomes. Prior to stepping off the curb, your assessment of the future contains an evaluation of all outcomes and their probabilities (although you might default by dismissing many as simply "unlikely," even though they are matters of life and death). Hence, your assessment of the future, even in this rather trivial example, contains many elements of the outcome space and your assessment of their probabilities of occurrence, and it is therefore unique to you.

***Basic concept 3*:** Prediction is about events.

When we speak about prediction, we are always speaking about the prediction of an event – the outcome of a game, the outcome of a bet, the outcome of an engineering design – and always an event that is in the future. Although the event can be complex, it is often couched in simple win/lose terms – a monetary gain or loss, for example. The outcome of an engineering design can be couched similarly, as a final profit or loss, for example, or one can look at it in much more detail (for example, on its third test flight, the airplane crashed and the project was abandoned with a loss of $5 million). But the addition of detail does not change the essence of the fact that the prediction is of an event and, in this case, an event is a "final" state resulting pursuant to a decision.

***Basic concept 4*:** All predictions begin with some feeling – a *prior* – about how a particular decision will influence the future.

In essence, what this is saying is that one cannot hope to make rational decisions unless one begins with some feeling about the outcome and what are the important factors in its determination. Thus, evidence alone does not determine the decision maker's final prediction at the time of a decision, rather it influences the decision maker's prior beliefs about the event.

***Basic concept 5*:** A decision maker holds a set of fundamental beliefs – $F = ma$, steel is hard, the sky is blue, etc. – and the object of prediction is to draw conclusions about the future that are consistent with these beliefs and any evidence that may be introduced.

The key here is consistency. If one truly holds a belief, that is, he or she is firm in that belief, then to make a decision contrary to that belief is to be irrational. So, the goal of a prediction is to forecast a future that is consistent with a set of beliefs (of course this cannot be done if one's fundamental beliefs contradict one another), and to allow the incorporation of evidence. For example, a finite element analysis enforces a set of conditions (beliefs), such as Hooke's Law, throughout a volume.

***Basic concept 6*:** In engineering, mathematical models are often used to provide evidence. The purpose of mathematical modeling is to provide a formal and rigorous framework to enable the simultaneous consideration of a set of fundamental beliefs held by a decision maker to better enable the decision maker to draw conclusions (and take actions) that are consistent with his or her beliefs.

These basic concepts form the core of a mathematical theory of predictive modeling.

5.3 The Dutch Book

The Dutch book is a mathematical argument that bears heavily on the theory of prediction. Although not totally clear, it is probably so named because of its origin, which is said to be the Dutch who were insuring shipping in the nineteenth century. These insurers noted that, based on inconsistencies in the assessment of risks by the shippers, they were able to structure their insurance so that it became irrelevant whether the ships arrived safely or not. They made money either way. The Dutch book argument (DBA) was formalized by Ramsey (1931) and de Finetti (1937). It provides a powerful philosophical justification for the Kolmogorov axioms of probability theory and Bayes Theorem. While, perhaps not totally airtight, it nonetheless is sufficiently compelling that anyone who desires to propose an alternative approach to thinking about the future (for example, fuzzy logic) should first provide an explanation why the DBA does not disqualify the proposed theory.[2]

The DBA rests on the notions that a decision maker (one who is about to enter into a wager, as is the case of an engineer doing engineering design) enters into a decision with a belief about the outcomes that could result from the various alternatives available and further would refuse to accept any wager that will result in a certain loss. Thus, the DBA connects degrees of belief held by a decision maker to the decision maker's willingness to wager. There is a rather extensive literature on the Dutch book, extending to lengthy and highly sophisticated derivations that put to rest many arguments against the procedure. We present here as an introduction to the DBA only a very cursory overview, and a rather simplistic derivation of the Dutch book conclusions.

We begin by presenting to a gambler[3] a small wager, say $1, for which we can assume that the gambler is risk neutral. By *risk neutral*,[4] we mean that, if presented with a large number of similar wagers, the gambler would bet in such a way as to maximize his return (or minimize his loss). The wagers will be presented by a bookie, whom we will refer to as the Dutch book. The Dutch book will seek to structure the wagers in such a way as to take advantage of any inconsistencies in the gambler's behavior. We begin the Dutch book argument with a simple wager that has only two possible events (or outcomes), E and $\leftarrow E$ (read E and not E). For example, in a wager on a coin flip, the events would be heads or tails, which we would instead refer to as heads and not heads simply to emphasize that we are considering only two events and that they are mutually exclusive and all inclusive (there are no other events that could occur).

The gambler holds a belief in E, which we will denote by $p(E)$, and a belief in $\leftarrow E$, denoted by $p(\leftarrow E)$. The gambler's beliefs in E and $\leftarrow E$ are strictly his own, and are not influenced in any way by the Dutch book. What the Dutch book controls are the stakes, or payouts, which we denote S_E and $S_{\leftarrow E}$. The Dutch book

[2] The reader who wants to explore the DBA in more depth should refer to the extensive literature on the subject. An impressive collection of papers on both sides is given by Alan Hájek ("Scotching Dutch Books," *Philosophical Perspectives*, 19, issue on Epistemology, ed. John Hawthorne, 2005). Additionally, there is an extensive treatment of Bayes Theorem by José M. Bernardo and Adrian F.M. Smith, (*Bayesian Theory*, John Wiley and Sons, 2000), which includes a discussion of the DBA. This book includes a 65-page list of references (none from the engineering literature) and a 46-page appendix on the faults in the major non-Bayesian approaches. The extensive engineering literature on predictive modeling is not referenced here because, as this paper shows, it is mostly flawed.

[3] Note here that the term *gambler* includes any engineer involved in any decision making process.

[4] Risk preference is formally defined by the shape of one's utility function. However, we choose to avoid this complication here.

is free to set S_E and $S_{\leftarrow E}$ to the values of his choice, positive or negative. The gambler's beliefs are defined in terms of what he would pay to place a bet on the stakes S_E or $S_{\leftarrow E}$. So, for example, for a bet on E, the gambler would pay any amount $b(E) \leq p(E)S_E$ (noting that, if the stake is negative, the Dutch book would have to pay the gambler to accept the wager), and he would be willing to sell the bet for any amount $b(E) > p(E)S_E$. Following the approach of Caves (2003), the gambler's gain should event E occur is the stake, given that E occurs, minus the money the gambler pays to enter the wager. Remember that the gambler is willing to bet, with appropriate odds, on both E and $\leftarrow E$, and the amount the gambler will bet on each outcome is determined by the stakes S_E and $S_{\leftarrow E}$, which are set by the Dutch book.

$$G_E = S_E - p(E)\mathrm{S_E} - p(\leftarrow E)S_{\leftarrow E}$$

And the gain should event $\leftarrow E$ occur is:

$$G_{\leftarrow E} = S_{\leftarrow E} - p(E)S_E - p(\leftarrow E)S_{\leftarrow E}$$

The Dutch book can always choose to set $S_{\leftarrow E} = 0$. Then the gains become:

$$G_E = (1 - p(E))S_E$$
$$G_{\leftarrow E} = -p(E)S_E$$

Noting that the Dutch book is free to set S_E and $S_{\leftarrow E}$ either positive or negative, the gambler can prevent both gains from being negative only by assuring that $(1-p(E)) \geq 0$ and $-p(E) \leq 0$ or that $(1 - p(E)) \leq 0$ and $-p(E) \geq 0$. Obviously, the second condition cannot be met, and so, to prevent a certain loss, the gambler must enforce the condition:

$$0 \leq p(E) \leq 1$$

Next, suppose the gambler knows with certainty that E will occur, and therefore also that $\leftarrow E$ will not occur. Then, his gain is:

$$G_E = S_E - p(E)S_E - p(\leftarrow E)S_{\leftarrow E} = (1 - p(E))S_E - p(\leftarrow E)S_{\leftarrow E}$$

By choosing S_E and $S_{\leftarrow E}$ appropriately, the Dutch book can set the gain at whatever level desired, positive or negative, unless $p(E) = 1$ and $p(\leftarrow E) = 0$. Thus, to prevent an assured loss, the gambler must assign $p(E) = 1$ to an event that he believes is certain to occur, and $p(E) = 0$ to an event that he believes is certain not to occur.

Now we turn to the arithmetic of probabilities. Consider two mutually exclusive events, E_1 and E_2, such that event E represents the occurrence of E_1 or E_2. We can write this as $E_1 \cup E_2 = E$, and we pose bets on E_1, E_2 and E with beliefs $p(E_1)$, $p(E_2)$ and $p(E)$ respectively and wagers of $p(E_1)S_{E1}$, $p(E_2)S_{E2}$ and $p(E)S_E$. Now the gains can be written:

$$G_{\leftarrow E} = -p(E)S_E - p(E_1)S_{E1} - p(E_2)S_{E2}$$

if neither E_1 nor E_2 occur,

$$G_{E1} = (1 - p(E))S_E + (1 - p(E_1))S_{E1} - p(E_2)S_{E2}$$

if E_1 occurs, and

$$G_{E2} = (1 - p(E))S_E - p(E_1)S_{E1} + (1 - p(E_2))p(E)S_{E2}$$

if E_2 occurs. This can be written in matrix form as $g = Ps$, where

$$g = \begin{pmatrix} G_{\neg\mathrm{E}} \\ G_{E1} \\ G_{E2} \end{pmatrix} \text{ and } s = \begin{pmatrix} S_E \\ S_{E1} \\ S_{E2} \end{pmatrix}$$

and the matrix P is

$$P = \begin{bmatrix} -p(E) & -p(E_1) & -p(E_2) \\ 1 - p(E) & 1 - p(E_1) & -p(E_2) \\ 1 - p(E) & -p(E_1) & 1 - p(E_2) \end{bmatrix}$$

Unless the determinant of P is zero, the Dutch book can again select the elements of s to assure that all gains are strictly negative. Ergo, to avoid a loss, the gambler must assure that

$$\det P = p(E) - p(E_1) - p(E_2) = 0$$

and this leads to the condition that probabilities of mutually exclusive events add.

At this point, we have, in a rather abbreviated way and without extreme rigor, shown that the axioms of Kolmogorov probability can be derived in a decision-making context based solely on the argument that a gambler will not accept a wager that results with certainty in a loss. We can extend the Dutch book argument to include Bayes Theorem. To do this, we must consider conditional wagers. Consider an event A that is conditioned on event B, that is, A is such that the decision maker's degree of belief in A depends upon the occurrence of event B. We write the joint occurrence of A and B as $A \cap B$, and the degrees of belief as $p(B)$ and $p(A \cap B)$. The Dutch book proposes wagers on the events $\leftarrow B$ (not B), $\leftarrow A \cap B$ (B but not A) and $A \cap B$ (both A and B):

$$G_{\leftarrow B} = -p(B)S_B - p(A \cap B)S_{A \cap B}$$

if B does not occur,

$$G_{\leftarrow A \cap B} = (1 - p(B))S_B - p(A \cap B)S_{A \cap B} - p(A|B)S_{A|B}$$

if B occurs but not A, where $p(A|B)$ is the degree of belief in A given that B has occurred, and

$$G_{A\cap B} = (1 - p(B))S_B - (1 - p(A\cap B))S_{A\cap B} + (1 - p(A|B))S_{A|B}$$

if both A and B occur. This can be written in matrix form as $g = Ps$, where

$$g = \begin{pmatrix} G_{\leftarrow B} \\ G_{\leftarrow A\cap B} \\ G_{A\cap B} \end{pmatrix} \text{ and } s = \begin{pmatrix} S_B \\ S_{A\cap B} \\ S_{A|B} \end{pmatrix}$$

and the matrix P is:

$$P = \begin{bmatrix} -p(B) & -p(A\cap B) & 0 \\ 1 - p(B) & -p(A\cap B) & -p(A|B) \\ 1 - p(B) & 1 - p(A\cap B) & 1 - p(A|B) \end{bmatrix}$$

Unless the determinant of P is zero, the Dutch book can again select the elements of s to assure that all gains are strictly negative. Ergo, to avoid a loss, the gambler must assure that

$$\det P = -p(A|B)p(B) + p(A\cap B) = 0$$

which is Bayes Theorem.

It is important to keep in mind that the gambler is in complete control of his assignment of belief to each event. All that the Dutch book does is to structure the wager in accordance with the beliefs of the gambler. The Dutch book, in fact, neither needs to have beliefs of his own with regard to the events, nor does he need to know anything about the events. Yet, unless the gambler structures his beliefs in complete accord with the Kolmogorov axioms of probability theory and Bayes Theorem, the Dutch book can structure wagers that the gambler would have to accept to be consistent with his beliefs, but which provide a guaranteed loss. Yet we wish to enforce the notion that a gambler will not accept a wager that he is certain of losing. So, to prevent certain loss, the only thing the gambler can do is to accept and abide by the axioms of Kolmogorov probability and Bayes Theorem.

It is further important to note that the axioms that underlie this result are few and simple. First, we accept arithmetic on real numbers (the Peano postulates) and, second, we state that the gambler will not accept a wager that is a certain loss. It seems quite reasonable that an engineer would accept both arithmetic and the notion that a decision that is a certain loss should be avoided. Further, the engineer has only three alternatives: (1) to accept Kolmogorov probability and Bayes Theorem as the only valid approach to prediction, (2) to agree never to use arithmetic again, or (3) to give all his money to the Dutch book. It would seem that alternative (1) is the only viable alternative for an engineer who wants to continue to practice engineering.

Another important concept that the Dutch book teaches us is that the correct interpretation of a *probability* is that it is the degree of belief held by the decision maker with respect to an event that could occur in a small wager. Probabilities are determined by the decision maker in terms of his willingness to pay for a wager. Thus, probabilities are always subjective, adjectives simply don't apply to the word *probability*, and the frequentist view of probability has no place in a theory of prediction. Indeed, an observed frequency is merely evidence that would be entered into the decision to modify the decision maker's beliefs.

Finally, we mention here only in passing that the Dutch book applies in the case of "small" wagers, that is, wagers for which the gambler is risk neutral. The extension of the Dutch book to the case of large wagers is the subject of utility theory. Utility theory enables the self-consistent evaluation of large, risky alternatives, but in no way invalidates the conclusions obtained by the Dutch book argument.

A brief example may help to understand how Dutch book gambling works. Suppose there is a game between the LA Lakers and the Chicago Bulls. John is a big Bulls fan, and believes strongly that they will win. He is willing to give 2:1 odds on bets against the Bulls. Sue is a Lakers fan, and she is willing to give 3:2 odds. You are the bookie. For each dollar you bet against John, you win \$2 if the Lakers win, and for each dollar you bet against Sue, you win \$1.50 if the Bulls win. So, if you bet $\$L$ on the Lakers and $\$B$ on the Bulls, your winnings would be $\$(2L - B)$ if the Lakers win, and $\$(1.5B - L)$ if the Bulls win. Assuming no risk at all, you can set these two wins equal, $2L - B = 1.5B - L$, or $B = 1.2L$. Then, no matter what the outcome of the game, you win $2L - B = 1.5B - L = 0.8L$ for a total bet of $1.2L + B = 2.2L$. Hence, for every dollar bet, you win about \$0.36 with probability one. In this example, you can play on the inconsistency between John and Sue. But, should John or Sue be inconsistent themselves in their assessment of the relative probabilities of the Lakers and Bulls winning, you could structure a bet with either alone and still win regardless of the outcome of the game.

5.4 The Use of Evidence in Prediction

As noted above in Basic concept 4, all prediction begins with a prior, that is, with a belief held before new evidence is provided. Evidence can come in the form of anything relevant to a decision. In the case of crossing a street, evidence may be in the form of seeing oncoming traffic. In an engineering decision, evidence may take the form of the result of a computation, for example, the result of a detailed finite element analysis. If the evidence is of sufficient relevance to the decision, the decision maker will want to take it into account in making the decision.

It is useful to take a moment to understand what evidence does in the mind of the decision maker. Remember, the decision maker has a prior before the evidence is provided. We noted that this prior can be delineated through a series of questions relating to possible wagers. Typically, however, a finite set of questions cannot define the prior perfectly. Think in terms of drawing a curve through a finite set of points. No single curve is unique in that it is the only curve that goes through the set of points (even, as required of a cumulative distribution, that the curve be everywhere monotonically increasing). Indeed, there is generally an infinite set of curves that pass precisely through a finite set of points. Thus, we typically are dealing with an infinite set of possible probability distributions, each of which precisely satisfies the conditions we have set. Now, we obtain evidence, and we place a belief function on the evidence. We normally would elicit this belief function in much the same way as we did for the prior, obtaining a finite set of points or conditions. Again, we could fit an infinite set of belief functions to our data. Now, when we take evidence into account, what we are really trying to do is to find a posterior belief that is entirely self-consistent with these two sets of beliefs. This sounds like an overwhelmingly daunting task. But, in the end, we don't have to do it perfectly. We just have to do it well enough to separate the alternatives posed in the decision at hand. In accordance with the results of the Dutch book proof, the only correct, that is, self consistent, way to do this is by updating the decision maker's prior in accord with Bayes Theorem. Interestingly, this result is not only sound mathematically, it is also sensible from several aspects.

First, not all evidence is equal. Suppose an engineer, who is also the decision maker for a design, needs to know the maximum stress in a part under a given load. Of course, the engineer has a gut feel for the stress to begin with (otherwise what basis would there have been for selecting a design for analysis). This is the engineer's prior. Now, three possible methods for obtaining evidence might be (1) perform a back-of-the-envelop analysis, (2) have the stress group do a detailed finite element analysis, or (3) go out on the street and ask a homeless person: "I'll give you $5 if you'll tell me what you think is the maximum stress in this part loaded thusly." It would be natural for the engineer to place a higher degree of belief in the result of the finite element analysis than the guess of the homeless person. And, of course, doing a Bayesian update of the engineer's prior demands a statement by the engineer of his belief in the evidence. This statement would normally take the form of a probability distribution around the evidence, for example, the result of the analysis. Accordingly, the guess of the homeless person might have little or no impact on the engineer's prior, while the results of the finite element analysis could significantly alter his prior.

Second, the order in which evidence is introduced has an impact on the extent to which the evidence changes the engineer's belief. For example, we would expect that a back-of-the-envelope calculation would impact the engineer's prior much more if it is introduced before the results of a finite element analysis are introduced than later, particularly if a high degree of belief is held in the finite element analysis result. This is equivalent to saying that the back-of-the-envelope analysis has more value to the engineer in the absence of the finite element analysis than it would in the presence of the finite element analysis result. In fact, the order in which evidence is introduced can even determine whether the decision maker finds it credible (that is, believable) or not.

Third, all models, from back-of-the-envelope to finite element, provide results that are "wrong" anyway, that is, no model produces results that precisely match reality. And the Bayesian update provides a mathematically rigorous approach for taking into account "wrong" evidence.

The extent to which "wrong" evidence can be important becomes clear with the following example. Suppose a gambler is to place bets on the flip of a coin. And suppose an engineer provides the gambler with a model that predicts the outcome of each flip of the coin. But the gambler tests the model over many flips of the coin and observes that each time the model predicts heads the coin lands tails, and each time the model predicts tails the coin lands heads. So, being no dummy, the gambler bets against the model and wins every time. The model provides wrong results every time, but the gambler, based on the model result, makes a winning wager on every flip. It doesn't get any better than this. So we certainly cannot say that the model is useless, and even the concept of "wrong" seems to be in jeopardy. We quickly observe that the output of the model is not what is important – indeed it is hardly relevant. What is important is the decision and how capable the gambler is of making good decisions given the availability of the model. This effect is properly accounted for through the Bayesian update. The Bayesian update includes both an assessment of the accuracy of the evidence and its relevance to the prediction.

We can take this example to another extreme. We have all seen analyses that were somehow done in error, that is, the results, due to some mistake, were just plain wrong. And we caught the error because the result, compared to our prior, was just not believable. So the prior also plays a role in the acceptance of the evidence.

Clearly, whatever method we adopt for the consideration of evidence, it must deal with the above issues in a way that is consistent with the overall prediction framework. The Bayesian update does this.

Now, let's consider an example to illustrate how the Bayesian update works to alter one's prior on a prediction. We shall consider a rather simple example so that we don't get bogged down in complex equations. An engineer must make a "design" decision, namely a wager on how many M&Ms there are in the jar pictured in Fig. 5.1.[5] A quick glance at the jar gives the engineer confidence that there are clearly more than 100 and clearly fewer than 1,000 M&Ms in the jar. Next, the engineer can assign his degree of belief to various numbers between these limits. This assignment can be obtained through a series of proposed wagers, "What would you wager against a dollar that the number of candies is fewer than m?" By varying m, we can plot a cumulative distribution directly from the responses to this question. A result might look like the plot in Fig. 5.2.

At this point, it is already clear that, should someone present to the engineer a model result that there are 27 M&Ms in the jar, the engineer would likely dismiss

[5] Note that this is indeed an engineering example. An engineer is asked to place a wager on a decision whose outcome can be estimated in part through the use of a physics-based model that produces an imprecise result – precisely what a design engineer must do. Indeed, the only distinction between this and a more complex engineering example is that the physics of the candy-jar model can be understood by almost any engineer.

Fig. 5.1 A jar full of M&Ms

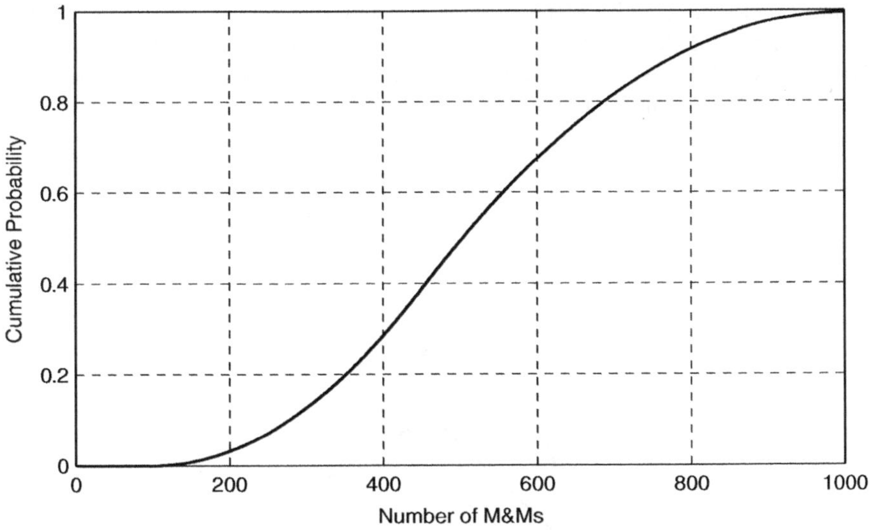

Fig. 5.2 A typical cumulative distribution for a particular decision maker

this result as utterly erroneous, or at least reexamine his prior distribution obtained above. But the importance of the prior goes well beyond this. Next, we introduce a model for the computation of the number of M&Ms in the jar. We will base our model on the simple notion that the jar contains candies and air, and the number of candies is the volume of the jar that is occupied by candies (not air), namely $V_j \mu$, where we can call μ the packing factor, divided by the average volume of a candy, v_c.

$$n = \frac{V_j}{v_c}\mu$$

The engineer can now estimate the input variables to the model and compute an estimate of the number of M&Ms in the jar. Let's say that this result is 380. The question is, given this model result, now what should the engineer think about how many candies are in the jar?

Obviously, if the engineer was certain that the model result was exact, he would believe that 380 is exactly the number of candies in the jar, and he would be willing to wager \$0.00 against a dollar that the number is fewer than 380 or any lower number, and \$1.00 that the number is fewer than 381 or any higher number. More likely, the engineer would use the model result as guidance, and he would place a degree of belief in the model. Again, the engineer could express his degree of belief in the model in terms of a series of proposed wagers of the form, "I would wager \$0.xx against \$1.00 that the model result is not high (or low) by more than y percent."

Let's assume for simplicity, that the engineer places a triangular distribution on his belief about the model, and that he is confident that the model result is true to within a factor of ± 2. This distribution is shown in Fig. 5.3 (note that, for simplicity, the cumulative distribution of Fig. 5.2 was constructed from a triangular density function, which is also shown in Fig. 5.3). We now use this distribution to update the prior of Fig. 5.2 using Bayes Theorem. The implementation of Bayes Theorem is as follows.

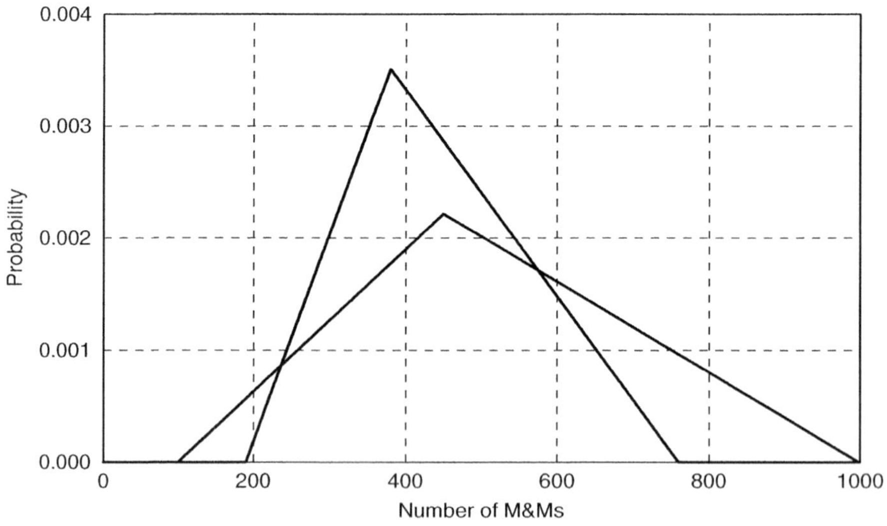

Fig. 5.3 Density functions representing the prior belief about the number of M&Ms in the jar and the belief in the model

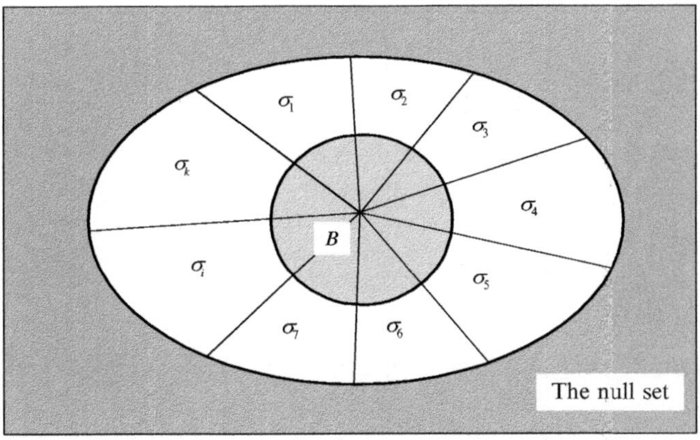

Fig. 5.4 A Venn diagram to assist in the use of Bayes theorem

Let $\sigma_1, \sigma_2, \sigma_3, \ldots, \sigma_k$ be feasible states of nature (in this case, how many M&Ms are actually in the jar), and let B denote the result of an experiment (in this case, the result of the model computation) such that $p(B \cap \sigma_i) > 0$. Taking the states of nature to be mutually exclusive and all inclusive, that is, they are the only possible states of nature, and referring to the Venn diagram of Fig. 5.4, we can write:

$$p(B) = p(B \cap \sigma_1) + p(B \cap \sigma_2) + p(B \cap \sigma_3) + \cdots + p(B \cap \sigma_k)$$

And we can then compute the posterior distribution $p(A|B)$ from the equation

$$\begin{aligned} p(\sigma_i|B) &= \frac{p(B \cap \sigma_i)}{p(B)} \\ &= \frac{p(\sigma_i)p(B|\sigma_i)}{\sum\limits_{j=1}^{k} p(B \cap \sigma_j)} \\ &= \frac{p(\sigma_i)p(B|\sigma_i)}{\sum\limits_{j=1}^{k} p(\sigma_j)p(B|\sigma_j)} \end{aligned}$$

The result is shown in Fig. 5.5.

We note here that this result is consistent with both the engineer's prior distribution $p(\sigma)$ and the distribution he put on the model result, $p(B)$. It is also interesting to apply Bayes Theorem to the case of the coin toss given above, where the model always produces the "wrong" result (that is, it always predicts tails, t, when the coin comes up heads, H, and heads, h, when the coin comes up tails, T). Thus, the terms $p(t|T) = p(h|H) = 0$ and $p(t|H) = p(h|T) = 1$, and Bayes Theorem

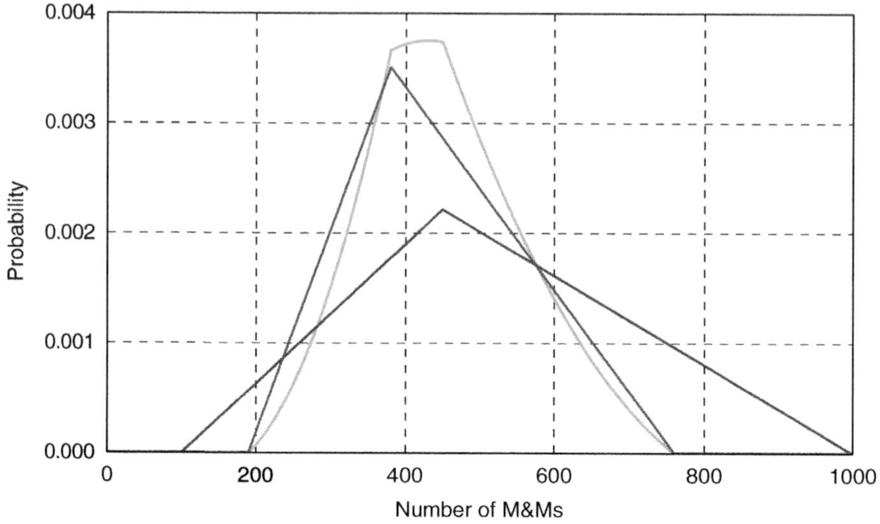

Fig. 5.5 The posterior distribution showing the belief of the decision maker after the introduction of evidence provided by a model

gives precisely the correct probability law, namely $p(T|t) = p(H|h) = 0$ and $p(T|h) = p(H|t) = 1$, on the outcome of the coin toss, even though the model gives the wrong result on every toss. And, of course, to make a good decision, we need a good probability law, not necessarily a good model result. The two are different as clearly demonstrated here. This case dramatically emphasizes the need to do a mathematically correct update of the prior distribution.

5.5 Stochastic Modeling

It might appear that we have solved all the problems of prediction with the analysis given above. But, unfortunately, these calculations alone leave considerable room for loss of consistency. Let's go back to the M&M prediction. The engineer used a deterministic model to compute his estimate of the number of M&Ms in the jar. As inputs to the model, he made estimates of V_j, μ and v_c. And, of course, he has a prior on the accuracy of each of these variables. So first, we need to assure that his prior on n is consistent with his priors on V_j, μ and v_c. But, second, we need to make sure that we don't make mathematical errors in the computation of n or the engineer's prior on n. To do this, we need to turn to stochastic modeling.

While the above equation for n might look appropriate, we need to remember that V_j, μ and v_c are random variables, as the engineer does not know them precisely. We say that they are uncertain. Now random variables are neither random nor are they variables. Rather they are mappings. For example, we can map a coin toss as heads = 0, tails = 1, and the resulting number 0 or 1 is called a random variable. In other words, a random variable is a mapping of events onto the real number line. In the case of events that are represented, that is, named, as numbers, for example, μ, which could take on values like 0.55, 0.56, 0.57, etc., it may be convenient to map these events onto the real number line such that the value assigned equals the name of the event. In other words, the event called 0.57 would map into the number 0.57 on the real number line. Nonetheless, the event is not a number, rather it maps onto the real number line as a number. Thus, random variables are mappings, they are not numbers, and it is not valid to perform arithmetical operations on them. It follows that, unless V_j, μ and v_c are known precisely, we cannot use the above equation for n. We could correct this equation by writing it in terms of expected values, which are numbers and, assuming that V_j, μ and v_c are independently distributed, we would get the equation

$$E\{n\} = E\{V_j\}\, E\left\{\frac{1}{v_c}\right\} E\{\mu\}$$

where $E\{.\}$ denotes the expectation. This would give us a correct expression for the expectation of n, and with some computational agility, we could also estimate the variance on this expectation.

Another method is to use Monte Carlo simulation to estimate the engineer's belief in n. The Monte Carlo method presumes that the variables in the equation are known precisely, and they are obtained by sampling their distributions repeatedly for many trials. Through repeated computations, a probability distribution(s) on the output variable(s) is obtained. This procedure corrects the mistake of treating the input variables as numbers.

As deterministic modeling is the norm for most engineering prediction, it is reasonable to ask how much error one is likely to incur by mistaking random variables for deterministic numbers? The answer is that, in the absence of a deliberate process to limit the error, it is unbounded. We can demonstrate this through a simple example. Suppose we are computing the ratio $x = y/z$. This would be the case with determination of a from the relationship $F = ma$, or the determination of I from the relationship $V = IR$. It's also the case in computing the number of M&Ms in the jar. Obviously, this is a common computation for an engineer. Let's take a very simple case for this computation in which y is known precisely, $y = 1$, and z is a random variable uniformly distributed on the interval 0 to 1 as shown in Fig. 5.6.

In the deterministic approach, the idea is to put one's best guess into the equation and compute the output. Given the distribution on z, it is reasonable to say that the best guess is $z = 0.5$, that is, its expected value. Thus, the computation yields $x = 1/0.5 = 2$. On the other hand, if we do the computation correctly, we obtain

$$E\{x\} = E\{y\}\, E\left\{\frac{1}{z}\right\} = 1\int_0^1 \frac{dz}{z} = \ln z|_0^1 = \infty$$

and so we have shown that the error can be completely unbounded.

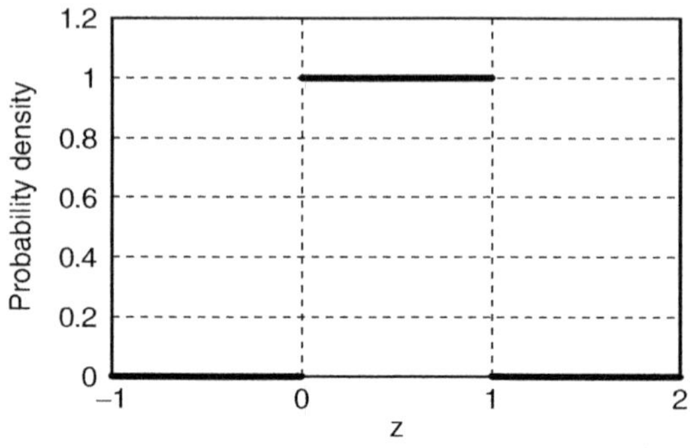

Fig. 5.6 A uniform distribution on the interval 0 to 1

Once we have corrected the error of performing arithmetic on things that are not cardinal numbers, we next need to recognize that there are two fundamentally different types of models. We shall call them definitional models and constitutive models. Definitional models are valid and precise by definition. For example, our equation for the number of M&Ms in the jar, $n = V_j \mu / v_c$, is a definitional model. By definition, the volume of the jar occupied by M&Ms is $V_j \mu$, and the average volume of an M&M is defined by the relationship $v_c = V_j \mu / n$. It follows, by definition, that if we pick values for V_j, μ and v_c that are precisely correct, we will compute exactly the correct value for n. Thus, in a definitional model, exclusive of computational errors, all of the uncertainty in the result is because of the uncertainty in the input variables. Generally, we can assume that computational errors are small (especially if we take care to keep them small, and we don't make mistakes). So, if we can quantify the uncertainty in the input variables, we can use a procedure such as Monte Carlo modeling to propagate that uncertainty into the result. And you may very well be quite surprised to see how much uncertainty results. In the M&M example given, and with very reasonable quantification of the uncertainty on the input variables, we find that the accuracy we can achieve by such a model may indeed be fairly well represented by the triangular distribution of Fig. 5.3, with error bounds of roughly a factor of ± 2. And this is a very simple case with only three variables, all relatively well known. Particularly when multiplications and divisions are present in models, uncertainty grows rapidly.

What is interesting here is that, if we correctly represent the uncertainty held by the decision maker on the input variables to the model, the result of the model, represented as a probability distribution on the output variable, is itself a prediction that is consistent with the beliefs of the decision maker. Of course, in this case, evidence would be introduced on the input variables, modifying their prior distributions in accord with Bayes Theorem. But the model properly aggregates these uncertainties into the final prediction. Hence, such a stochastic model is a predictive model, whereas we recognize that, taken as a deterministic model, it merely produces evidence to be incorporated into the decision maker's final prediction using Bayes Theorem.

In contrast to definitional models, constitutive models typically make use of laws of nature to predict the future. But neither do we know the laws of nature with precision and certainty, nor can we implement them precisely. Most often, the models we generate to represent a system are based on a set of simplifying assumptions, neglecting phenomena that we think are not significant. And the solutions to these models are themselves approximations. Thus, in the case of constitutive models, in addition to uncertainty in the input variables, there is a substantial component of uncertainty in the result that is attributable to the fundamental inaccuracy of the model itself and the algorithm by which the model is solved. This uncertainty has been addressed by Hazelrigg (2003).

Suppose we have a model of the form $y = f(x)$, which we know to be imprecise. Assuming that $f(x) \neq 0$, we could imagine a term $\boldsymbol{\phi}(x)$ by which, if we had the wisdom to determine it, we could multiply by $f(x)$ to obtain the true value of y, namely $y = \boldsymbol{\phi}(x) f(x)$. Of course, it is hopeless to think of actually obtaining $\boldsymbol{\phi}(x)$. But we can determine the statistical properties of $\boldsymbol{\phi}(x)$, and that is exactly what we need to estimate the total uncertainty in our model. And, as with all things probabilistic, $\boldsymbol{\phi}(x)$ is determined subjectively. It is the decision maker's belief in the accuracy of the model.

Now, if we formulate a stochastic constitutive model, in which we treat input variables as uncertain, then we can, using a method such as Monte Carlo, determine the uncertainty in the result that obtains from the uncertainty in the input variables. To this we must add the uncertainty that the decision maker holds with respect to the model itself and its solution algorithm. Again, this computation is done using Bayes Theorem

$$
\begin{aligned}
p(y) &= p(\phi|f)P(f) \\
&= \sum_f p\left(\frac{y}{f}\middle| f\right) p(f) \\
&= \sum_f p\left(\frac{y}{f}\right) p(f)
\end{aligned}
$$

where, because of the multiplicative form chosen for $\boldsymbol{\phi}(x)$, we have assumed that it is independent of $f(x)$. For example, the decision maker might feel that the model is accurate to $\pm 10\%$ over its entire range. The effect of adding model uncertainty to input variable uncertainty is to spread the probability distribution wider and reduce the peaks of the distribution.

5.6 Conclusions

The Dutch book argument presented in the context of prediction for engineering decision making provides us with many important insights. First, it emphasizes that probability theory is all about the future, and all about prediction for the purpose of decision making. There really is no other reason to be concerned about the future, at least from the engineering point of view.

Second, the DBA provides us with a clear definition of *probability*: it is defined in terms of the beliefs held by a gambler in the context of a small wager. We gain a clear distinction between events that will occur in the future and past events together with the frequentist view of probabilities. We now view observed frequencies as evidence, and we support the notion that a probability is a belief, no more, no less, and that adjectives such as *objective* and *subjective* simply do not apply. All beliefs are subjective.

Third, we see that the axioms of Kolomogorov probability theory and Bayes Theorem provide a rigorous and philosophically sound framework for thinking about the future. But more than this, the DBA provides an argument that is sufficiently compelling to all but dismiss outright any alternative framework.

Fourth, we now understand the notion of predictive modeling to mean only stochastic models. We see the notion of deterministic "predictive models" to be rather naive, and we see that, instead of predicting the future, deterministic models merely provide evidence that may be incorporated into the predictive process. This was made quite clear by the example in which the model gave exactly the wrong answer every time, leading the gambler to make the best wager every time.

Fifth, it is now clear that validation of a model is not even a relevant concept. Rather we seek to validate a decision. So we now look at a typical deterministic engineering model only in the context of providing evidence for prediction within the context of a specific decision with a specific decision maker.

Sixth, we have emphasized the importance of using stochastic models. Stochastic models correct errors present in deterministic models, particularly the error of doing arithmetic on random variables, and they provide an estimate of the accuracy of the model that is consistent with the beliefs of the decision maker as regards the inputs to the model.

Seventh, we see that properly formulated stochastic models can be predictive and provide a prediction of the future complete with an assessment of the accuracy of the prediction.

Finally, it is worth reemphasizing that the reason for any mathematic is to provide a framework for consistency. The mathematics of prediction provide a framework for thinking rationally, that is, in a way that is entirely self-consistent, about the future. Thus, this mathematic provides a framework for getting the most out of evidence. Failure to follow this logic is very likely to lead to significant losses in engineering decision making.

The notions presented in this paper are not new. A significant debate on probability theory took place among the top mathematicians during the early half of the twentieth century. The above theories have been heavily vetted in mathematical circles. Yet these mathematics have not been embraced by engineering, and much of what is taught in engineering is considerably at odds with the theory presented here. It is important to distinguish the mathematics of applied science from the mathematics of engineering, the latter having a substantive base in probability theory (not statistics),[6] and it is important that engineers learn these mathematics in order that they can be good decision makers.

References

Caves, C.M. (2003) "Betting probabilities and the Dutch book," originally written June 24, 2000, revised May 24, 2003. Unpublished manuscript.

de Finetti, B. (1937) "La Prévision: Ses Lois Logiques, ses Sources Subjectives," *Annales de lyInstitut Henri Poincaré* 7, 1–68. Translated into English by Henry E. Kyburg Jr., "Foresight: Its Logical Laws, its Subjective Sources." In Henry E. Kyburg Jr. & Howard E. Smokler (1964, Eds.), *Studies in Subjective Probability*, Wiley, New York, 53–118; 2nd edition 1980, Krieger, New York.

Hazelrigg, G.A. (2003) "Thoughts on Model Validation for Engineering Design," Paper DETC2003/DTM-48632, Proceedings of the ASME 2003 Design Engineering Technical Conference, September 2–6, 2003, Chicago, Illinois.

Ramsey, F.P. (1931) "Truth and Probability," in "*The Foundations of Mathematics and other Logical Essays*," 156–198, Routledge and Kegan Paul, London. Reprinted in Henry E. Kyburg Jr. & Howard E. Smokler (1964, Eds.), *Studies in Subjective Probability*, 61–92 Wiley, New York; 2nd edition 1980, Krieger, New York.

Thurston, D.L. (1999) "Real and Perceived Limitations to Decision Based Design," Paper DETC99/DTM-8750, Proceedings of the 1999 ASME Design Engineering Technical Conferences, September 12–15, 1999, Las Vegas, Nevada.

[6] A useful view is that probability theory is a framework for thinking logically about the future. Statistics is a framework for thinking logically about the past. These are quite different things. Science has a focus on explaining observations (data), which are always about the past. Designers must think about the future, they must predict it.

Chapter 6
An Exploratory Study of Simulated Decision-Making in Preliminary Vehicle Design

Joseph A. Donndelinger, John A. Cafeo, and Robert L. Nagel

Abstract Results are presented from a simulation in which three engineers were independently tasked with choosing a vehicle subsystem design concept from a set of fictitious alternatives. The authors acted as analysts and responded to the decision-makers' requests for information while also observing their information collection and decision-making processes. The results of this study serve as a basis for comparing and contrasting common decision-making practices with established theories of normative decision analysis, cognition, and psychological type.

Keywords Decision-making · New product development · Cognitive fit theory · Information visualization · Myers-Briggs Type Indicator

6.1 Introduction

Although there has been a considerable amount of research into decision-making in engineering design, there is not yet a consensus within the engineering design community as to the definition of a decision. Conceptually, Herrmann and Schmidt (2006) view decisions as value-added operations performed on information flowing through a product development team, while Carey et al. (2002) view decisions as strategic considerations that should be addressed by specific functional activities at specific points in the product development process to maximize the market success of the product being developed. Matheson and Howard (1989) offer this definition:

> "A decision is an irrevocable allocation of resources, in the sense that it would take additional resources, perhaps prohibitive in amount, to change the allocation."

J.A. Donndelinger (✉) and J.A. Cafeo
General Motors R&D Center, Mail Code 480-106-256, 30500 Mound Road,
Warren, MI 48090, USA
e-mail: joe.donndelinger@gm.com

R.L. Nagel
Missouri University of Science and Technology, G-4B Interdisciplinary Engineering Building,
1870 Miner Circle, Rolla, MO 65409, USA

N.R.S. Raghavan and J.A. Cafeo (eds.), *Product Research: The Art and Science Behind Successful Product Launches*, DOI 10.1007/978-90-481-2860-0_6,

Commonly, it is understood that a decision is a selection of one from among a set of alternatives after some consideration. This is illustrated well by Hazelrigg (2001) in his discussion of the dialogue between Alice and the Cheshire cat in Alice in Wonderland. He notes that in every decision, there are alternatives. Corresponding to these alternatives are possible outcomes. The decision-maker weighs the possible outcomes and selects the alternative with the outcomes that he or she most prefers. Although apparently simple, this discussion contains several subtle but powerful distinctions. First, there is a clearly designated decision-maker who will select an alternative and will experience the consequences of that selection. Second, the decision is made according to the preferences of the decision-maker – not those of the decision-maker's stakeholders, or customers, or team members, or for that matter anyone's preferences but the decision-maker's. Third, the decision-maker's preferences are applied not to the alternatives, but to the outcomes. Finally, and perhaps most importantly, the decision is an action taken in the present to achieve a desired outcome in the future.

The intent of this work is to compare and contrast theory and practice in the area of decision-making. This begins with an overview of normative decision analysis and some additional aspects of the cognitive thought process. Next, the results from several observations of simulated decision-making simulations are presented. Following these observations a discussion of the connections and mismatches between theory and practice is presented that hopefully sets the stage for further connecting the theory and practice of decision-making.

6.2 Prior Work

6.2.1 Decision Analysis

Decision analysis is the practical application of normative decision theory concerned with how a decision ought to be made. A good decision is based on the information, values and preferences of a decision-maker and is one that is favorably regarded by a decision-maker. Good decisions, however, may yield either good or bad outcomes from the perspective of the decision maker.

Implicit in this view is that different decision-makers will make different decisions (i.e., make different choices from the same set of alternatives) due to differences in background and experiences, beliefs about the future, presentation of information and risk attitude. Exploration and development of alternatives is typically an integral part of the decision-making process. Often times this is not a separate step but is intertwined in an iterative decision analysis cycle. Beliefs about the future and assessments of risk can be and often are informed by evidence external to a person's experience. This evidence can be drawn from past experience, models of events, statistical analyses, and simulation of future possibilities. Gathering evidence is an iterative and interactive process in which decision-makers will likely request additional information beyond what is originally presented. Finally, it is important to understand that aligning the presentation of information with the structure of the decision at hand facilitates the decision-maker's natural cognitive processes (Speier 2006), potentially leading to more accurate interpretation of information and faster execution of the decision analysis cycle. A simulation of decision-making for design concept selection on a new vehicle program is presented to fully develop these points.

To appreciate the context for decisions in product development for the case study presented here, it is instructive to review the basic ideas around the vehicle development process (VDP). The VDP is the series of actions taken to bring a vehicle to market. For many vehicle manufacturers, the VDP is based on a classical systems engineering approach to product development. The initial phase of the VDP is focused on identifying customer requirements and then translating them into lower level specifications for various functional activities, such as product planning, marketing, styling, manufacturing, finance, and various engineering activities. Work within the VDP then proceeds in a highly parallel fashion. Engineers design subsystems to satisfy the lower level requirements; the subsystems are then integrated to assess their compatibility with one another and to analyze the vehicle's conformance to the customer requirements. Meanwhile, other functional staffs work to satisfy their own requirements: product planning tracks the progress of the VDP to ensure that the vehicle program is adhering to its schedule and budget, marketing ensures that the vehicle design will support its sales and pricing goals, finance evaluates the vehicle design to ensure that it is consistent with the vehicle's established cost structure, manufacturing assesses the vehicle design to ensure that it is possible to build within the target assembly plant, and so on. This is typically the most complex, and the most iterative phase of the VDP, as literally thousands of choices and trade-offs are made. Finally, the product development team converges on a compatible set of requirements and a corresponding vehicle design. Engineers then release their parts for production and the vehicle proceeds through a series of pre-production build phases culminating in the start of production.

The VDP is typically preceded by a preliminary phase (hereinafter P-VDP) that is analogous to the VDP in its structure and content, but conducted at a higher level and on a smaller scale due to the high levels of uncertainty involved. The P-VDP is a structured phase-gate product development process in which a fully cross-functional product development team rigorously evaluates the viability of a new vehicle concept. Through the course of the P-VDP, a concept begins as an idea and matures to a preliminary design for a vehicle that could be produced and incorporated into the product portfolio. The process encompasses a broad scope of activities, including determining the marketing requirements for the vehicle, determining a manufacturing strategy, designing the vehicle's appearance, assessing the engineering design feasibility, and developing a business case to assess the vehicle's financial viability. At the end of the process, the results are presented to a decision board that decides whether development of the vehicle will continue through the VDP or will be suspended with the results of the study saved for future consideration. The benefit of executing the P-VDP is that potential impasses in the development of the vehicle may be determined at a very early stage before significant investments of human resources and capital have been made.

6.2.2 Decision Analysis Cycle

Decision-makers in vehicle development are, by definition, executives or senior managers who have the authority to allocate an organization's resources. There is some evidence (Donndelinger 2006) to suggest that they make decisions using the Decision Analysis Cycle (DAC) shown in Fig. 6.1 and described completely in Matheson and Howard (1989).

The discussion of the phases in the DAC contains several precisely defined terms from the language of formal decision analysis (Matheson and Howard 1989). The term "Value" is used to describe a measure of the desirability of each outcome. For a business, the value is typically expressed as some measure of profit. The term "Preferences" refers to the decision-makers' attitude toward postponement or uncertainty in the outcomes of his decision.

The three phases of the Decision Analysis Cycle that precede the Decision are: (1) Deterministic Phase – the variables affecting the decision are defined and related, values are assigned, and the importance of the variables is measured without consideration of uncertainty. (2) Probabilistic Phase – probabilities are assigned for the important variables, and associated probabilities are derived for the values. This phase also introduces the assignment of risk preference, which provides the solution in the face of uncertainty. (3) Informational Phase – the results of the previous two phases are reviewed to determine the economic value of eliminating uncertainty in each of the important variables of the problem. A comparison of the value of information with its cost determines whether additional information should be collected.

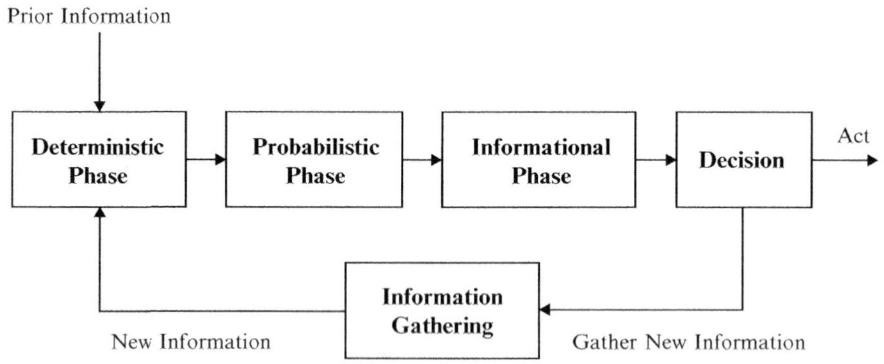

Fig. 6.1 The decision analysis cycle (Matheson and Howard 1989)

6.2.3 Human Aspects

Decisions-making is a human activity, not the result of executing a mathematical program. This is necessarily true because the models underlying mathematical programs are limited abstractions of reality. Even if the models scopes are broad enough to cover all factors relevant to the decision they supports (which rarely occurs), there still remains in every decision an element of aleatory uncertainty that models cannot address. The role of the human in the decision-making process is to supplement outputs from models with their understanding of the problem, experiences, beliefs about the future and risk attitude and ultimately to commit to a course of action. It may help to assume that people are rational when developing methods and tools to support decision-making; however, rationality is a complex notion. An understanding or at least an appreciation for the human side of decision-making is important when considering what an assumption of rationality implies.

6.2.3.1 State of Information

The state of information is an important concept as it is an important part of the basis that a person will use to frame the decision at hand. In order to properly frame a decision, three elements must be considered: the alternatives available (and perhaps creatable), the state of information, and the decision-maker's preferences on outcomes. A decision-maker's state of information is based both on experience and on the specific information that is collected for a particular decision. The author's believe that there is an interaction between these two contributions as experience will influence what type of information is sought and how it is used in the decision.

It is clear from this study that past experiences heavily influence the decision making process. Experience, loosely defined is the collection of events and activities from which an individual may gather knowledge, opinions, and skills. Experience may influence the decision-making process in several ways. It may be applied to reduce uncertainty when similar or analogous circumstances from the past can be used to project outcomes for the situation at hand. Analogies are powerful tools in problem understanding. Experience can provide confidence (or perhaps false confidence) that success is possible even when facing a great deal of uncertainty. Confidence based on past experiences both informs decision-makers and shapes their risk attitudes. Experience can provide a template for weeding through the overabundance of information to find the key pieces of information needed for reducing uncertainty.

6.2.3.2 Cognition

The manner in which information is visualized or presented to the intended audience plays a key role in the decision-making process and can either aid or hamper the decision-making process. The importance of using effective visualizations in the decision-making process is recognized in many texts (Card 1999; Tegarden 1999;

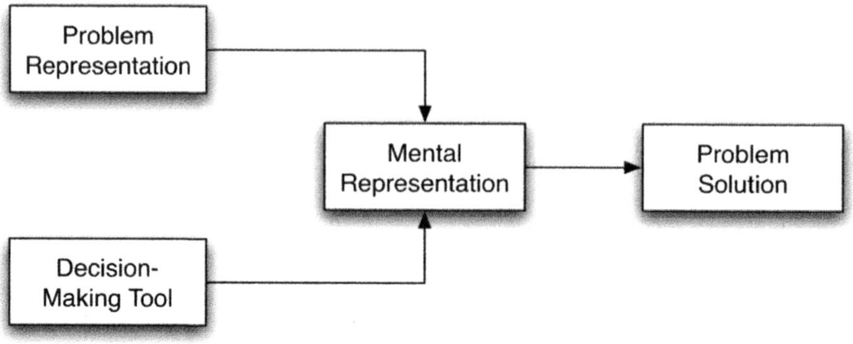

Fig. 6.2 Cognitive Fit Theory (Card 1999)

Ware 2004; Tufte 2006). In Tufte (2006) a number of visualization techniques are explored to demonstrate design strategies for effective visualizations. Cognitive fit theory, schematically shown in Fig. 6.2, addresses the importance of matching the representation of information with the decision making task being performed (Vessey 1991), and "suggests that decision performance improves when the task representation and problem representation (e.g., the format in which information is presented) match as this allows decision-makers to develop a more accurate mental representation of the problem" (Speier 2006). In Vessey (1991), cognitive fit theory is applied to simple symbolic (tabular) and spatial (graphical) tasks with paired data sources demonstrating that when the information provided is matched to the decision making task provided, the speed and accuracy of performance is improved. Cognitive fit theory has since been expanded to complex tasks, where it was found that as tasks became more complex, the time required to make decisions grew more sensitive to the alignment between information format and problem structure (Speier 2006). Other extensions to cognitive fit theory include: geographic tasks using geographical information representations (Denis and Carte 1998), representations based on objected-oriented and process-oriented methodologies (Agarwal et al. 1996), and the application of cost–benefit analysis to further influence the data representation chosen for a task (Vessey 1994).

This concept manifests itself within many large corporations. To a large extent, the availability of software in large corporations is determined by a corporate need for common tools that meet most general requirements at reasonable costs. The ubiquitous applications for storing and sharing information are the spreadsheet, word processing, and presentation applications in an office productivity suite. These well-known tools each have unique capabilities but can also be used for many of the same purposes. Indeed, some newer versions of these tools are structured such that the underlying data are stored in common formats and are interchangeable. Over the long term, it has been observed that spreadsheets have become very capable and heavily used engineering tools; they have well known interfaces that seem to resonate with many engineers. Their tabular format and simple graphing capabilities facilitate many of the discussions that take place throughout the product development process. Spreadsheets also provide adaptive short term knowledge capture and generation. Once consensus is reached, it is common practice to prepare a presentation to communicate the end result. These files, particularly some key presentation slides, can reside for years and often become the official knowledge repositories within some domains of the corporation. It has been argued by Tufte (2006) that some presentation programs reflect the organizational structure from which they were created and hence promote a certain cognitive style that may not be conducive to serious information exchange.

6.2.3.3 Personality

Personality loosely defined is comprised of the enduring patterns of perceiving, relating to, and thinking about the environment and oneself. Personality traits are prominent aspects of personality that are exhibited in a wide range of important social and personal contexts (Heuer 2007). In this study, the Myers-Briggs Type Indicator (MBTI) is used to frame discussions around the effect of personality on decision-making. This was chosen because it is a convenient, well known framework for discussing the effect of personality. It should be noted that this is strictly a United States (US) study. The participants were all US natives. There is much research and discussion about adapting the MBTI for use in other cultures [http://www.ndu.edu/library/ic6/93S86.pdf].

The typology model originated by Jung (and further developed by Briggs and Myers) regards personality type as similar to left or right handedness: individuals are either born with, or develop, certain preferred ways of thinking and acting. The MBTI sorts some of these psychological differences into four opposite pairs. None of these types is “better” or “worse”; however, Briggs and Myers posited that everyone naturally matches best to one overall combination of psychological type. In the same way that writing with the left hand is hard work for a right-hander, so people tend to find using their opposite psychological preferences more difficult, even if they can become more proficient (and therefore behaviorally flexible) with practice and development (Myers 1998).

The MBTI identifies preferences on the following four scales, with two opposite preferences defining the extremities or poles of each scale with a resulting 16 possible psychological types (Myers 1998):

1. Where an individual prefers to focus his/her attention and get energy (**E**xtraversion or **I**ntroversion)
2. The way an individual prefers to take in information (**S**ensing or **IN**tuition)
3. The way an individual prefers to make decisions (**T**hinking or **F**eeling)
4. The way an individual orients himself/herself to the external world (**J**udging or **P**erceiving)

Preference characteristics for these four dimensions are stated in Vessey (1994). An excellent description of MBTI applied to engineering project teams can be found in Culp and Smith (2001).

6.3 Methodology

Herein, observations are presented of a decision process where three decision-makers are independently presented with a fictitious new technology introduction. During this experiment, real people working in General Motors' product development organization, are enlisted as decision-makers, while the authors act as analysts, responding to the decision-makers requests for information. Observations around this human-in-the-loop simulation about a fictitious new technology decision allow the development of some understanding about how "real" decisions fit into the theoretical perspectives discussed previously. It should be noted, that in the real product development process, it is more common to employ a team approach to decision-making.

6.3.1 Method

At the beginning of the study, each subject was presented with detailed instructions covering the roles of the various agents in the study, the frame for the decision to be made, and processes to be used for decision-making and collecting information. The two types of agents in this study were decision-makers and analysts. There were three decision-makers, each an engineer selected from within General Motors' Advanced Vehicle Development Center. Each decision-maker was tasked to choose a design concept for a door seal on a new vehicle program from a set of fictitious alternatives. All three decision processes began on the same day and proceeded in parallel. Each decision-maker was aware that exercise was being conducted with others, but worked independently and was instructed not to discuss this exercise with each other until every decision-maker had selected a concept. There were two analysts in this study, each a researcher from the Vehicle Development Research Laboratory at the General Motors Research & Development Center. Both analysts participated in all three decision processes. The analysts also observed and recorded the questions, comments, and actions of the decision-makers. It should be noted that no attempt was made in this study to account for the effects of group dynamics between the decision-makers and the analysts.

In the initial meeting with each decision-maker, the decision was described as a choice between two fictitious design concepts (discussed in Section 6.3.2) for a new luxury coupe. The decision-makers were given no explicit direction on decision framing; they were not instructed to question the decision frame, nor were they instructed not to question it. The decision-makers were also given no explicit direction to use or not to use any specific decision-making techniques; they were only informed that when they made their decision, they would be required to discuss the process they used to reach their decision and the manner in which the information they collected was used in this process. The decision-makers were presented with hard decision deadlines 21 days from their initial session at which point information collection would cease and each decision-maker would be required to choose a design concept. It was also made clear at the beginning of the study that the decision-makers were free to make a decision before the deadline if prepared to do so.

The information collection process began for each decision-maker during the initial meeting. Each decision-maker was presented with descriptions of two fictitious door seal design concepts; these are shown exactly as they were presented to the decision-makers in Section 6.3.2. At the initial meetings, each decision-maker was also instructed that follow-up sessions would be scheduled at his discretion for the purposes of requesting additional information and clarifying interpretations of information provided in response to requests. No restrictions were imposed on information requests; the decision-makers were instructed to request whatever information they believed to be necessary for making this decision. This process of collecting and providing information gave rise to a distinction between "information requests" (questions asked about the alternatives by the decision-makers) and "information objects" (responses to the decision-makers' questions provided by the analysts in the form of memos, graphs, tables of data, and so forth). It is noteworthy that although the exact forms, wordings, and even numbers of "information requests" varied widely across the three decision-makers, there were much higher degrees of structure and similarity in the "information objects" that were prepared for the decision-makers and subsequently accepted as responses to their "information requests". Therefore, in this work the authors have chosen to use "information objects" rather than "information requests" as a basis for analyzing similarities and differences in the information collection processes across decision-makers.

6.3.2 Problem Statement

A concept selection is required on door seals for a vehicle that will be marketed as a luxury sports coupe. There are two potential door sealing options:

1. A traditional, passive sealing approach which utilizes three seals placed around the exterior of the door. This traditional sealing approach, shown in Fig. 6.3 (*left*), utilizes two inner belt seals and a single gap-closeout seal.

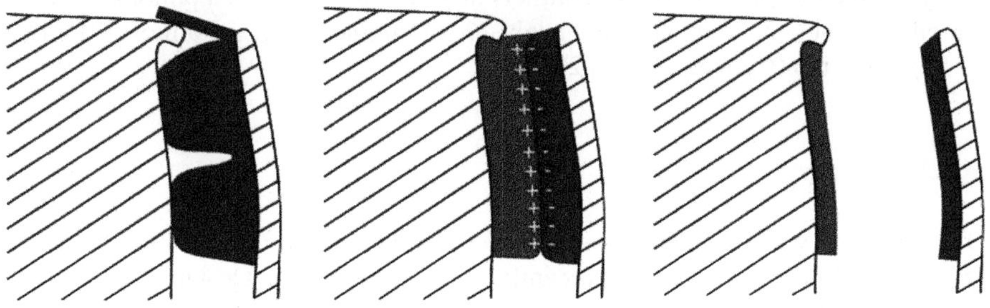

Fig. 6.3 Traditional, passive sealing approach with two inner belt seals and a gap-closeout seal (left), biology-inspired, active sealing approach in the sealed position (middle), biology-inspired, active sealing approach in the open position (right)

2. A new, biology-inspired, active sealing approach proposed by a small technology startup company is based on adhesion properties of the feet of many insects and some small animals such as tree frogs and lizards. The biological sealing or gripping techniques of the insects and animals work in one of two unique ways: (1) using specialized hairs forming thousands of small attachments points relying on Van der Waal forces for stiction or (2) using a specialized material structure that allows the shape of the feet to reform and perfectly match the surface (Creton and Gorb 2007). The biology-inspired door seal takes inspiration from both techniques and consists of two oppositely charged elastic filaments – one around the door ring and the other around the door. When the elastic filaments come into contact, the filaments morph and stretch together to seal the gap between the door and the frame (shown in Fig. 6.3 (*middle*)). The new door sealing technology has the potential to provide a number of benefits to the customer such as significantly reduced closing and opening effort through door opening and closing assist, active acoustic tuning, vibration damping, and improved water leak performance. The cost of the new door seals is currently projected to be $10/vehicle higher than the cost of the conventional seal; however, the supplier has a cost reduction plan that, if completely successful, could bring the cost of the new door sealing system below the cost of a conventional triple-seal system.

6.3.3 Description of Decision-Makers

6.3.3.1 Jim

Jim is an experienced systems engineer. He has spent the vast majority of his 35 years of professional experience working in various systems engineering roles and is very knowledgeable in the areas of representing the Voice of the Customer and developing vehicle-level requirements at the earliest stages of the vehicle development process. Jim's MBTI is ISFP. According to Myers (1998), ISFPs may be characterized as quiet, friendly, sensitive and kind. They enjoy the present moment and like to have their own space and to work within their own time frame. They are loyal and committed to their own values and to people who are important to them. They dislike disagreements and conflicts and do not force their opinions or values on others. They make up approximately 2% of engineers but are almost 9% of the U.S. population (Culp and Smith 2001).

6.3.3.2 Terry

Terry is a knowledgeable and well-respected structural engineer with 25 years of professional experience. He has recently transferred into the Design for Six Sigma (DFSS) activity and in this capacity frequently participates in gate reviews for DFSS projects. Over the last ten years, Terry has consistently tested as an ISTP on MBTI. According to Myers (1998) an ISTP is tolerant and flexible. They are quiet observers until a problem appears, then act quickly to find workable solutions. They analyze what makes things work and readily get through large amounts of data to isolate the core practical problems. They are interested in cause and effect and organize facts using logical principles. Finally, they value efficiency. They make up approximately 6% of the engineers and are almost 6% of the U.S. population (Culp and Smith 2001).

6.3.3.3 Glenn

Glenn is an energetic, young engineer in the systems engineering department. He has 6 years of professional experience, predominantly in the automotive sector with some additional experience in the defense sector. He has spent much of his professional career developing empirical models to support activities ranging from component design to product planning strategies. His mathematical skills made him a natural fit in the DFSS activity, where he met Terry and initiated a mentor-protégé partnership with him. Glenn has recently transferred to Systems Engineering to broaden his professional experience and works closely with Jim on a high-profile project. His MBTI is ESTJ. According to Myers (1998) an ESTJ is practical, realistic and matter-of-fact. They are decisive and mover quickly to implement decisions. They organize projects and people to get things done and focus on getting results in the most efficient way possible. They take care of routine details and have a clear set of logical standards that they systematically follow and want others to do the same. They are forceful in implementing their plans. They make up approximately 8% of engineers and are almost 9% of the U.S. population (Culp and Smith 2001).

6.4 Results

As expected, there were substantial differences in the processes, information requests, and decisions of the three decision-makers; however, there were also a number of noteworthy similarities. In this section, the common elements across all three decision processes are reviewed, followed by an account of each individual's decision process with an emphasis on its unique elements.

6.4.1 Common Elements

Overall, the area of greatest commonality between the three decision-makers was in their decision-making process. Each decision-maker employed a multi-criteria decision-making process in which criteria were defined (or at least disclosed) towards the end of the information collection process. Each decision-maker used a flat list of criteria, with; no provisions for aggregating or hierarchically organizing criteria or for differentiating means criteria from ends criteria. None of the decision-makers assigned weights or relative importances to any of their criteria. All three decision-makers accepted the decision frame as originally presented; none attempted to define additional alternatives based on the information they collected. Each of the decision-makers discussed risk in semantic terms only; risks were identified for both alternatives on some of the criteria, but there were no attempts made to quantify them.

Most of the remaining commonalities were in the information collection processes of the three decision-makers. They shared several information objects in common, focused on the biology-inspired seal and arguably in the most obvious areas. Two examples of these are the company background for the supplier, shown in Fig. 6.4, and test results showing the temperature sensitivity of the seal material, shown in Fig. 6.5. Other information objects common to all three decision-makers were detailed explanations of the principles of operation, measures of sealing performance during usage by the end customer, and results from an accelerated durability test. There were very few temporal differences in information requests; when two or more decision-makers requested the same information, 19 out of 20 times the requests were separated by 1 meeting or less. Finally, all decision-makers exhibited a tendency to challenge information that was presented in unconventional formats; in this exercise, this included 2 graphs with non-standard usage of axes and a Failure Mode and Effects Analysis (FMEA) that did not conform to the standard prescribed by the Automotive Industry Action Group (2008), but instead conformed to the Function-Failure Design Method (Stone et al. 2005).

Technologies Biomimétiques
La nature nous pousse
63 57W Avenue Ouest
Blainville, QC, Canada
(450) 971-8080

Company Profile

Technologies Biomimétiques is a leading developer, manufacturer and provider of biology inspired technologies. The goals of the company are two fold: (1) provide cutting edge biology inspired design consulting to industry leading companies and (2) manufacture and provide quality biology inspired products. The company's design consulting portfolio is diverse and includes: active door sealing technology based on the adhesion principles of insects and animals, pumping systems based on the esophageal contractions, contaminate neutralizations technology based on the stomach and intestinal tracks of cows, and personal body armor based on armadillo scutes. Technologies Biomimétiques is currently the leading manufacturer and provider of its biology inspired active door sealing technology to the aircraft industry.

Technologies Biomimétiques was founded in 2002 and is based in Blainville, Quebec just 35 minutes from Montréal. Their headquarters includes state-of-the-art manufacturing and testing facilities as well as an engineering and technical center.

Over the last couple years, Technologies Biomimétiques has experienced exponential growth with last year having higher than expected sales as the result of a new contract where approximately 100 active sealing systems were installed on business class jet airliners. Technologies Biomimétiques is currently expanding its sealing technology into new sectors and expanding its manufacturing capabilities to meet demand.

Fig. 6.4 Company profile and background information object

Technologies Biomimétiques
La nature nous pousse
63 57W Avenue Ouest
Blainville, QC, Canada
(450) 971-8080

Seal Life: Temperature Impact

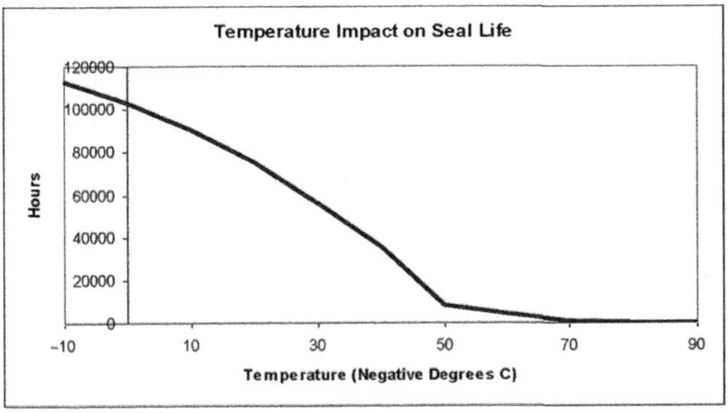

Test for effect of temperature on longevity of the biology inspired active seal. Test performed in Cryogenic Test Chamber #1 at Technologies Biomimétiques. Test performed with samples of 5 specimens each in temperature range 0 C to -70 C in 10 C increments. Accelerated life test performed with electrical current applied at 10x nominal levels. Specimens failed when seal attractive force declined below minimum value of 1.8 N/ cm^2.

Fig. 6.5 Test results for the temperature sensitivity of the seal material

6.4.2 *Jim's Decision*

Jim focused on the biology-inspired seal almost exclusively throughout his decision-making process. The scope of Jim's request was the broadest of all three decision-makers, covering the performance of the seal in use as well as life cycle issues from manufacturing to end-of-life processing with noteworthy emphases on regulatory requirements, environmental issues, and supplier relationships. He made by far the most information requests in his initial meeting with a total of 16. He asked only a few follow-up questions at the first follow-up meeting concerning physical and functional interactions of the biology-inspired seal system with other vehicle systems and clarifications of the function-based failure mode analysis provided by the supplier. At the second follow-up session, Jim seemed satisfied that all of his questions had been addressed. When prompted, Jim paused momentarily and declared somewhat abruptly that he was prepared to make his decision.

Jim chose the biology-inspired seal concept. He presented his rationale orally without referring to any of the information provided or to any decision aids. As the most important source of information, he cited the prior application of the biology-inspired seal in the aircraft industry. Jim had a high degree of respect for aircraft applications due to the extreme environmental conditions faced by aircraft. He also cited opportunities for favorable reviews from customers and media by using the biology-inspired seal in a luxury coupe application; it is a visible, "advertisable", and marketable application of new technology. Finally, Jim cited the responsiveness of the biology-inspired seal supplier in addressing his information requests as a key criterion in his decision process, commenting that relative to other requests for quotations that he had reviewed in the past, this supplier's response was very timely and well-prepared. Jim acknowledged some risk that the biology-inspired seal might prove to be sufficiently durable for automotive application due to differences in operation of aircraft and automobiles over their respective lifecycles, but also recognized a need for strategic risk-taking in new product development and believed that in a luxury coupe application where quietness and new product features are key competitive elements, the biology-inspired seal offered a greater net benefit than conventional door sealing technology.

6.4.3 Terry's Decision

Terry initially focused his information gathering on the biology-inspired seal and later shifted his focus to the conventional seal to provide for a side-by-side comparison of the two sealing concepts. The scope of Terry's information requests was slightly narrower than Jim's, with less emphasis on exogenous factors but with greater depth in the needs of end users and internal stakeholders. His requests were specific and quantitative in nature, especially in the areas of marketing and functional performance of the door seal. He made several unique requests, including queries about the serviceability of the biology-inspired seal, the electromagnetic interactions of the biology-inspired seal with other vehicle systems, and the differences in package space requirements between the conventional seal and the biology-inspired seal.

Terry made a total of 24 information requests, six more than either Jim or Glenn. Of his 24 requests, 20 were distributed evenly over the first two sessions. He made the remainder of his requests in the third and fourth sessions; most of his time in these sessions was spent reviewing previously collected information and asking confirmatory questions to check his interpretations. The fifth session with Terry lasted more than 2 h. He began by opening a spreadsheet template for a Pugh Selection Matrix (Otto and Wood 2001) and defining 14 decision criteria. After this he sorted the available information into piles corresponding to his 14 decision criteria. He then filtered and analyzed each pile of information to identify a preferred alternative for each of the 14 criteria. He spent much of this time cross-checking information from different sources that pertained to the same criteria; for example, he simultaneously analyzed variation in test data and frequencies of problems reported by customers to form hypotheses on cause-effect relationships in the data.

Terry chose the biology-inspired seal concept. He used his completed Pugh Selection Matrix as a visual aid while discussing the rationale for his decision. Based on the supplier's cost reduction plan and test data, he found the biology-inspired seal to be favorable in terms of cost and performance and equivalent to the conventional seal in terms of durability. He expressed several concerns over the biology-inspired seal, including availability of service parts, increased complexity of the vehicle's electrical and electronic systems, and impact on final assembly procedures. He expressed some concern that his decision to employ the biology-inspired seal might be "overruled by manufacturing". He described these as "risks that would need to be carefully managed" and commented that managing these risks would be much simpler if the packaging requirements for the biology-inspired seal did not preclude the use of a conventional sealing system. He concluded his review by stating that strictly in terms of the information collected, he was indifferent between the two seals and that he favored the biology-inspired seal because he believed it would be necessary to implement new technology on the luxury coupe program to succeed in the marketplace. Terry was silent on the subject of previous aerospace applications of the biology-inspired seal.

6.4.4 Glenn's Decision

Glenn initially focused his information collection efforts on the biology-inspired seal, stating that there would be "loads of data" available on conventional seals; he later shifted his information collection efforts to conventional seals to prepare for a direct comparison of the two seals. The scope of his information requests was relatively narrow, mostly covering issues in the marketing and engineering domains. Glenn's peak number of information requests occurred relatively late in his information collection process; he made 10 of his 18 information requests in the second meeting, with seven requests before the second session and one afterwards.

Glenn exhibited a relatively high sensitivity to the formats used for presentation of information. He requested revisions on two of the graphs presented as responses to information requests. One was a plot of seal durability vs. temperature on which the X-axis was reversed, reading positive to negative from left to right; Glenn asked to have this graph regenerated with a conventional scale on the X-axis. Another was a graph of changes in aircraft cabin pressure and contaminant ingress over time for the biology-inspired seal with two different sets of units used on the Y-axis for measurements of pressure (psi/10) and contaminant ingress (parts per million). He struggled to interpret this graph and initially misinterpreted it as a main effects plot from a designed experiment. He later commented that he found the format of the graph confusing and would prefer to see graphs with no more than one units used on each axis. In contrast to this, he cited results from a durability test of the biology-inspired seal in a tabular format and said he found that document to be "very clear".

Glenn began forming preferences relatively early in his information collection process. He concluded his second session with the statement that if he were to choose a seal that day, it would be the conventional seal, but that he would use the remaining 15 days before the decision deadline to review additional data. He began filtering data in the third session, sorting information into three categories: obsolete (prior versions of documents that were later revised), insignificant (information that was either irrelevant or of low enough impact to be safely ignored), and significant (information that would be used to differentiate the alternatives in decision-making). He also began cross-validating the significant information in the third session and requested additional information to resolve discrepancies he identified between variation in test data and rates of customer complaints. At the fourth session, Glenn sketched a scorecard on a piece of scratch paper and expressed three decision-criteria in question form: (1) How well does it perform? (2) Will it fail? (3) How much does it cost?

In terms of performance, Glenn considered both seal concepts equivalent. In terms of failure, Glenn considered the conventional seal to be favorable to the biology-inspired seal, citing concerns over the long-term failure rates of the biology-inspired seal and differences in ambient temperature and numbers of door opening and closing cycles between aerospace and automotive applications. In terms of cost, Glenn based his decision on the supplier's initial price quote and therefore favored the conventional seal. On Glenn's scorecard, the conventional seal weakly dominated the biology-inspired seal and on this basis he selected the conventional seal.

6.5 Discussion

Admittedly the small sample size in this experiment poses challenges in explaining differences observed between the decision-makers. However, a number of interesting issues have been, identified and a number of hypotheses have been formed based on the observations. These ideas are presented and discussed below.

6.5.1 State of Information

The two most salient observations of the decision-makers' states of information were in pairwise similarities of requested information and cadence of information delivery. Jim and Terry shared six information objects, a relatively high number but not a surprising one given that both are experienced automotive engineers. Terry and Glenn shared nine information objects, an unusually high number considering that half of the information provided to Glenn was shared with Terry. This could be explained by a combination of two factors. The first is similarity in background; both have experience in DFSS and, as previously discussed, both made multiple requests for sensitivity analyses and confidence bounds which would be typical in DFSS project execution. The second factor is the mentor-protégé relationship between Terry and Glenn. The large overlap of Glenn's information objects with Terry's, suggest that Glenn may be inheriting information collection paradigms through his relationship with Terry. The same two factors may explain why Jim and Glenn only share one information object in common. Glenn has recently transferred into Systems Engineering, an area in which Jim has worked for the bulk of his career; consequently there have been few opportunities for Jim and Glenn to share professional experiences or to develop social bonds.

There were pronounced differences in the cadence of information collection between the three decision-makers. Jim's information collection process was heavily front-loaded, Glenn's was somewhat back-loaded, and Terry's proceeded at a relatively even pace. These differences could be attributable to a number of factors. One potential factor is overall level of experience; Glenn may have initially requested less information than the other decision-makers because he began the exercise with less background knowledge on which to base his information requests. Another potential factor is the type of experience; it is very likely that Jim applied knowledge of typical subsystem requirements from his extensive background in systems engineering to front-load his information collection process.

6.5.2 Cognition

A wide variety of cognitive issues surfaced in the course of this exercise. Cognitive fit issues were most apparent for Glenn, who commented toward the end of the exercise that he could now "empathize with the need to summarize information for management." His sensitivity to information format and his struggles in interpreting unconventional graphs may likely be traced to his relatively low level of experience in complex decision-making situations. A more widespread cognitive issue is a tendency towards tabular thinking in all three decision-makers. Both Terry and Glenn used tabular scorecards as decision-making aids; Jim presented his decision rationale orally in a very similar row/column style. Glenn also commented that he found data presented in a tabular format to be "very clear." It is fascinating to speculate to what extent the decision-makers thought processes have been shaped by the ubiquity of spreadsheet tools and how their decision-making processes may have unfolded without the influence of spreadsheets. One noteworthy perceptual issue also surfaced; toward the end of the experiment, one of the decision-makers disclosed that he was colorblind and asked for clarification in interpreting a graph with several lines shown in various shades of blue and purple. When prompted, the decision-maker assured the analyst that there had been no prior issues with interpreting information presented using colors and that trends communicated through the use of color may be inferred by context. This is nonetheless a noteworthy issue that may have introduced artifacts into the data as well as a key factor to be considered in future experiments.

6.5.3 Prior Knowledge

The effects of prior knowledge were evident in this exercise. This was pronounced in Jim's decision-making process; his selection of the biology-inspired seal was based heavily on prior knowledge of the impact of new technology features on sales of luxury vehicles and of the rigorous systems engineering practices employed in the aerospace sector. Interestingly, similarities or differences with Jim's prior knowledge correlate strongly to similarities or differences with Jim's decision. Terry's prior knowledge concerning the impact of new technology features on luxury vehicle sales was very similar to Jim's, influencing him to also choose the biology-inspired seal over the conventional seal. In contrast, Glenn's prior knowledge of systems engineering practices in the aerospace and automotive sectors was very different than Jim's, leading him to scrutinize the durability test data for the biology-inspired seal and ultimately to choose the conventional seal largely based on concerns over long-term durability.

6.5.4 Personality

Considering personality from the perspective of Myers-Briggs type, effects were most pronounced for the thinking/feeling and perceiving/judging functions. Glenn's almost exclusive focus on analysis and test data suggests that he strongly favors the thinking function, whereas Jim's consideration of the supplier's responsiveness to information requests shows his feeling function at work. Terry's completion of the data collection process prior to preference formation is characteristic of the perceiving function; likewise Glenn's early announcement that he favored the conventional seal is typical of the judging function. Jim's ISFP type indicates a predilection towards the perceiving function; however, his practically exclusive focus on the biology-inspired seal throughout the decision process suggests judging-type behavior in this situation. Interestingly, Jim commented at the conclusion of the exercise on his rapid and focused decision-making process, saying that he is "usually not like this" in decision-making situations. In terms of experimental design, the personality types of the decision-makers in this exercise may have introduced some noise in the experiment. All three of the decision-makers are inclined towards sensing over intuition; this may have introduced some artificial consistency into the observations and results. Also, because requests for information were processed offline on an as-needed basis in this work, the process of information may have favored introverted types over extraverted types.

6.5.5 Decision-Analytic Principles

The decision processes followed by each of the decision-makers were remarkably similar; their implementation of decision-analytic principles, however, varied widely by topical area. There is clear evidence supporting iterative application of prior knowledge in the decision process. Glenn was perhaps the closest of the three to practicing an iterative decision process, including definition of objectives and formation of preferences as well as information in the decision analysis cycle. Terry clearly deferred the definition of objectives and formation of preferences until the conclusion of his iterative, information collection process. Jim's decision process was completely implicit and it is therefore unknown which elements in it were included in the iterative cycle. Prior knowledge factored much more heavily in Jim's and Terry's decision-making process than in Glenn's, presumably because Jim and Terry have had much more time to accumulate prior knowledge.

There is much less evidence for either the disciplined representation of uncertainty or the explicit identification of values. Both Terry and Glenn requested various forms of bounds on data, such as confidence intervals and sensitivity analyses, showing an appreciation for the impact of variation. However, none of the decision-makers attempted to quantify uncertainty related to outcomes of future events or to ambiguities in available data. There was no explicit mention of conditional probability; to the extent that the decision-makers considered variation and uncertainty, they seemed to do so under an implicit assumption of independence. At best, it could be claimed that Terry and Glenn factored uncertainties into their formation of preferences at the level of individual criteria. As for value functions, none of the three decision-makers explicitly expressed a value function or made any attempt to distinguish means objectives from ends objectives. In using the Pugh selection method, both Terry and Glenn implicitly expressed a simple additive value function; Jim's value function was completely implicit. Intriguingly, the decision-maker who disclosed the least information about his value function (Jim) expressed the greatest confidence in his decision at the end of the exercise while the decision-maker who disclosed the most information about his value function (Terry) expressed the least confidence in his decision.

6.5.6 Evaluation of Decisions

A natural question to ask at the conclusion of this exercise is: Who made the right decision? Because the decision is an action taken in the present to achieve an outcome in the future that cannot be known with certainty, it is impossible to answer this question. However, a related question can be addressed, "Who made a good decision." Recall the normative decision viewpoint, that a good decision is one that is well framed, takes into account the decision makers preferences and values. Not one explicitly stated their value function. From this standpoint, one could argue that none of the decision makers made a good decision. However, each decision-maker chose the alternative that he believed to offer the greatest net benefit. Each decision was based on a unique subset of information and supplemented with unique sets of prior knowledge and experiences. This leads the authors to speculate on the need for providing tools to support the human issues involved in the quantitative view of normative decision analysis.

6.6 Conclusions

The experiment conducted in this work illustrates several characteristics of decision-making in preliminary vehicle development. It is iterative largely due to its dynamic processes for collection and reduction of information. It is also an activity that can be sensitive to formats in which information is presented – particularly when these formats do not conform to established norms. This work features noteworthy distinctions between sets of information collected by decision-makers and the alternatives selected by the decision-makers. Jim and Terry collected distinctly different sets of information and selected the same alternative; Glenn collected a set of information very similar to Terry's yet selected a different alternative. Additional observations that were not anticipated but are intriguing and useful are the similarities in the multi-criteria decision-making process employed by all three decision-makers and the tendency to build confidence in information through cross-validation of information collected from different sources.

The foremost recommendation for future work is to repeat this experiment (or one similar to it) with a much larger sample of decision-makers to provide a realistic basis for confirming or rejecting hypotheses developed through observation. In this experiment, the biographical and psychographic profiles of the decision-makers were collected at the conclusion of the experiment and any knowledge of sensory or cognitive issues was discovered accidentally. In future work, this information should be collected beforehand; ideally, subjects for the experiment would be recruited according to their biographical, psychographic, sensory, and cognitive profiles to provide a balanced sample for the experiment. There are additional opportunities in exploration of the effects of pooling information or decision-makers; it would be interesting to observe any changes in concept selections if each decision-maker were provided all of the information generated in the exercise or if the decision-makers were allowed to discuss their interpretations of the information with each other. In its current state, the information pool is of limited usefulness because it has been developed on a demand-pull basis by a group of three decision-makers and it likely lacks several important types of information. For example, no engineering drawings of the biology-inspired seal concept exist because these decision-makers requested none. If, the experiment were repeated, however, with additional decision-makers until the pool of requested information stabilized, this case study could be developed into a testbed problem for further decision-making research.

As stated at the outset, it is the authors' hope that further research will be conducted into further incorporating human aspects into decision-making specifically in the product development domain. From this work it is clear that inherent differences between decision-makers lead to substantial differences in information and decision-making processes. It therefore seems that there is tremendous potential for increasing the efficiency and the effectiveness of product development processes by designing flexibility into them to accommodate many types of decision-makers. This would enable firms to leverage both the power of structured decision-making methods and the strengths of their diverse product development organizations. Developing a sound understanding of the effects of human aspects on product development decision-making is a necessary precursor to realizing these benefits.

References

Agarwal, R., Sinha, A., and Tanniru, M., 1996, "Cognitive Fit in Requirements Modeling: A Study of Object and Process Methodologies," *Journal of Management Information Systems*, **13(2)**: 137–162.

Automotive Industry Action Group, 2008, *Potential Failure Mode and Effects Analysis*, Fourth Edition.

Card, S., 1999, *Readings in Information Visualization – Using Vision to Think*, Morgan Kaufmann.

Carey, H. et al., 2002, "Corporate Decision-Making and Part Differentiation: A Model of Customer-Driven Strategic Planning," *Proceedings of the ASME Design Engineering Technical Conferences*, Montreal, Canada.

Creton, and Gorb, S., Eds., 2007, "Sticky Feet: From Animals to Materials." *MRS Bulletin*, **32**: 466–71.

Culp, G. and Smith, A., 2001, "Understanding Psychological Type to Improve Project Team Performance," *Journal of Management in Engineering*, **17(1)**: 24–33.

Denis, A. and Carte, T., 1998, "Using Geographical Information Systems for Decision Making: Extending Cognitive Fit Theory to Map-Based Presentations," *Information Systems Research*, **9(2)**: 194–203.

Donndelinger, J. 2006, "A Decision-Based Perspective on the Vehicle Development Process," *Decision Making in Engineering Design*, Ch. 19, ASME Press.

Hazelrigg, G., 2001, "The Cheshire Cat on Engineering Design," submitted to *Journal of Mechanical Design*.

Herrmann, J. and Schmidt, L., 2006, "Viewing Product Development as a Decision Production System," *Decision Making in Engineering Design*, Ch. 20, ASME Press.

Heuer, R., 2007, "Glossary of Selected Psychiatric Terms," *Adjudicative Desk Reference: Background Resources for Personal Security Adjudicators, Investigators, and Managers*, V 3.1, www.rjhresearch.com/ADR/ psychconditions/glossarypsychterms.htm.

Matheson, J. and Howard, R., 1989, "An Introduction to Decision Analysis," *Readings on the Principles and Applications of Decision Analysis*, Strategic Decisions Group.

Myers, I.B., 1998, *Introduction to Type*, Consulting Psychologies Press.

Otto, K. and Wood, K., 2001, Product Design: *Techniques in Reverse Engineering, Systematic Design, and New Product Development*, Prentice-Hall, 493–500.

Stone, R., Tumer, I., and Van Wie, M., 2005, "The Function-Failure Design Method," *Journal of Mechanical Design*, **127(3)**: 397–407.

Speier, C., 2006, "The Influence of Information Presentation Formats on Complex Task Decision-making Performance," *International Journal of Human-Computer Studies*, **64**: 1115–1131.

Tegarden, D., 1999, "Business Information Visualization," *Communications of the Association for Information Systems*, **1(4)**: 1–38.

Tufte, E., 2006, *Beautiful Evidence*, Graphics Press.

Vessey, I., 1991, "Cognitive Fit: A Theory-Based Analysis of the Graph Versus Tables Literature," *Decision Sciences*, **22(2)**: 219–240.

Vessey, I., 1994, "The Effect of Information Presentation on Decision making: A Cost-Benefit Analysis," *Information and Management*, **24**: 103–119.

Ware, C., 2004, *Information Visualization – Perception for Design*, Morgan Kaufmann.

Chapter 7
Dempster-Shafer Theory in the Analysis and Design of Uncertain Engineering Systems

S.S. Rao and Kiran K. Annamdas

Abstract A methodology for the analysis and design of uncertain engineering systems in the presence of multiple sources of evidence based on Dempster-Shafer Theory (DST) is presented. DST can be used when it is not possible to obtain a precise estimation of system response due to the presence of multiple uncertain input parameters. The information for each of the uncertain parameters is assumed to be available in the form of interval-valued data from multiple sources implying the existence of large epistemic uncertainty in the parameters. The DST approach, in conjunction with the use of the vertex method, and the evidence-based fuzzy methodology are used in finding the response of the system. A new method, called Weighted Dempster Shafer Theory for Interval-valued data (WDSTI), is proposed for combining evidence when different credibilities are associated with different sources of evidence. The application of the methodology is illustrated by considering the safety analysis of a welded beam in the presence of multiple uncertain parameters. The epistemic uncertainty can be modeled using fuzzy set theory. In order to extend the mathematical laws of crisp numbers to fuzzy theory, we can use the extension principle, which provides a methodology that fuzzifies the parameters or arguments of a function, resulting in computable fuzzy sets. In this work, an uncertain parameter is modeled as a fuzzy variable and the available evidences on the ranges of the uncertain parameter, in the form of basic probability assignments (bpa's), are represented in the form of membership functions of the fuzzy variable. This permits the use of interval analysis in the application of a fuzzy approach to uncertain engineering problems. The extension of DST in the decision making of uncertain engineering systems based on using different combination rules such as Yager's rule, Inagaki's extreme rule, Zhang's center combination rule and Murphy's average combination rule is also presented with an illustrative application.

Keywords Dempster-Shafer Theory · Engineering design · Evidence · Fuzzy

---

S.S. Rao (✉) and K.K. Annamdas
Department of Mechanical and Aerospace Engineering, University of Miami,
Coral Gables, FL 33124-0624, USA
e-mail: srao@miami.edu

N.R.S. Raghavan and J.A. Cafeo (eds.), *Product Research: The Art and Science Behind Successful Product Launches*, DOI 10.1007/978-90-481-2860-0_7,

7.1 Introduction

7.1.1 Background

Uncertainty can be considered as the lack of adequate information to make a decision. It is important to quantify uncertainties in mathematical models used for design and optimization of nondeterministic engineering systems. In general, uncertainty can be broadly classified into three types (Bae et al. 2004; Ha-Rok 2004; Klir and Wierman 1998; Oberkampf and Helton 2002; Sentz 2002). The first one is aleatory uncertainty (also referred to as stochastic uncertainty or inherent uncertainty) – it results from the fact that a system can behave in random ways. For example, the failure of an engine can be modeled as an aleatory uncertainty because the failure can occur at a random time. One cannot predict exactly when the engine will fail even if a large quantity of failure data is gathered (available). The second one is epistemic uncertainty (also known as subjective uncertainty or reducible uncertainty) – it is the uncertainty of the outcome of some random event due to lack of knowledge or information in any phase or activity of the modeling process. By gaining information about the system or environmental factors, one can reduce the epistemic uncertainty. For example, a lack of experimental data to characterize new materials and processes leads to epistemic uncertainty. The third one is numerical uncertainty (also known as error) – it is present, for example, when there is numerical error due to round-off and truncation errors as in the case of the numerical solution of ordinary and partial differential equations.

The estimation of uncertainty of engineering systems is sometimes referred to as the simulation of nondeterministic systems. The mathematical model of the system, which includes the influence of the environment on the system, is considered non-deterministic in the sense that: (i) the model can produce non-unique system responses because of the existence of uncertainty in the input data of the model, or (ii) there are multiple alternative mathematical models of the system.

Conventionally, probability theory has been used to characterize both aleatory and epistemic uncertainties. However, recent developments in the characterization of uncertainty reveal that traditional probability theory provides an inadequate model to capture epistemic uncertainty (Alim 1988; Beynon et al. 2000; Parikh et al. 2001; Tanaka and Klir 1999). It is recognized that probability theory is best suited to deal with aleatory uncertainty. The present study uses Dempster-Shafer theory (DST) as the framework for representing uncertainty and investigates the various aspects of combining evidence.

Depending on the nature and extent of uncertainty involved in an engineering system, three different approaches can be used for the analysis given by Rao and Sawyer (1995). The fuzzy or imprecise information may be present in the geometry, material properties, external effects, or boundary conditions of a system. Uncertainties, usually, are present in the form of interval values (i.e., the values of the parameters are known to lie between two limits, but the exact values are unknown) with specific confidence levels. The present study also uses evidence-based fuzzy approach as a framework for representing uncertainty and investigates the various aspects of combining evidence.

7.1.2 Review of Dempster Shafer Theory

The DST (Shafer 1976), also known as evidence theory, is a branch of mathematics that concerns with the combination of empirical evidence in order to construct a coherent picture of reality. When the evidence is sufficient enough to permit the assignment of probabilities to single events, the DST model reduces to the traditional probabilistic formulation. One of the most important features of DST is that the model is designed to cope with varying levels of precision regarding the information and no further assumptions are needed to represent the information (Agarwal et al. 2004; Butler et al. 1995; Ferson et al. 2003).

Three important functions are defined in DST: the *basic probability assignment (or mass) function,* the *Belief* function (*Bel*), and the *Plausibility* function (*Pl*). The basic probability assignment (bpa) or mass is a primitive of evidence theory. It is customary in DST to think about the degree of belief in evidence as similar to the mass of a physical object i.e., mass of evidence supports a belief. Generally, the term "basic probability assignment" does not refer to probability in the classical sense. The bpa, m, defines a mapping of the power set to the interval between 0 and 1, where the bpa of the null set is 0 and the sum of the bpa's of all the subsets of the power set is equal to 1. The value of the bpa for a given set A, $m(A)$, expresses the proportion of all relevant and available evidence that supports the claim that a particular element of Θ (the universal set) belongs to the set A but to no particular subset of A. The value of $m(A)$ pertains only to the set A i.e., portion of total belief assigned exactly to the proposition A and makes no additional claims about any subsets of A. Any further evidence on the subsets of A would be represented by other bpas, for example, if $B \subset A$, $m(B)$ denotes the bpa of the subset B. Formally, the description of m is given by the following three equations:

$$\mathrm{m}(A) \geq 0 \text{ for any } A \in 2^{\Theta} \tag{7.1}$$

$$\mathrm{m}(\phi) = 0 \tag{7.2}$$

$$\sum_{A \in 2^{\Theta}} m(A) = 1 \tag{7.3}$$

From the basic probability assignment, the upper and lower bounds of an interval can be defined. This interval contains the precise probability of a set of interest (in the classical sense) and is bounded by two nonadditive continuous measures termed Belief and Plausibility. The lower bound, *Belief,* for a set A is defined as the sum of all the basic probability assignments of the proper subsets B of the set of interest $A(B \subseteq A)$. The upper bound, *Plausibility*, is defined as the sum of all the basic probability assignments of the sets B that intersect the set of interest $A(B \cap A \neq \emptyset)$. Thus

$$Bel(A) = \sum_{B|B \subseteq A} m(B) \tag{7.4}$$

$$Pl(A) = \sum_{B|B \cap A \neq \phi} m(B) \tag{7.5}$$

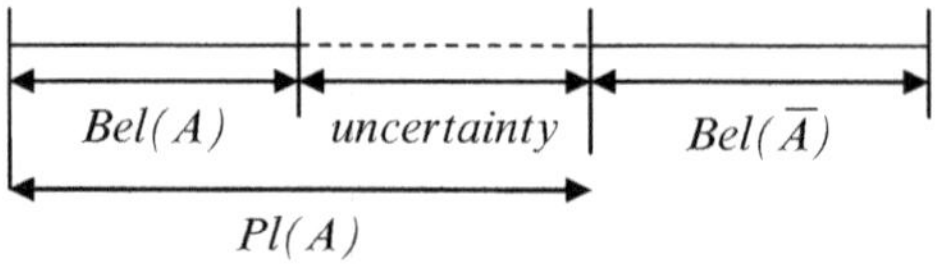

Fig. 7.1 Belief (*Bel*) and Plausibility (*Pl*)

The belief and plausibility are related as

$$Pl(A) = 1 - Bel(\bar{A}) \tag{7.6}$$

where $\bar{A}$ is the classical complement of A. When the available information is inadequate, it is more reasonable to present bounds to quantify the uncertainty as opposed to representation by a single value of probability. The precise probability of an event (in the classical sense), $P(A)$, lies within the lower and upper bounds given by the belief and plausibility functions:

$$Bel(A) \leq P(A) \leq Pl(A) \tag{7.7}$$

Thus the total degree of belief in the proposition A is expressed in the form of bounds, [*Bel(A), Pl(A)*], which lie in the unit interval [0,1], as shown in Fig. 7.1.

Dempster's rule of combination: Originally, Dempster's rule of combination was introduced to permit computation of the orthogonal sum of given evidences from multiple sources. Dempster's rule combines multiple belief functions through their bpas (m). These belief functions are defined on the same frame of discernment, but are based on independent arguments or bodies of evidence. One of the important areas of research in DST is the effect of independence on the bodies of evidence when combining evidence. Dempster's rule involves a conjunctive operation (AND) for combining evidence (Kozine and Utkin 2004). Two bpa's m_1 and m_2 can be combined to yield a new bpa m_{12}, denoted as $m_{12} = m_1 \oplus m_2$, as follows:

$$m_{12}(A) = \frac{\sum_{B \cap C = A} m_1(B) m_2(C)}{1 - k} \quad \text{where} \quad A \neq \phi \tag{7.8}$$

$$m_{12}(\phi) = 0 \tag{7.9}$$

$$\text{where } k = \sum_{B \cap C = \phi} m_1(B) m_2(C) \tag{7.10}$$

The denominator in Dempster's rule, $1 - k$, in Eq. (7.8) is called the normalization factor. This will not only ignore conflict but also attribute bpa associated with conflict to the null set. In addition, this normalization will yield counterintuitive results when significant conflict is present in certain contexts. Thus, the rule is not suitable for cases where there is considerable inconsistency in the available evidence. However, it is applicable, when there is some degree of consistency or sufficient agreement among the opinions of different sources. In this work, we assume that there is enough consistency among the given sources of evidence so that the Dempster's rule can be applied.

7.2 Vertex Method

When the uncertain parameters are described by interval values, the vertex method can be used for computing the response efficiently (Dong and Shah 1987). When $y = f(x_1, x_2, x_3, \ldots, x_n)$ is continuous in the n-dimensional rectangular region with no extreme points in the region (including the boundaries), then the interval value of the function can be obtained as

$$Y = f(X_1, X_2, \ldots\ldots, X_n) = \left[\min_j \left(f\left(c_j\right)\right), \max_j \left(f\left(c_j\right)\right)\right], \; j = 1, 2, \ldots, N \tag{7.11}$$

where c_j denotes the ordinate of the j-th vertex. If m extreme points $E_k (k = 1, 2, \ldots, \mathrm{m})$, exist among the vertices, then Eq. (7.11) is to be modified as

$$Y = \left[\min_{j,k} \left(f\left(c_j\right), f\left(E_k\right)\right), \max_{j,k} \left(f\left(c_j\right), f\left(E_k\right)\right)\right] \tag{7.12}$$

The vertex method is based on the α -cut concept and interval analysis. The α-cut is a discretization technique on membership value domains of uncertain variables instead of on the domains of the variables themselves. The vertex method reduces the widening of the function value set due to multi-occurrences of a variable when interval analysis is implemented on the expression of the function.

7.2.1 Computational Aspects of the Vertex Method

The following step-by-step procedure is used to implement the vertex method for determining the belief and plausibility functions:

1. Initialize the interval ranges of the uncertain parameters as $X_1, X_2, \ldots, X_n$ and the corresponding bpa's as $Y_1, Y_2, \ldots, Y_n$ respectively where n is the number of uncertain parameters.
2. Construct the bpa product table using $Y_1, Y_2, \ldots, Y_n$ and store the result in a matrix A where n represents the dimensionality of A.
3. Each element of A, corresponding to the interval ranges of $X_1, X_2, \ldots, X_n$ represents an n-sided hyper cube where each vertex denotes a different combination of the values of the uncertain parameters.
4. The belief is calculated as the sum of the bpa's (or elements of A) corresponding to the n-sided hyper cube where the response function evaluated at each vertex is less than the limit value of the response (or output) function.

5. The plausibility is calculated as the sum of all those bpa's from the matrix A that not only correspond to the belief but also any n-sided hyper cube for which any of its vertices has a function value less than the limit value of the output/response function.
6. The number of function evaluations can be optimized (minimized) in the computation of plausibility in identifying the function values at the vertices of the n-sided hyper cube that has values less than the limit value of the response function.

7.3 Analysis of a Welded Beam

The failure/safety analysis of a beam of length L with cross-sectional dimensions t and b that is welded to a fixed support, as shown in Fig. 7.2, is considered. The weld length is l on both top and bottom surfaces and the beam is required to support a load P. The weld is in the form of a triangle of depth h. The maximum shear stress developed in the weld, τ, is given by

$$\tau = \sqrt{(\tau')^2 + 2\tau'\tau'' \cos\theta + (\tau'')^2} \tag{7.13}$$

where

$$\tau' = \frac{P}{\sqrt{2}.h.l} \tag{7.14}$$

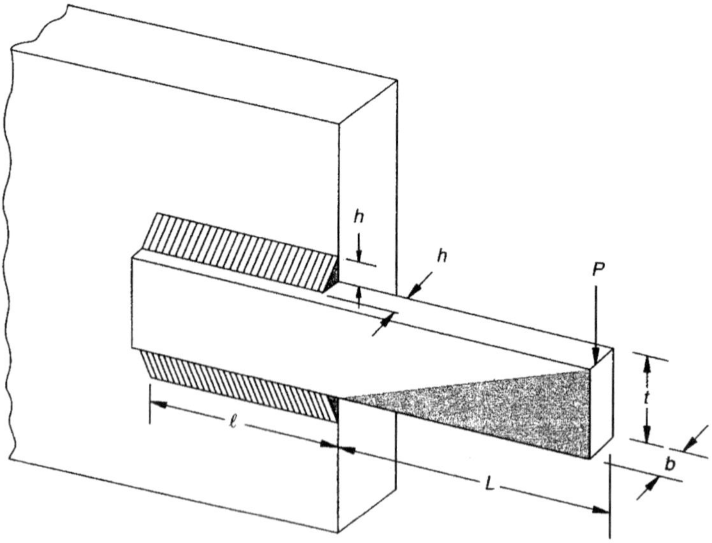

Fig. 7.2 A welded beam

$$\tau'' = \frac{MR}{J} = \left[P\left(L + \frac{l}{2}\right)\right] \frac{\sqrt{\left[\frac{l^2}{4} + \left(\frac{h+t}{2}\right)^2\right]}}{2\left(\sqrt{2}hl\left[\frac{l^2}{12} + \left(\frac{h+t}{2}\right)^2\right]\right)} \tag{7.15}$$

$$\cos\theta = \frac{l}{2\sqrt{\left[\frac{l^2}{4} + \left(\frac{h+t}{2}\right)^2\right]}} \tag{7.16}$$

The beam is considered unsafe if the maximum shear stress in the weld is greater than 9066.67 $\mathrm{lb/in^2}$. The nominal (input) data is $P = 6000\,\mathrm{lb}$, $L = 14$ in., $E = 30x10^6$ psi, $\sigma_{\max} = 30{,}000$ psi, $\delta_{\max} = 0.25\,in$, $\tau_{\max} = 13{,}600$ psi, $h = 0.2455$ in., $l = 6.196$ in. and $t = 8.273$ in. The analysis of the welded beam is conducted for two cases. In the first case, the beam is assumed to have two uncertain parameters while four uncertain parameters are assumed in the second case. The nondeterministic character of the system is due to the presence of uncertainty in the parameters embodied in the mathematical model of the welded beam.

Figure 7.3(a) shows a three-dimensional representation of the shear stress induced in the weld, τ, over the range of possible values of x_1 and x_2 with several level curves, or response contours, of τ. The rectangle defined by $0.7 \le \mathrm{x}_1 \le 1.3$ and $0.8 \le \mathrm{x}_2 \le 1.3$ is referred to as the input product space. Figure 7.3(b) shows the limit shear stress contour in the $\mathrm{x}_1 - \mathrm{x}_2$ plane for $\tau = 9066.67\,\mathrm{lb/in^2}$ in the weld.

7.3.1 Analysis with Two Uncertain Parameters

The length of the weld (l) and the height of the weld (h) are treated as the uncertain parameters. Let x_1 and x_2 be the multiplication factors that denote the uncertainty of the parameters l and h, respectively. It is assumed that two sources of evidence (experts) provide the possible ranges (intervals) of x_1 and x_2 along with the corresponding bpa's as indicated in Table 7.1. The belief and plausibility are computed as follows:

Step 1: Find the intersections of intervals for the uncertain variable x_1 from the two sources of evidence (experts) as shown in Table 7.2(a).

Step 2: Find the bpa for the interval ranges calculated in Step 1 as indicated in Table 7.2(b).

Step 3: Find the normalization factor for x_1 as shown in Table 7.2(c) and scale the bpa's to obtain $\sum m = 1$. Use a similar procedure for x_2 and combine the evidences for both x_1 and x_2 as shown in Table 7.2(d).

Step 4: Since there are two uncertain parameters in the problem, the dimensionality of the bpa product table is 2. Apply DST for the evidences obtained for x_1 and x_2 to find the bpa product table of order 7×6 as shown in Table 7.2(e).

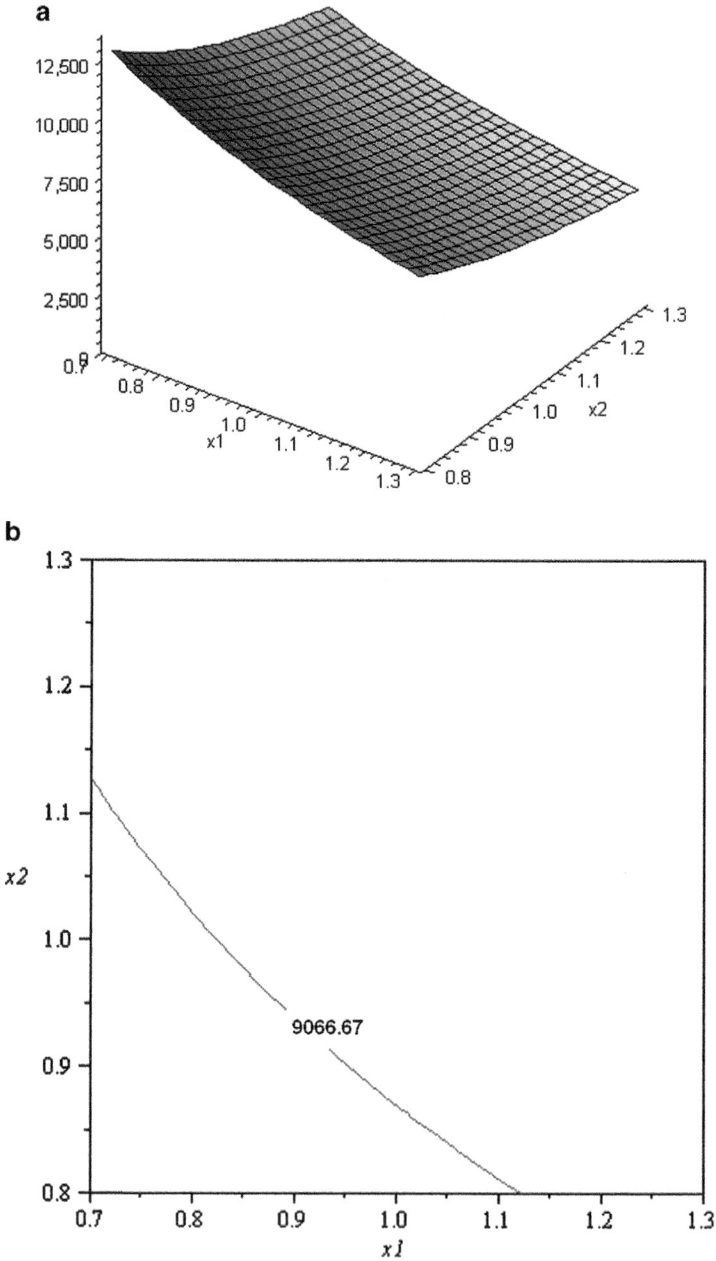

Fig. 7.3 (**a**) Representation of shear stress (τ) in the weld in the range $0.7 \leq x_1 \leq 1.3$ and $0.8 \leq x_12 \leq 1.3$. (**b**) Contour of the shear stress in the weld (τ) in the range $0.7 \leq x_1 \leq 1.3$ and $0.8 \leq x_2 \leq 1.3$

Table 7.1 Evidences for the uncertain parameters x_1 and x_2

x_1	Expert1 (Evidence1)	Interval	[0.7,0.8]	[0.8,1.1]	[1.0,1.2]	[1.2,1.3]
	(S1)	Bpa	0.1	0.4	0.4	0.1
	Expert2 (Evidence2)	Interval	[0.7,0.9]	[0.8,1.0]	[1.0,1.2]	[1.1,1.3]
	(S2)	Bpa	0.1	0.4	0.3	0.2
x_2	Expert1 (Evidence1)	Interval	[0.8,0.9]	[0.9,1.1]	[1.0,1.2]	[1.2,1.3]
	(S1)	Bpa	0.1	0.4	0.4	0.1
	Expert2 (Evidence2)	Interval	[0.7,0.9]	[0.9,1.0]	[1.0,1.2]	[1.1,1.3]
	(S2)	Bpa	0.2	0.4	0.2	0.2

Table 7.2(a) Combined interval ranges of x_1

	Expert 2				
	Interval	[0.7,0.9]	[0.8,1.0]	[1.0,1.2]	[1.1,1.3]
Expert 1	[0.7,0.8]	[0.7,0.8]			
	[0.8,1.1]	[0.8,0.9]	[0.8,1.0]	[1.0,1.1]	
	[1.0,1.2]			[1.0,1.2]	[1.1,1.2]
	[1.2,1.3]				[1.2,1.3]

Table 7.2(b) Combined bpa values of x_1

		Expert 2				
		Interval	[0.7,0.9]	[0.8,1.0]	[1.0,1.2]	[1.2,1.3]
	Interval	Bpa (m)	0.1	0.4	0.3	0.2
Expert 1	[0.7,0.8]	0.1	0.01			
	[0.8,1.1]	0.4	0.04	0.16	0.12	
	[1.0,1.2]	0.4			0.12	0.08
	[1.1,1.3]	0.1				0.02

Table 7.2(c) Normalization factor of x_1

x_1	
Interval	m
[0.7,0.8]	0.01
[0.8,0.9]	0.04
[0.8,1.0]	0.16
[1.0,1.1]	0.12
[1.0,1.2]	0.12
[1.1,1.2]	0.08
[1.2,1.3]	0.02
$\sum m$	0.55

Table 7.2(d) Combined normalized evidence (bpas) from experts 1 and 2 of x_1 and x_2

x_1		x_2	
Interval	m	Interval	m
[0.7,0.8]	0.01818	[0.8,0.9]	0.04545
[0.8,0.9]	0.07272	[0.9,1.0]	0.3636
[0.8,1.0]	0.29091	[1.0,1.1]	0.1818
[1.0,1.1]	0.21818	[1.0,1.2]	0.1818
[1.0,1.2]	0.21818	[1.1,1.2]	0.1818
[1.1,1.2]	0.14545	[1.2,1.3]	0.04545
[1.2,1.3]	0.03636		

Table 7.2(e) Bpa product table of x_1 and x_2

		x_2					
x_1		[0.8,0.9]	[0.9,1.0]	[1.0,1.1]	[1.0,1.2]	[1.1,1.2]	[1.2,1.3]
Interval	m	0.04545	0.3636	0.1818	0.1818	0.1818	0.04545
[0.7,0.8]	0.01818	0.00083	0.00661	0.00331	0.00331	0.00331	0.00083
[0.8,0.9]	0.07272	0.00331	0.02644	0.01322	0.01322	0.01322	0.00331
[0.8,1.0]	0.29091	0.01322	0.10577	0.05289	0.05289	0.05289	0.01322
[1.0,1.1]	0.21818	0.00992	0.07933	0.03967	0.03967	0.03967	0.00992
[1.0,1.2]	0.21818	0.00992	0.07933	0.03967	0.03967	0.03967	0.00992
[1.1,1.2]	0.14545	0.00661	0.05289	0.02644	0.02644	0.02644	0.00661
[1.2,1.3]	0.03636	0.00165	0.01322	0.00661	0.00661	0.00661	0.00165

Table 7.2(f) Bpa product table of x_1 and x_2 represent matrix A

		x_2					
x_1		[0.8,0.9]	[0.9,1.0]	[1.0,1.1]	[1.0,1.2]	[1.1,1.2]	[1.2,1.3]
Interval	m	0.04545	0.3636	0.1818	0.1818	0.1818	0.04545
[0.7,0.8]	0.01818	A(1,1)	A(1,2)	**A(1,3)**	**A(1,4)**	**A(1,5)**	*A(1,6)*
[0.8,0.9]	0.07272	A(2,1)	**A(2,2)**	**A(2,3)**	**A(2,4)**	*A(2,5)*	*A(2,6)*
[0.8,1.0]	0.29091	**A(3,1)**	**A(3,2)**	**A(3,3)**	**A(3,4)**	*A(3,5)*	*A(3,6)*
[1.0,1.1]	0.21818	**A(4,1)**	*A(4,2)*	*A(4,3)*	*A(4,4)*	*A(4,5)*	*A(4,6)*
[1.0,1.2]	0.21818	**A(5,1)**	*A(5,2)*	*A(5,3)*	*A(5,4)*	*A(5,5)*	*A(5,6)*
[1.1,1.2]	0.14545	**A(6,1)**	*A(6,2)*	*A(6,3)*	*A(6,4)*	*A(6,5)*	*A(6,6)*
[1.2,1.3]	0.03636	*A(7,1)*	*A(7,2)*	*A(7,3)*	*A(7,4)*	*A(7,5)*	*A(7,6)*

Step 6: Use the vertex method to handle the two uncertain parameters represented by interval numbers to obtain an assessment of the likelihood of the maximum induced shear stress not exceeding the specified limit state value of 9066.67 lb/in^2 (a matlab program developed in this work).

Belief is calculated as the sum of all bpas of matrix A whose corresponding hyper cube satisfies all the design constraints and maximum induced shear stress less than 9066.67 lb/in^2 using the procedure described in Section 7.2. These bpas are indicated by italic letters in Table 7.2(f). Similarly, plausibility is calculated as the sum of all the bpas of the matrix for which at least one vertex of the corresponding hyper cube satisfies the condition that the maximum induced shear stress is less than $9066.67\,\mathrm{lb/in^2}$. These bpas are represented by both bold and italic letters in Table 7.2(f). This procedure is implemented in a matlab program to obtain belief and plausibility values for the safety of the welded beam.

If the program is not optimized for the number of function evaluations, the required number of function evaluations using the vertex method would be $(2^2)\,(7)\,(6) = 168$ to find the belief and plausibility for realizing the maximum induced shear stress to be less than the limit value $9066.67\,\mathrm{lb/in^2}$. In the optimized program, when computing plausibility, we do not evaluate function values (maximum induced shear stress) at all other vertices of the hypercube once the program finds at least one vertex that corresponds to a function value less than $9066.67\,\mathrm{lb/in^2}$ and at least one vertex with a function value (τ) greater than $9066.67\,\mathrm{lb/in^2}$. When the current procedure/program is optimized, the number of function evaluations required is 166. Thus, a slight reduction in function evaluations is achieved by the program. This reduction increases with an increase in the number of uncertain input parameters and/or an increase in the number of interval data for the uncertain input parameters. The numerical results indicate a belief of 0.67515 and a plausibility of 0.98927 (for the maximum shear stress less than $9066.67\,\mathrm{lb/in^2}$). Thus the degree of plausibility, 0.98927, is the maximum limit state violation for the design while there is at least 0.67515 belief for a safe design. The belief and plausibility are nothing but lower and upper bounds on the unknown probability. The DST thus indicates that the probability of a safe design with $\tau < 9066.67\,\mathrm{lb/in^2}$ will be as low as 0.67515 and as high as 0.98927 with the given body of evidence.

7.4 DST Methodology when Sources of Evidence Have Different Credibilities

When the credibilities of the various expert opinions are different, a modified DST, proposed in this section, can be used for combining evidences. Let c_i be the weighting/credibility factor for the source of evidence i where $0 \le \mathrm{c_i} \le 1$. The combined bpa with all the available evidence is determined as:

$$
\begin{aligned}
m_{12..n} = {} & m_1(S1)\, m_2(S2) \ldots .m_n(Sn) + (1-\mathrm{c}_1)\,(1-m_1(S1)) \\
& m_2(S2) \ldots .m_n(Sn) \\
& + (1-\mathrm{c}_2)\, m_1(S1)\,(1-m_2(S2))\, m_3(S3) \ldots \ldots m_n(Sn) \\
& + \ldots\ldots\ldots\ldots\ldots \\
& + (1-\mathrm{c_n})\, m_1(S1)\, m_2(S2)\, m_3(S3) \ldots \ldots m_{n-1}(Sn{-}1) \\
& (1-m_n(Sn))
\end{aligned} \tag{7.17}
$$

where $m_1(S1)$, $m_2(S2) \ldots m_n(Sn)$ are the bpa's for a particular interval range from sources $1, 2, \ldots\ldots \mathrm{n}$ (S1, S2, ..., Sn), respectively, and $m_{12\ldots n}$ is the combined bpa obtained from the DST rule for the same interval range. For example, if the number of sources is two (n = 2), Eq. (7.17) gives

$$m_{12} = m_1(S1)\,m_2(S2) + (1 - c_2)(1 - m_2(S2))\\ m_1(S1) + (1 - c_1)(1 - m_1(S1))\,m_2(S2) \tag{7.18}$$

where $m_1(S1)$ and $m_2(S2)$ are the bpa's for a particular interval range from sources 1 and 2, respectively, and m_{12} is the bpa obtained from the DST rule for the same interval range.

Note that the degree of uncertainty, $m(\Theta)$, itself is not multiplied by the weighting factor and Eq. (7.18) reduces to Eq. (7.8) when all the credibility factors are equal to 1. Also, when $c_1 = 0$ or $c_2 = 0$, Eq. (7.17) reduces to the formula corresponding to the case with only n – 1 sources of evidence. Once the modified bpa is determined, the DST can be used to combine evidences. The procedure, termed the Weighted Dempster Shafer Theory for Interval-valued data (WDSTI), is outlined below. The procedure of WDSTI is developed so as to make the results obtained with different credibilities (for different sources of evidence) converge to those obtained using the standard Dempster-Shafer Theory when all the credibilities are identical.

7.4.1 Solution Procedure with Weighted Dempster Shafer Theory for Interval-Valued Data (WDSTI)

The procedure to determine the belief and plausibility functions using WDSTI is indicated in the following steps:

1. Use Dempster's rule to combine the evidences from interval valued input data.
2. Find the bpa's using Eq. (7.18) for the product of evidences.
3. Let the sum of all the bpa's be n. Using the normalization factor 1/n, multiply each of the bpa's by 1/n.
4. The number of uncertain parameters used to find the combined evidence determines the dimensionality of the product table of bpa.
5. Calculate the belief and plausibility functions using the vertex method as described earlier.

7.4.2 Analysis of a Welded Beam

The safety analysis of the welded beam described in Section 7.3 is considered again. The maximum shear stress induced in the welded beam can be calculated using Eqs. (7.13)–(7.16). The problem is solved with two uncertain parameters. The length of weld (l) and the height of the weld (h) are considered to be uncertain. The problem is solved using data from two sources of evidence. The beam is considered unsafe if the maximum induced shear stress in the weld is greater than

9066.67 lb/in^2. If the credibilities of the sources of evidence 1 and 2 are 1 and c $(0 \le c \le 1)$, respectively, the belief and plausibility can be determined as follows:

(i) The intervals are combined in the usual manner (as in the case of equal credibilities for all sources of evidence) except that the bpa's are calculated using Eq. (7.17).
(ii) The Dempster's rule is used to combine the evidences as per the WDSTI method to compute the combined bpa values and the resulting values are normalized as indicated earlier.
(iii) The belief and plausibility functions are computed using the vertex method.

7.4.3 Numerical Results

The methodology of WDSTI is illustrated for the analysis of a welded beam considered earlier.

The two sources of evidence are assumed to provide possible ranges or intervals of x_1 and x_2 (or l and h) along with the corresponding bpa's as given in Table 7.3. The values of belief and plausibility computed for different values of the credibility (c) of source 2, with $(0 \le c \le 1)$, are shown in Fig. 7.4.

Table 7.3 Evidences for the uncertain parameters x_1 and x_2

	x_1		x_2	
Interval	Expert1 (Evidence1) (S1) Bpa	Expert2 (Evidence2) (S2) Bpa	Expert1 (Evidence1) (S1) Bpa	Expert2 (Evidence2) (S2) Bpa
[0.85,0.86]	0.02	0.03	0.03	0.04
[0.86,0.87]	0.04	0.03	0.05	0.05
[0.87,0.88]	0.04	0.04	0.04	0.05
[0.88,0.89]	0.05	0.05	0.05	0.04
[0.89,0.90]	0.06	0.05	0.04	0.04
[0.90,0.91]	0.06	0.07	0.08	0.07
[0.91,0.92]	0.08	0.08	0.08	0.08
[0.92,0.93]	0.09	0.1	0.1	0.1
[0.93,0.94]	0.11	0.1	0.09	0.08
[0.94,0.95]	0.09	0.09	0.1	0.11
[0.95,0.96]	0.06	0.05	0.08	0.08
[0.96,0.97]	0.05	0.05	0.05	0.05
[0.97,0.98]	0.06	0.06	0.06	0.06
[0.98,0.99]	0.04	0.04	0.04	0.04
[0.99,1.00]	0.05	0.05	0.05	0.05
[1.00,1.01]	0.04	0.04	0.03	0.02
[1.01,1.02]	0.04	0.05	0.02	0.03
[1.02,1.03]	0.02	0.02	0.01	0.01

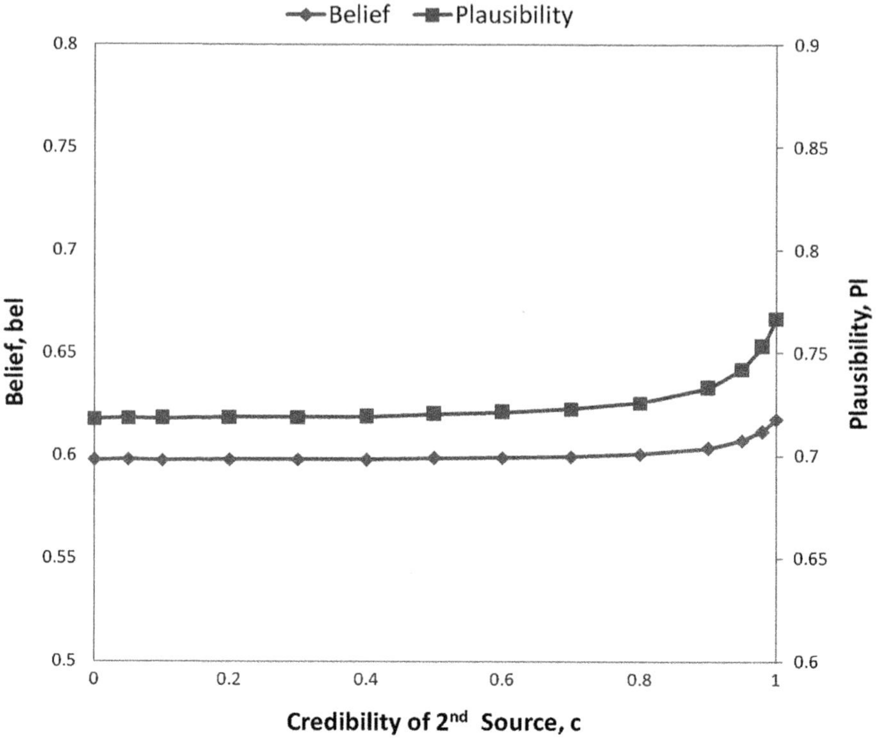

Fig. 7.4 Variations of belief and plausibility with credibility (c) of source 2

7.5 Evidence-Based Fuzzy Approach

When the uncertain parameters are described as fuzzy quantities by different sources of evidence, the approach described in this section can be used (Zadeh 1965).

7.5.1 α-Cut Representation

In general, when a fuzzy set is discretized, the number of elements in the set could become quite large. Thus, in numerical computations, it is convenient to express fuzzy numbers as sets of lower and upper bounds of a finite number of α-cut subsets as shown in Fig. 7.5. Corresponding to a level of α (α-cut), the value of x is extracted in the form of an ordered pair with a closed interval $[x_l, x_u]$. The α-cut can be taken anywhere ranging from $\alpha = 0$ (total uncertainty) to $\alpha = 1$(total certainty). An interval number is represented as an ordered pair $[x_l, x_u]$ where $[x_l \leq x_u$. In case $x_l = x_u$, the interval is called a fuzzy- point interval, e.g. [a, a]. Thus membership functions are constructed in terms of intervals of confidence at several levels

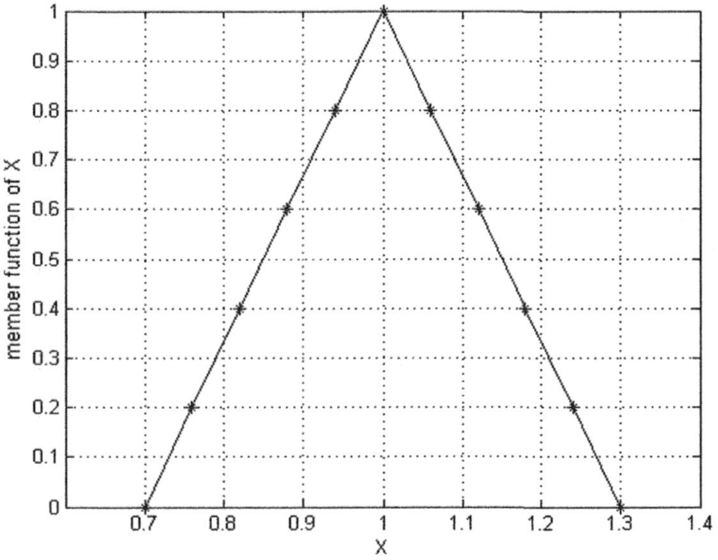

Fig. 7.5 Typical fuzzy number, **X**

of α-cuts. The level of α, $\alpha \in [0, 1]$, gives an interval of confidence $\mathbf{X}_\alpha$, defined as

$$\mathbf{X}_\alpha = \{x \in \mathbf{R}, \mu_{\mathbf{X}}(x) \geq \alpha\}. \tag{7.19}$$

where $\mathbf{X}_\alpha$ is a monotonically decreasing function of α, that is

$$(\alpha_1 < \alpha_2) \Rightarrow (\mathbf{X}\alpha_2 \subset \mathbf{X}\alpha_1) \tag{7.20}$$

or

$$(\alpha_1 < \alpha_2) \Rightarrow \left[a_1^{\alpha 2}, a_2^{\alpha 2}\right] \subset \left[a_1^{\alpha 1}, a_2^{\alpha 1}\right] \text{for every } \alpha_1, \alpha_2 \in [0, 1] \tag{7.21}$$

The fuzzy numbers thus defined are known as intervals. Once the intervals or ranges of a fuzzy quantity corresponding to specific α – cuts are known, the system response at any specific α – cut, can be found using interval analysis. Thus in the application of a fuzzy approach to uncertain engineering problems, interval analysis can be used.

7.5.2 *Fuzzy Approach for Combining Evidences (Rao and Annamdas 2008)*

In Dempster-Shafer theory, evidence can be associated with multiple or sets of events. By combining evidence from multiple sources, Dempster-Shafer theory provides the lower and upper bounds, in the form of belief and plausibility, for the probability of occurrence of an event. In this work, uncertain parameters are denoted as fuzzy variables and the available evidences on the ranges of the uncertain parameters, in the form of basic probability assignments (bpa's), are represented in the form of membership functions of the fuzzy variables. The membership functions constructed from the available evidences from multiple sources are added as multiple fuzzy data to find the combined membership of the uncertain or fuzzy parameter. The resulting combined membership function of the fuzzy parameter is then used to estimate the lower and upper bounds of any response quantity of the system.

7.5.3 Computation of Bounds on the Margin of Failure/Safety

If the margin of failure (or margin of safety) is indicated as shown in Fig. 7.6 (or Fig. 7.7), the lower and upper bounds on the margin of failure (or margin of safety) greater than zero can be expressed as

For margin of failure:

$$\text{Lower bound} = 0, \text{ Upper bound} = \frac{A_2}{A_1 + A_2} \text{ with } A_1 \gg A_2 \tag{7.22}$$

For margin of safety:

$$\text{Lower bound} = \frac{A_2}{A_1 + A_2}, \text{ Upper bound} = 1 \text{ with } A_1 \ll A_2 \tag{7.23}$$

The bounds on the margin of safety and margin of failure of the welded beam described in Section 7.3 are computed by representing the uncertain parameters in

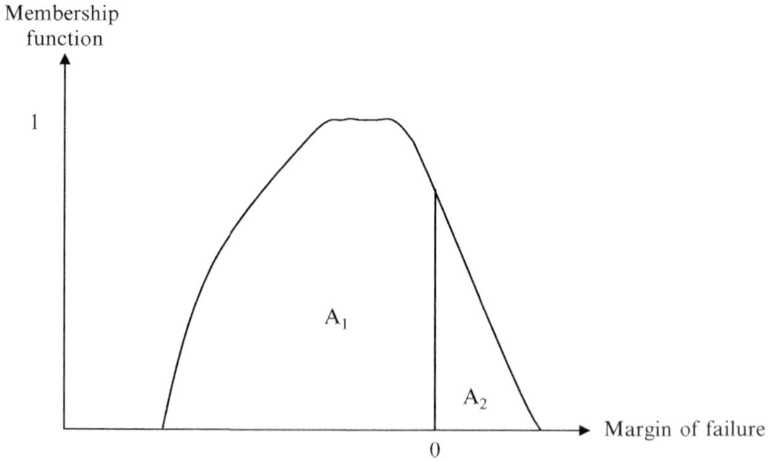

Fig. 7.6 General membership function of the margin of failure

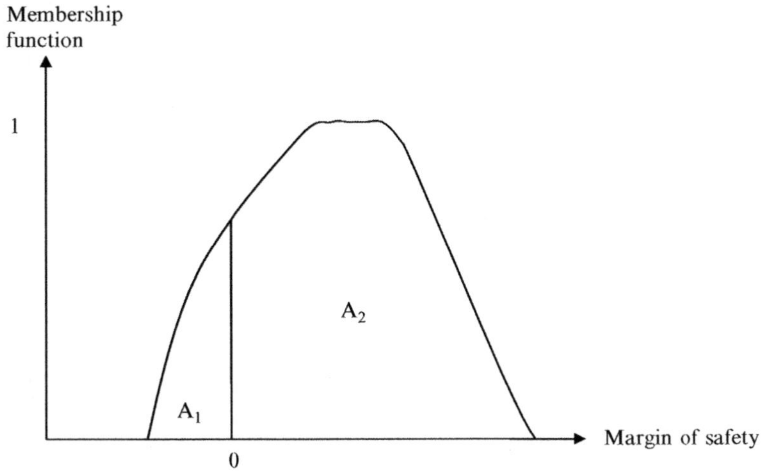

Fig. 7.7 General membership function of the margin of safety

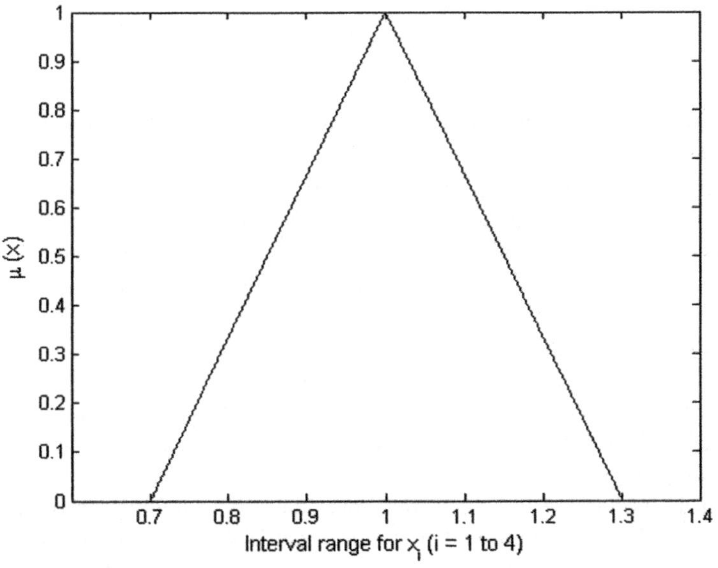

Fig. 7.8 Fuzzy membership functions of x_i ($i = 1$ to 4)

the form of triangular membership functions. The depth of the cantilever (t), length of the cantilever (L), length of the weld (l) and height of the weld (h) are treated as four uncertain parameters with x_1, x_2, x_3 and x_4 indicating their corresponding multiplication factors. The membership functions of the uncertain variables are shown in Fig. 7.8. A Matlab program is developed to implement the fuzzy arithmetic required to handle both the two and the four uncertain parameters and obtain

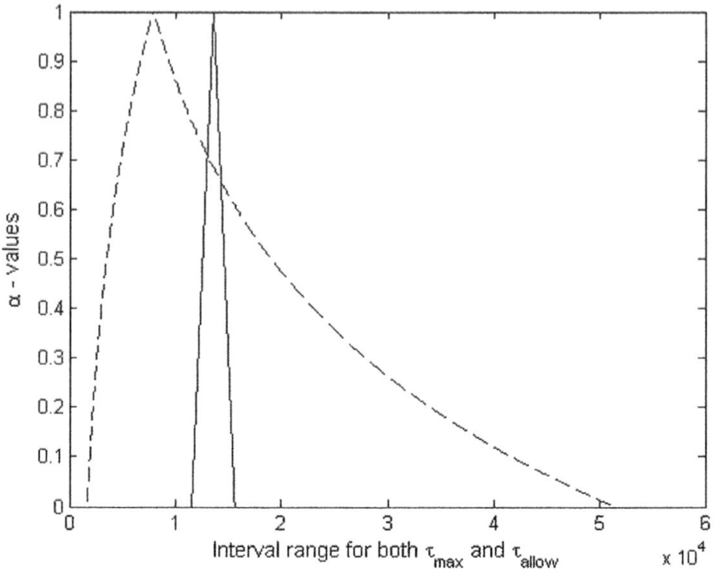

Fig. 7.9 Fuzzy membership functions of τ_{max} and $\tau_{allowable}$

an assessment of the likelihood of the maximum induced shear stress exceeding the specified permissible value.

The procedure indicated earlier is used to find the likelihood of the maximum induced shear stress exceeding the specified permissible/allowable value, $\tau_{allowable}$, which is considered to be a fuzzy quantity with a triangular membership function with a range of $\pm$ 15% of $\tau_{max}(0.85\ \tau_{max}$ psi to $1.15\ \tau_{max}$ psi) with $\tau_{max} = 13{,}600$ psi as shown by the solid line in Fig. 7.8. The dotted curve in Fig. 7.9 shows the fuzzy description of the maximum shear stress induced in the beam.

The margin of safety, calculated as the fuzzy difference between $\tau_{allowable}$ and τ_{max}, i.e., $\tau_{allowable} - \tau_{max}$ and is shown in Fig. 7.10. Similarly, the margin of failure, calculated as the fuzzy difference between τ_{max} and $\tau_{allowable}$, i.e., $\tau_{max} - \tau_{allowable}$, is shown in Fig. 7.11.

Using the procedure indicated, the lower and upper bounds on the margins of safety and failure are computed from the membership function curves shown in Figs. 7.10 and 7.11, respectively, to obtain the results indicated in Table 7.4.

7.6 Other Combination Rules

The use of different combination rules for combining evidences is illustrated in this section through an automobile example. An automobile is assumed to have two states: safe and unsafe. The frame of discernment consists of these states,

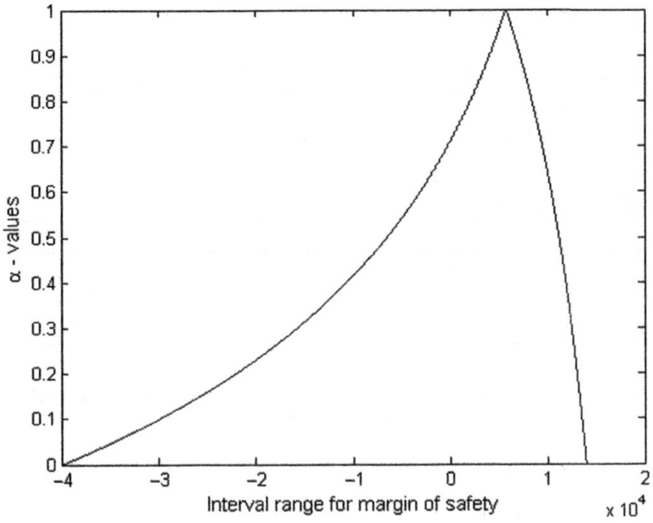

Fig. 7.10 Fuzzy membership function of the margin of safety

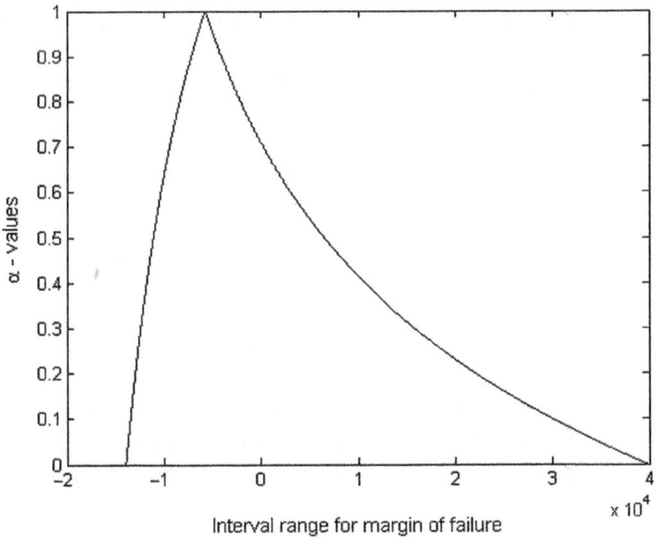

Fig. 7.11 Fuzzy membership function of the margin of failure

$\Theta = \{I, II\}$, where $\{I\}$means that automobile is safe and $\{II\}$ means unsafe to drive. The evidences (E_1 and E_2) are as follows.

$$E_1 : m\{I\} = 0.5,\ m\{II\} = 0.3,\ m\,(\Theta) = 0.2$$
$$E_2 : m\{I\} = 0.6,\ m\{II\} = 0.1,\ m\,(\Theta) = 0.3$$

Table 7.4 Lower and upper bounds on the margins of failure and safety

Response quantity	Lower bound	Upper bound
Margin of failure	0	0.50311
Margin of safety	0.49689	1

Table 7.5 Combination of bpa values for sources E_1 and E_2

E_1E_2	$m\{I\} = 0.5$	$m\{II\} = 0.3$	$m\{\Theta\} = 0.2$
$m\{I\} = 0.6$	0.30	0.18	0.12
$m\{II\} = 0.1$	0.05	0.03	0.02
$m\{\Theta\} = 0.3$	0.15	0.09	0.06

The safety analysis of this automobile can be conducted using different rules of combination as indicated below and the results are summarized in Table 7.10.

7.6.1 Dempster's Rule

This rule is described in Section 7.1 (Eqs. (7.8)–(7.10)).

Using Dempster's rule, the combined bpa values are calculated as shown in Table 7.5.

By using Eq. (7.10), we obtain $k = 0.18 + 0.05 = 0.23$. Using DST combination rule (Eq. (7.8)), we obtain

$$m\{I\} = \frac{0.30 + 0.12 + 0.15}{1 - 0.23} = 0.74026$$

$$m\{II\} = \frac{0.03 + 0.02 + 0.09}{1 - 0.23} = 0.1818$$

$$m\{\Theta\} = \frac{0.06}{1 - 0.23} = 0.07792$$

The belief values are calculated as $Bel\{I\} = m\{\text{I}\} = 0.74026$, and $Bel\{II\} = m\{\text{II}\} = 0.1818$. The plausibility values are calculated by using Eq. (7.6) as $Pl\{\text{I}\} = 1 - Bel\{II\} = 0.81818$, and $Pl\{II\} = 0.25974$.

7.6.2 Yager's Rule (Yager 1987)

The combined ground probability mass assignment is defined as follows

$$q(A) = \sum_{B \cap C = A} m_1(B)\, m_2(C) \tag{7.24}$$

where $q(A)$ denotes the ground probability assignment associated with A. Equation (7.24) is same as Eq. (7.8) with no denominator. We convert the ground probabilities to the basic probability assignment of the universal set $m^Y(\Theta)$ as

$$m^Y(\Theta) = q(\Theta) + q(\phi) \tag{7.25}$$

where $q(\phi)$ and $q(\Theta)$ are the ground probability assignments of the null set and universal set, respectively. In Yager's rule, there is no normalization or scaling to the resulting combined evidence and mass associated with conflict, $m(\phi)$ will appear as ignorance in the belief interval. When there is any contradiction among the evidence from different sources, the Yager's combination rule treats that contradiction as coming from ignorance. Using Eq. (7.21) in conjunction with Table 7.5, we obtain $q\{I\} = 0.57$, and $q\{II\} = 0.14$. The ground probability mass assignments to the null set and the universal set are calculated exactly the same manner as in the case of Dempster's rule except for normalization to obtain $q\{\phi\} = 0.23$ and $q\{\Theta\} = 0.06$. Equation (7.25) gives $m^Y\{\Theta\} = 0.29$. The belief values are calculated as $Bel\{I\} = q\{I\} = 0.57$ and $Bel\{II\} = q\{II\} = 0.14$. The plausibility values are given by Eq. (7.6) as $Pl\{I\} = 1 - Bel\{II\} = 0.86$ and $Pl\{II\} = 0.43$.

7.6.3 Inagaki's Extreme Rule

Inagaki's (1991) unified combination rule (denoted m^U) is defined in the following way:

$$m^U(C) = \left[1 + \tilde{k}q(\phi)\right] q(C),\ where\ C \neq \Theta, \phi \tag{7.26}$$

$$m^U(\Theta) = \left[1 + \tilde{k}q(\phi)\right] q(\Theta) + \left[1 + \tilde{k}q(\phi) - \tilde{k}\right] q(\phi) \tag{7.27}$$

$$\text{where } 0 \leq \tilde{k} \leq \frac{1}{1 - q(\phi) - q(\Theta)} \tag{7.28}$$

With Inagaki's rule, Dempster's rule is obtained by setting $\tilde{k} = [1 - q(\phi)]^{-1}$ and Yager's rule can be realized with $\tilde{k} = 0$ in Eq. (7.26)–(7.28). Since $\tilde{k}$ is continuous-valued, the unified rule of combination results in an infinite number of rules of combination, as shown in Fig. 7.12.

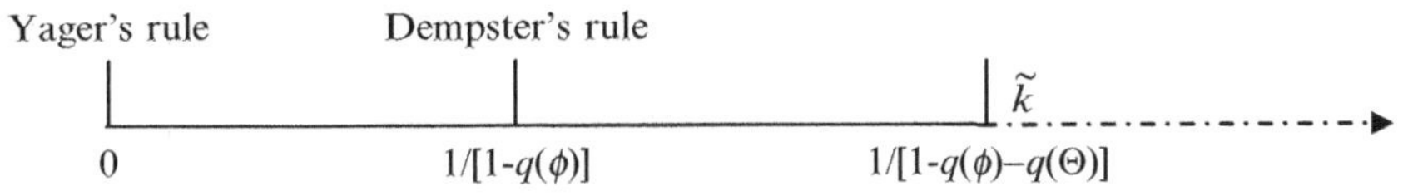

Fig. 7.12 Rules of combination with different values of the parameter $\tilde{k}$

The Inagaki's extreme rule (also called "the extra rule") can be obtained by setting the value of $\tilde{k}$ equal to the upper bound indicated in Eq. (7.28).

$$m^U(C) = \frac{1 - q(\Theta)}{1 - q(\phi) - q(\Theta)} q(C), \; where \; C \neq \Theta \tag{7.29}$$

$$m^U(\Theta) = q(\Theta) \tag{7.30}$$

The ground probability mass assignment to the null set, $q\{\phi\}$, and the universal set, $q\{\Theta\}$, are calculated exactly the same manner as in the case of Yager's rule to find $q\{\phi\} = 0.23$ and $q\{\Theta\} = 0.06$. Denoting $p = \frac{1-q(\Theta)}{1-q(\phi)-q(\Theta)} = 1.3239$, Eq. (7.29) gives, using the values of $q\{I\}$ and $q\{II\}$ from Yager's rule for $q(C)$, as $m^U\{I\} = 0.57p = 0.7546$ and $m^U\{II\} = 0.06p = 0.0805$. The belief values are calculated as $Bel\{I\} = m^U\{I\} = 0.7546$ and $Bel\{II\} = 0.0805$. The plausibility values are calculated by using Eq. (7.6) as $Pl\{I\} = 0.81465$ and $Pl\{II\} = 0.24535$.

7.6.4 Zhang's Rule

Zhang (1994) introduced a combination rule as an alternative to the Dempster's rule. In this rule, a two frame interpretation of DST is offered. Let S and T denote two frames of discernment; these could be the opinions of two experts. We are concerned with the truth in T but the probability P is available for the truth in S. For the two frame model, we define a subset C_s of T for each $s \in S$ by compatibility relation $C_s = \{t | (s, t) \in C\}$. Thus, the bpa over T can be defined by

$$m(A) = \sum \{P(C_s) | s \in S \text{ such that } C_s = A\} \tag{7.31}$$

The belief function for A as calculated using Eq. (7.31) as

$$Bel(A) = \sum \{P(s) | C_s \subset A\} \tag{7.32}$$

and plausibility function as

$$Pl(A) = \sum \{P(s) | C_s \cap A \neq 0\} \tag{7.33}$$

where bpa of $P(s)$ is assigned to C_s. We use a moderate approach to include only a portion of $P(s)$ in the Eq. (7.32) or (7.33) which leads to a center probability P_c of

$$P_c(A) = \sum_s \frac{|A \cap C_s|}{|C_s|} P(s) \tag{7.34}$$

where $|.|$ represents cardinality. Zhang's rule of combination introduces a measure to the intersection of the two sets A and B assuming them to be finite. The measure of intersection is the ratio of the cardinality of the intersection of the two sets divided by the product of the cardinality of the individual sets. Thus the measure of intersection of the sets A and B, denoted $r(A, B)$, is given by

$$r(A, B) = \frac{|A \cap B|}{|A|\,|B|} = \frac{|C|}{|A|\,|B|} \tag{7.35}$$

where $A \cap B = C$. The use of Eq. (7.35) in the combination rule reduces to the moderate approach given by Eq. (7.34). Zhang's combination rule scales the products of the bpa's of the intersecting sets $(A \cap B = C)$ with r(A, B), given by Eq. (7.35). This is repeated for all intersecting pairs that give C. These scaled products of all bpa's whose intersection equals C are added and then multiplied by a factor $\hat{k}$ where $\hat{k}$ denotes a renormalization factor which is independent of C, m_1, and m_2. This renormalization factor makes the summation of the bpa's equal to 1:

$$m(C) = \hat{k} \sum_{A \cap B = C} \left[\frac{|C|}{|A|\,|B|} m_1(A) m_2(B) \right] \tag{7.36}$$

The product of bpa values is same as the one obtained in Table 7.5. The measures of intersections are shown in Table 7.6. The products of the entries of Table 7.5 and the corresponding r-values of Table 7.6 (as indicated by Eq. (7.36)) are shown in Table 7.7. Re-normalizing the combined bpa values yield Table 7.8.

Using Zhang's rule Eq. (7.36), we obtain $m\{I\} = 0.7909$ and $m\{II\} = 0.1545$, and $m\{\Theta\} = 0.0545$. The belief values are calculated as $Bel\{I\} = m\{II\} = 0.79091$ and $Bel\{II\} = m\{II\} = 0.15455$. The plausibility values are calculated by using Eq. (7.6) as $Pl\{I\} = 1 - Bel\{II\} = 0.84545$, and $Pl\{II\} = 0.20909$.

Table 7.6 Values of intersection measure (r) for combined bpa values from sources E_1 & E_2

E_1E_2	$m\{I\} = 0.5$	$m\{II\} = 0.3$	$m\{\Theta\} = 0.2$
$m\{I\} = 0.6$	1	0	0.5
$m\{II\} = 0.1$	0	1	0.5
$m\{\Theta\} = 0.3$	0.5	0.5	0.5

Table 7.7 Combined bpa values from sources E_1 and E_2using Zhang's rule

E_1E_2	$m\{I\} = 0.5$	$m\{II\} = 0.3$	$m\{\Theta\} = 0.2$
$m\{I\} = 0.6$	0.3	0	0.06
$m\{II\} = 0.1$	0	0.03	0.01
$m\{\Theta\} = 0.3$	0.075	0.045	0.03

Table 7.8 Combined bpa values from sources E_1 and E_2 using Zhang's rule after normalization

E_1E_2	$m\{I\} = 0.5$	$m\{II\} = 0.3$	$m\{\Theta\} = 0.2$
$m\{I\} = 0.6$	0.5454	0	0.10909
$m\{II\} = 0.1$	0	0.05454	0.01818
$m\{\Theta\} = 0.3$	0.1363	0.08181	0.05454

Table 7.9 Combination of bpa values for sources E_1 and E_2 using Murphy's rule

E_1E_2	$m\{I\} = 0.55$	$m\{II\} = 0.2$	$m\{\Theta\} = 0.25$
$m\{I\} = 0.55$	0.3025	0.11	0.1375
$m\{II\} = 0.2$	0.11	0.04	0.05
$m\{\Theta\} = 0.25$	0.1375	0.05	0.0625

7.6.5 Murphy's Rule

Murphy's rule (2000) uses the average beliefs in the combination rule. Thus, the initial averaging of bpas is followed by combining the evidence, using DST, 'n – 1' times where n is the number of evidences. The belief functions are calculated exactly the same manner as in the case of Dempster's rule from the product bpas given in Table 7.9. This gives the belief values as $Bel\{I\} = m\{I\} = 0.740385$ and $Bel\{II\} = m\{II\} = 0.179487$. The plausibility values are calculated by using Eq. (7.6) as $Pl\{I\} = 1 - Bel\{II\} = 0.820513$ and $Pl\{II\} = 0.259615$.

7.6.5.1 Observations on the Results of the Automobile Safety Problem

1. It is clear that all combination rules indicate that the automobile is safe but differ in the matter of degree of the belief assigned to I.
2. The belief assigned to the automobile to be safe (I) can be predicted intuitively to be around 0.75 based on the given initial evidence as the sum of the evidences given by E_1 and E_2.
3. Zhang's rule gives the maximum belief to safe condition (I) (i.e., more than 0.75). Hence, this rule may not be applicable for this example.
4. For the safety analysis of an automobile, the combination rule that gives the lowest belief to safety (I) can be taken as the best one. Thus Yager's rule is best suited for this example.
5. For the failure analysis of an automobile, the rule that gives highest belief to the unsafe condition (II) can be taken as the best one. Thus, Inagaki's extreme rule is best suited in this case (Table 7.10).

Table 7.10 Results of various combination rules for an automobile example

Combination rule E_1 & E_2	Belief Interval for *I*	Belief Interval for *II*	$m(\Theta)$
DST	[0.7402,0.8181]	[0.1818,0.2597]	0.0779
Yager's	[0.57,0.86]	[0.14,0.43]	0.29
Inagaki's	[0.7546,0.8146]	[0.1853,0.2453]	0.06
Zhang's	[0.7909,0.8454]	[0.1545,0.2090]	0.0545
Murphy's	[0.7403,0.8205]	[0.1794,0.2596]	0.0801

7.7 Conclusion

When evidences on several uncertain parameters are available from a number of sources (experts) in an engineering problem, and if the response of the system involves mathematical expressions, DST can be used. In the case of the safety analysis of the welded beam, evidences of two and four uncertain parameters are combined to find the maximum shear stress induced in the weld. A modified DST method is also presented to combine evidences from multiple sources, when different sources of evidence have different credibilities. The variations in the belief and plausibility have been observed to depend on the number of interval ranges for the interval-valued data used for combining the evidence. When the evidences on the uncertain parameters are available in the form of fuzzy membership functions, the method outlined in the work can be used to find the response, such as the margin of safety, of the system. The application of different methods of combining evidences is also illustrated using an automobile example. An essential feature of knowledge representation in engineering applications is the ability to represent and manage imprecise concepts and the examples presented in this work represent an initial step in this direction.

References

Agarwal, H., Renaud, J. E., Preston, E. L., and Padmanabhan, D. (2004), "Uncertainty quantification using evidence theory in multidisciplinary design optimization", *Reliability Engineering and System Safety*, Vol. 85, pp. 281–294.

Alim, S. (1988), "Application of Dempster-Shafer theory for interpretation of seismic parameters", *Journal of Structural Engineering*, Vol. 114, No. 9, pp. 2070–2084.

Bae, H. R., Grandhi, R. V. and Canfield, R. A. (2004), "Epistemic uncertainty quantification techniques including evidence theory for large scale structures", *Computers and structures*, Vol. 80, pp. 1101–1112.

Beynon, M., Curry, B. and Morgan, P. (2000), "The Dempster-Shafer theory of evidence: An alternative approach to multicriteria decision modeling", *Omega*. Vol. 28, pp. 37–50.

Butler, A. C., Sadeghi, F., Rao, S. S. and LeClair, S. R. (1995), "Computer-aided design/ engineering of bearing systems using Dempster-Shafer theory", *Artificial Intelligence for Engineering Design, Analysis and Manufacturing*, Vol. 9, pp. 1–11.

Dong, W. M., and Shah, H. C. (1987), "Vertex method for computing functions of fuzzy variables", *Fuzzy Sets and Systems*, Vol. 24, pp. 65–78.

Ferson, S., Kreinovich, V. Ginzburg, L., Myers, D. S. and Sentz, K. (2003). "Constructing probability boxes and Dempster-Shafer Structures", *SAND Report*.

Ha-Rok, B. (2004), "Uncertainty quantification and optimization of structural response using evidence theory", PhD Dissertation, Wright State University.

Inagaki, T. (1991), "Interdependence between Safety-Control Policy and Multiple-Sensor Schemes Via Dempster-Shafer Theory", IEEE Transactions on Reliability 40(2):182–188.

Klir, G. J. and M. J. Wierman (1998), *Uncertainty-Based Information: Elements of Generalized Information Theory*, Physica-Verlag, Heidelberg.

Kozine, I. O. and Utkin, L. V. (2004), "An Approach to Combining Unreliable Pieces of Evidence and their Propagation in a System Response Analysis," *Reliability Engineering and System Safety*, Vol. 85, No. 1–3, pp. 103–112.

Murphy, C. K. (2000), "Combining belief functions when evidence conflicts", *Decision Support Systems* 29, pp. 1–9.

Oberkampf, W. L., and Helton J. C., "Investigation of Evidence Theory for Engineering Applications," *Non-Deterministic Approaches Forum*, Denver, CO, April-2002, Paper No. AIAA-2002-1569.

Parikh, C. R., Pont, M. J. and Jones, N. B. (2001), "Application of Dempster- Shafer Theory in condition monitoring systems: A case study", *Pattern Recognition Letters*, Vol. 22, No. 6–7, pp. 777–785.

Rao, S. S., and Sawyer, J. P., "A fuzzy element approach for the analysis of imprecisely-defined systems," AIAA Journal, Vol. 33, No. 12, 1995, pp. 2364–2370.

Rao, S. S. and Annamdas, K. K. (2008), "Evidence-Based Fuzzy Approach for the Safety Analysis of Uncertain Systems", *AIAA Journal*, Vol. 46, No. 9, pp. 2383–2387.

Sentz, K. (2002), "Combination of evidence in Dempster-Shafer theory", PhD Dissertation, Binghamton University.

Shafer, G. (1976), *A Mathematical Theory of Evidence*, Princeton University Press, Princeton, NJ.

Tanaka, K. and G. J. Klir (1999), "A design condition for incorporating human judgement into monitoring systems", *Reliability Engineering and System Safety*, Vol. 65, No. 3, pp. 251–258.

Yager, R. (1987a), "On the Dempster-Shafer framework and new combination rules", *Information Sciences* 41, pp. 93–137.

Zadeh, L., "Fuzzy sets," *IEEE Information Control*, Vol. IC-8, 1965, pp. 338–353.

Zhang, L. (1994), "Representation, independence, and combination of evidence in the Dempster-Shafer theory", *Advances in the Dempster-Shafer Theory of Evidence* (Eds. R. R. Yager, J. Kacprzyk and M. Fedrizzi), John Wiley, New York, pp. 51–69.

Chapter 8
Role of Robust Engineering in Product Development

Rajesh Jugulum

Abstract Principles of Robust Engineering (some times referred to as Taguchi Methods) have been successfully applied in many applications across the globe and have resulted in hundreds of thousands of dollars in savings besides having robust products/service. Robustness can be defined as designing a product/service in such a way that its performance is similar across all customer usage conditions. In this paper we will describe the applicability of robust engineering principles in total product/service development with emphasis on the robustness aspects. We discuss several robustness strategies with case examples.

Keywords Robust engineering · Taguchi Methods · Signal to Noise Ratio · Design of Experiments · Optimal Design

8.1 Introduction to Robust Engineering

Robustness can be defined as designing a product in such a way that the level of its performance at various customer usage conditions is same as that at the nominal conditions. Robust engineering methods (also known as Taguchi Methods) are intended as cost-effective methods to improve the performance of a product by reducing its variability in the customer usage conditions. Because they are intended to improve companies' competitive position in the market, these methods have attracted the attention of many industries and academic community across the globe. The methods of Robust Engineering are the result of research efforts led by Dr. Genichi Taguchi.

Quality, in the context of Robust Engineering (Taguchi et al. 2004) can be classified into two types:

1. Customer Driven Quality
2. Engineered Quality

R. Jugulum (✉)
Bank of America and MIT, Boston, MA 02110, USA
e-mail: rajesh_jugulum@yahoo.com

N.R.S. Raghavan and J.A. Cafeo (eds.), *Product Research: The Art and Science Behind Successful Product Launches*, DOI 10.1007/978-90-481-2860-0_8,

Customer quality leads to the size of the market segment. It includes product features such as color, size, appearance and function. The market size becomes bigger as the customer quality gets better. Customer quality is addressed during the product planning stage. Customer quality is extremely important to create the new market. On the other hand, engineered quality includes the defects, failures, noise, vibrations, pollution, etc. While the customer quality defines the market size, the engineered quality helps in winning the market share within the segment.

All problems of engineered quality are caused by three types of controllable factors (called as noise factors).

I. Various usage conditions
   - Environmental conditions

II. Deterioration and wear
   - Degradation over time

III. Individual difference
   - Manufacturing imperfection. Robust engineering (RE) methods or Taguchi methods (TM) aim at improving the engineered quality

While designing a new product, optimization for robustness can be achieved in the following three stages:

1. Concept Design
2. Parameter Design
3. Tolerance Design

Although there are three stages, most of the RE applications focus on parameter design optimization and Tolerance design. It is widely acknowledged that the gains in terms of robustness will be much more if one starts designing a process with a robust concept selected in the concept stage. Techniques like Pugh concept selection, Theory of Inventive Problem Solving (TRIZ), and Axiomatic Design and P-diagram strategies developed for conceptual robustness through Ford-MIT collaboration by Taguchi et al. (2004) can be used to achieve robustness at the concept level.

The methods of robust engineering are developed based on the following five principles.

1. Measurement of function using energy transformation
2. Take Advantage of interactions between control and noise factors
3. Use of orthogonal arrays and signal-to-noise ratios
4. Two-step optimization
5. Tolerance design using quality loss function and on-line quality engineering

Taguchi (1987) and Phadke (1989) provided a detailed discussion on the principles of TM. These principles have been successfully applied in many engineering applications to improve the performance of the product/process. They are proved to be extremely useful and cost effective. A brief illustration of these principles is given below:

1. *Measurement of function using energy transformation*
   The most important aspect of Taguchi methods (TM) is to find a suitable function (called ideal function) that governs the energy transformation in the system from input signal to output response. It is important to maximize the energy transformation by minimizing the effect of uncontrollable or noise factors. In Taguchi approach, we measure the functionality of the system to improve the product performance (quality). The energy transformation is measured in terms of signal to noise (S/N) ratios. Higher S/N ratio means better is the energy transformation and hence functionality of the system.
2. *Take advantage of interactions between control and noise factors*
   In TM, we are not interested in measuring the interaction among the control factors. We are interested in the interaction between the control and noise factors since the objective is to make the design robust against the noise factors.
3. *Use of orthogonal arrays (OA's) and signal-to-noise ratios (S/N ratios)*
   OA's are used to minimize the number of runs (or combinations) needed for the experiment. Usually, many people are of the opinion that the application of OA is TM. However, it should be noted that the application of OA's is only a part of TM. S/N ratios are used as a measure of the functionality of the system.
4. *Two-step optimization*
   After conducting the experiment, the factor-level combination for the optimal design is selected with the help of a two-step optimization. In two-step optimization, the first step is to minimize the variability (maximize S/N ratios). In the second step, the sensitivity (mean) is adjusted to the desired level. It is easier to adjust the mean after minimizing the variability.
5. *Tolerance design using quality loss function and on-line quality engineering*
   While the first four principles are related to parameter design, the fifth principle is related to the tolerance design and on-line quality engineering. Having determined the best settings using parameter design, the tolerancing is done with the help of quality loss function. If the performance deviates from the target, there is a loss associated with the deviation. This loss is termed as the loss to the society. This loss is proportional to the square of the deviation. It is recommended to design the safety factors using this approach. On-line QE is used in monitoring the process performance and detect the changes in the process.

8.2 Concepts of Robust Engineering

8.2.1 Parameter Diagram (P-Diagram)

This is a block diagram, which is often quite helpful to represent a product or process or a system. In Fig. 8.1, the energy transformation takes place between input signal

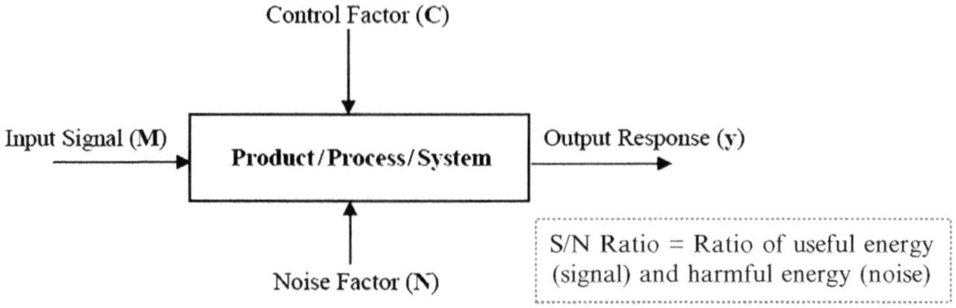

Fig. 8.1 Parameter diagram or P-diagram

(**M**) and the output response (**y**). The goal is to maximize the energy transformation by adjusting control factors (**C**) settings in the presence of noise factors (**N**).

1. **Signal Factors (M)**: These are the factors, which are set by the user/operator to attain the target performance or to express the intended output. For example, the steering angle is a signal factor for the steering mechanism of an automobile. The signal factors are selected by the engineer based on the engineering knowledge. Sometimes more than one signal factor is used in combination, for example, one signal factor may be used for coarse tuning and one for fine-tuning.
2. **Control Factors (C)**: These are the product parameters specification whose values are the responsibility of the designer. Each of the control factors can take more than one value, which will be referred to as levels. It is the objective of the design activity to determine the best levels of these factors that will make the product insensitive to noise factors or robust against noise factors. Robustness is the insensitivity to the noise factors.
3. **Noise Factors (N)**: Noise factors are the uncontrollable factors. They influence the output **y** and their levels change from one unit of the product to another, from one environment to another and from time to time. As mentioned before, noise factors can be one or a combination of the following three types. (1) Various usage conditions (2) Deterioration and wear and (3) Individual difference.

8.2.2 Experimental Design

Experimental design consists of a set of techniques and body of knowledge to help the investigator to conduct experiments in a better way, analyze results of experiments and find optimal parameter combinations for design to achieve the intended objectives. In experimental design, various factors affecting the product/process performance are varied (by changing their values called levels) in a systematic fashion to estimate their effects on product/process variability in order to find optimal factor combination for a robust design. There is an extensive literature on this subject. A typical experimental design cycle is shown in Fig. 8.2.

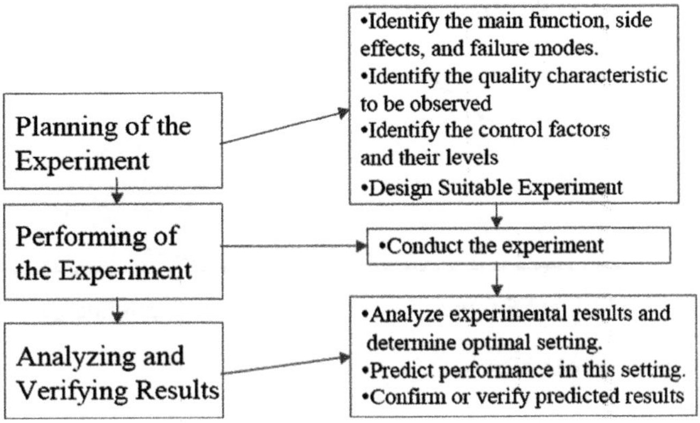

Fig. 8.2 Experimental design cycle

8.2.2.1 Types of Experiments

There are typically two types of experiments, full factorial experiments and fractional factorial experiments. In full factorial experiments, all combinations of factors are studied. All main effects and interaction effects can be estimated with results of such an experiment. In fractional factorial experiments, a fraction of total number of experiments is studied. This is done to reduce cost, material and time. Main effects and selected interactions can be estimated with such experimental results. Orthogonal array is an example of this type. In robust engineering, the main role of OA's is to permit engineers to evaluate a product design with respect to robustness against noise, and cost involved by changing settings of control factors. OA is an inspection device to prevent a "poor design" from going "down stream".

8.2.3 Signal to Noise (S/N) Ratios

The term S/N ratio means signal to noise ratio. S/N ratio tries to capture the magnitude of true information (i.e., signals) after making some adjustment for uncontrollable variation (i.e., noise). From Fig. 8.1, we can say that a system consists of a set of activities or functions that are designed to perform a specific operation and produce an intended or desired result by minimizing functional variations due to noise factors. In engineering systems the input energy is converted into an intended output through laws of physics.

These engineered systems are designed to deliver specific results as required by customers. All Engineered Systems are governed by an ideal relationship between the input and the output. This relationship is referred to as Ideal Function. Robust engineering approach uses this relation and brings the system closer to the ideal state (Fig. 8.3).

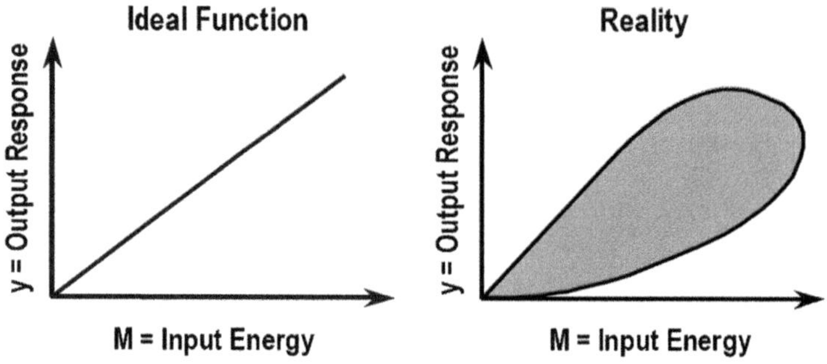

Fig. 8.3 Ideal function and reality

If all the input energy is converted into output, then there will be no energy losses. As a result there would be no squeaks, rattles, noise, scrap, rework, quality control personnel, customer service agents, complaint departments, warranty claims, etc. Unfortunately it does not happen because reality is much different from ideal situations.

In reality there is energy loss when input is transformed to an output. This loss occurs because of variability and noise factors. This energy loss will create unintended functions. Further, the bigger the energy loss, the bigger will be the problems.

The Signal-to-Noise Ratio (S/N ratio) is a measure of robustness as it measures energy transformation that occurs within a design. S/N ratio is expressed as:

S/N Ratio = Ratio of Energy (or power) that is transformed into intended output and Energy (or power) that is transformed into unintended output.

= Ratio of useful energy and harmful energy

= Ratio of work done by signal and work done by the noise

Higher the S/N ratio, more robust the system's function is.

8.2.4 Simulation Based Experiments

In the past, the experiments were performed as hardware experiments. But, today it is possible to conduct simulation based experiments. Computer simulations are quite important in robust engineering because:

- Simulated experiments play a significant role in reducing product development time because there is no need to conduct all hardware experiments.
- Simulated based robust engineering can serve as a key strategy for research and development.
- They help to conduct functionality based analysis.
- Simulation experiments are typically inexpensive, less time consuming, more informative and many control factors can be studied.

Once a simulation method is available, it is extremely important to optimize the concept for its robustness. The results of simulated experiments are analyzed by calculating SN ratios and sensitivities. However, the optimal combination needs to be tried using hardware experimentation to confirm the results. This book focuses only on simulation-based experiments. While using simulated based experiments, it is important to select suitable signals, noise factors and output response.

The examples given in Section 8.3 explain how to conduct simulation based experiments. These examples also explain the difference between Taguchi methods of robust engineering with other methods of design of experiments.

8.3 Case Examples

8.3.1 Circuit Stability Design

Consider a circuit stability design as a simple example with a theoretical equation.

The output current, y, in amperes of an alternating-current circuit with resistance, R, and self-inductance, L, is given by the equation below,

$$y = \frac{V}{\sqrt{R^2 + (2\pi fL)^2}} \tag{8.1}$$

where:

V: Input alternating – current voltage (V)
R: Resistance (Ω)
f: Frequency of input alternating current (Hz)
L: Self-inductance (H) and, $\omega = 2\pi f$

The target value of output current y is 10 amps. If there is a shift of at least 4 amps during use by the consumer, the circuit no longer functions. The target value of y is 10 amps and the tolerance is 10 ± 4 amps.

The purpose of parameter design is to determine the parameter levels (normal values and types) of a system of elements. For example, it determines whether the nominal value of resistance, R, in the circuit is to be 1 or 9. The purpose of tolerance design is to determine the tolerance around the nominal value of the parameter.

8.3.1.1 Classification of Factors: Control Factors and Noise Factors

Those variables in the equation whose central values and levels that the designer can select and control are the control factors.

The input voltage is 100 V AC in the case of a household source, and because the designer cannot alter this, it is not a control factor. Neither is the frequency, f, a control factor.

Only R and L are control factors. The designer sets the resistance value at $1\Omega, 3\Omega$, or 9Ω. The same is true with regard to self-inductance, L. It is obvious that since the target value of the output current is given as 10 amps, in this case, once one decides on the central value of the resistance, R, the central value of the inductance, L, will be determined by the equation.

When there are three or more control factors, it is best not to consider such limiting conditions. The reason for this is that it is more advantageous to adjust the output to the target value by changing the levels of those control factors that have little influence on stability after an optimum design for stability has been determined. Factors that are favorable for adjustment are factors that can be used to change the output to the target value. These factors are sometimes termed signal factors. We can perform parallel analyses for the S/N Ratio, and for the mean, which are measures of stability and sensitivity, we can adjust it once the variability has been reduced. In the present case, we choose three levels of R and L independently, as follows.

$$R_1 = 0.5\,(\Omega) \quad R_2 = 5.0\,(\Omega) \quad R_3 = 9.5\,(\Omega)$$
$$L_1 = 0.010\,(H)\ L_2 = 0.020\,(H) L_3 = 0.030 (H)$$

An orthogonal array is used if the control factors are more numerous, but since there are only two here, we can use a two-way layout with full factorial experiment. An orthogonal array to which control factors have been assigned is termed an inner orthogonal array or control orthogonal array. One should use as many control factors as possible, take a wide range of levels and assign only the main effects.

Next, let us examine the noise factors. As described earlier, a noise factor is a general term for a cause that disperses the target characteristic from the ideal value.

Dispersion by the environment consists of the following two factors in this example, and their levels are taken as follows:

Voltage of input source (V) 90 100 110

Frequency f(Hz) 50 60

The environmental temperature, etc., also changes the value of the resistance, R, and the coil inductance, L, although only slightly. However, changes due to deterioration of the resistance and coil are greater than the effects of environmental changes. After having considered changes due to dispersion of the initial-period value, let us decide that the resistance, R, and coil inductance, L, are to have the following three levels:

First level	Normal value × 0.9
Second level	Nominal value
Third level	Nominal value × 1.1

Table 8.1 Noise factors and levels

Noise factors and levels			
	First level	Second level	Third level
V	90	100	110 (V)
R′	–10	0	+10(%)
F	50	55	60 (Hz)
L′	–10	0	+10(%)

The levels of the noise factors are as given in Table 8.1 It should be noted that a prime has been affixed to symbols for the noise factors, R and L. In this instance, it means there is no error with respect to the noise factors V, R′, and L′ when they are at the second level. Frequency f is 50 Hz or 60 Hz, depending on the location in a country like, say, Japan. If one wishes to develop a product that can be used in both conditions, it is best to design it so that the output meets the target when the frequency is at 55 Hz, midway between the two. Sometimes one uses an intermediate value that has been weighted by the population ratio.

But now, as is evident from equation 1.1, the output becomes minimum with $V_1 R'_3 f_3 L'_3$ and maximum with $V_3 R'_1 f_1 L'_1$. If the direction of output changes is known when the noise factor levels are changed, all the noise factors can be compounded into a single factor. If the Compound factor is expressed as N, it has two levels. In this case:

$$N_1 = V_1 R'_3 f_3 L'_3 \text{ Minus side minimum value}$$
$$N_2 = V_3 R'_1 f_1 L'_1 \text{ Plus side maximum value}$$

When we know which noise factor levels cause the output to become large, it is a good strategy to compound them so as to obtain one factor with two levels, or one factor with three levels, including the central level. When we do this, our noise factor becomes a single compounded factor, no matter how many factors are involved. Since one can determine the tendencies of all four noise factors in this example, they have been converted essentially into a single compounded noise factor, N, with two levels. Thus, we need merely investigate the value of the outgoing current for these two levels.

Compounding of noise factors is even more important when there is no theoretical equation. This is because with a theoretical equation, the calculation time is short, usually no more than several seconds. Even if noise factors are assigned to the columns of a separate orthogonal array, there is little improvement in the efficiency of calculation even with compounded noise factors of two levels as in our example. Compounding was necessary in the past when computers were not available. However, when there are too many noise factors, the orthogonal array becomes large and much time and expense is required, even with a computer. In such cases, it is still useful either to reduce the number of noise factors to a few important ones or to compound noise factors.

8.3.1.2 Parameter Design

Design calculations are performed (experiments are performed when there is no theoretical equation) for all combinations of the inner orthogonal array to which control factors R and L have been assigned (two-way layout, here) and the noise factor assignment (two-level noise factor here). This is the logical method for parameter design. The assignment is shown in the left half of Table 8.2.

The S/N Ratio is the reciprocal of the square of the relative error (also termed the coefficient of variation, $\frac{\sigma}{m}$). Although the equation for the S/N Ratio varies, depending on the type of quality characteristic, all S/N Ratios have the same properties. When the S/N Ratio becomes 10 times larger, the loss due to dispersion decreases to one-tenth.

The S/N Ratio, η, is a measure for optimum design, and sensitivity, S, is used to select one (sometimes two or more) factor(s) to later adjust the mean value of the output to the target value, if necessary.

To compare the levels of control factors, we construct a table of mean values for the S/N Ration, η, and sensitivity, S. For example, in the case of R_1, $\eta = 7.5$ and $S = 24.0$.

In Table 8.3 the mean values are given for the other control factor levels.

Directing our attention to the measure of stability, the S/N Ratio, in Table 8.3, we can see that the optimum level of R is R_3, and the optimum level of L is L_1. Therefore, the optimum design is R_3L_1. This combination gives a mean value of 10.1 which is very close to the target value. If there is no difference between

Table 8.2 Calculations of SN ratios and sensitivities

Assignment of factors						
			Compound noise		S/N ratio	
No.	R	L	N_1	N_2	η	Sensitivity S
1	1	1	21.5	38.5	7.6	29.2
2	1	2	10.8	19.4	7.5	23.2
3	1	3	7.2	13.0	7.4	19.7
4	2	1	13.1	20.7	9.7	24.3
5	2	2	9.0	15.2	8.5	21.4
6	2	3	6.6	11.5	8.0	18.8
7	3	1	8.0	12.2	10.4	20.0
8	3	2	6.8	10.7	9.6	18.6
9	3	3	5.5	9.1	8.9	17.0

Table 8.3 Average factorial effects

	η	S
R_1	7.5	24.0
R_2	8.7	21.5
R_3	9.6	18.5
L1	9.2	24.5
L2	8.5	21.1
L3	8.1	18.5

the mean value and the target value, we might consider this as optimal design and therefore, we don't have to adjust control factors based on sensitivities. If there is a difference, it is required to compare the influence of the factors on the S/N Ratio and on the sensitivity, and use a control factor or two whose effect on sensitivity is great compared with its effect on the S/N Ratio in order to adjust the output to the target.

8.3.2 Robust Parameter Design of Brake System

This case study was conducted in a well-known automotive industry in Japan. When a system is very large, it is usually more effective to conduct Robust Design at a subsystem level. Since a brake system is so large, this section will focus on the Pad/Rotor (friction couple) subsystem because it is a critical subsystem producing significant energy transformation. Figure 8.4 illustrates how customer requirements can be translated into perceived output.

The energy transformation in this subsystem works in the following way: brake line sends pressure into the caliper. Because of this activity,

- The pressure displaces the pad against the rotor.
- The pad generates friction.
- The friction generates heat which then dissipates.
- The friction generates braking torque.

The Input signal for this subsystem is the line pressure and the output response is the braking torque (Fig. 8.5). The response may be friction, heat, or heat dissipation, depending on how you would define the boundary of the subsystem.

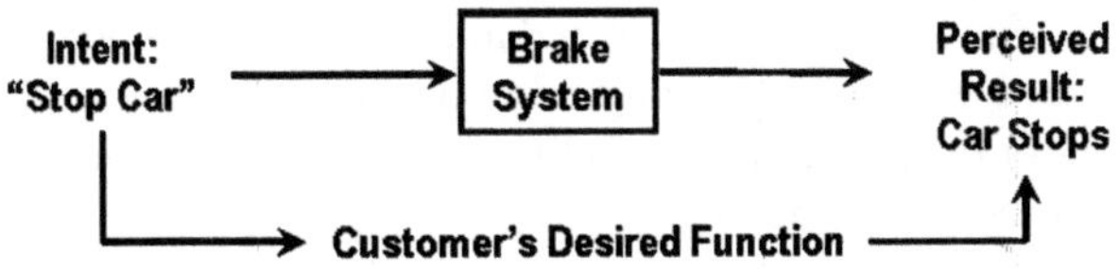

Fig. 8.4 Brake System

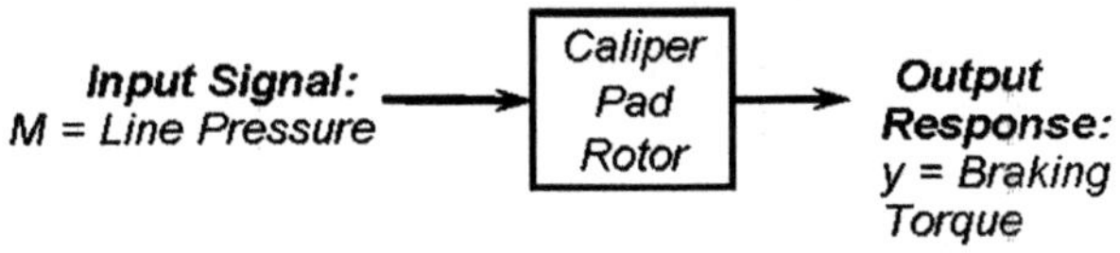

Fig. 8.5 Input and output relation for Pad Rotor

8.3.2.1 Signal Factor and Levels

After determining the scope of an experiment and ideal function, the next step is to decide how to vary the signal and noises experimentally. Often it is more effective to select signal factor levels so that the range covers all usage conditions including the entire family of present and future products.

The signal levels selected for this brake example are:

M = Brake Line Pressure

$M_1 = 0.008 \quad M_2 = 0.016 \quad M_3 = 0.032 \quad M_4 = 0.064$

8.3.2.2 Noise Factors and Noise Strategy

The noise factors corresponding to conditions of usage, environment and aging are considered. These three factors were compounded into one single noise factor. For environment, the pad temperature was considered, for usage, the condition of the wheels (dry or wet) was considered and for aging, the percentage of wear was considered.

For this case, two noise conditions, labeled N_1 and N_2 are determined by compounding these three noise factors. These two levels are:

N_1 = Noise condition when the response tends to be lower
N_2 = Noise condition when the response tends to be higher

The noise factors are compounded into two levels are:

N_1 = Low Temp/ Wet/ 80% Worn Pad
N_2 = Nominal Temp/ Dry/ 10% Worn Pad. This noise factor (N) represents noise conditions of usage, environment and aging. The most Robust Design against noise (N) should be the most robust against all other noises as well. In addition to these noises, the torque generated needs to remain consistent during braking is also considered as another noise factor.

8.3.2.3 Control Factor and Levels

The following control factors with different levels as shown in the Table 8.4 have been considered for the purpose of optimization:

8.3.2.4 Experimental Details

With these factors and levels, the following experimental layout was considered for robust optimization (Table 8.5):

The results of the experiments are given in Table 8.6.

Table 8.4 Control factors and levels

Control factor	Level-1	Level-2	Level-3
A: Pad material	Type-1[a]	Type-2	–
B: Pad shape e/w	4	5[a]	6
C: Pad curve profile	Type-1	Type-2[a]	Type-3
D: Pad additive	Low	Med[a]	High
E: Rotor material	Gray	Cast[a]	Steel
F: Pad taper	Low	Med[a]	High
G: Tapering thickness	Low	Med[a]	High
H: Rotor structure	Type-1	Type-2[a]	Type-3

[a]Indicates the initial design level.

Table 8.5 Experimental layout for robust optimization

Outer Array

Inner Array

	A	B	C	D	E	F	G	H	M_1: 0.008				M_2: 0.016				M_3: 0.032				M_4: 0.064			
									N_1		N_2		N_1		N_2		N_1		N_2		N_1		N_2	
	1	2	3	4	5	6	7	8	Q_1	Q_2	Q_1	Q_2	Q_1	Q_2	Q_1	Q_2	Q_1	Q_2	Q_1	Q_2	Q_1	Q_2	Q_1	Q_2
1	1	1	1	1	1	1	1	1																
2	1	1	2	2	2	2	2	2																
3	1	1	3	3	3	3	3	3																
4	1	2	1	1	2	2	3	3																
5	1	2	2	2	3	3	1	1																
6	1	2	3	3	1	1	2	2																
7	1	3	1	2	1	3	2	3																
8	1	3	2	3	2	1	3	1																
9	1	3	3	1	3	2	1	2																
10	2	1	1	3	3	2	2	1																
11	2	1	2	1	1	3	3	2																
12	2	1	3	2	2	1	1	3																
13	2	2	1	2	3	1	3	2																
14	2	2	2	3	1	2	1	3																
15	2	2	3	1	2	3	2	1																
16	2	3	1	3	2	3	1	2																
17	2	3	2	1	3	1	2	3																
18	2	3	3	2	1	2	3	1																

Response: Torque (Y)

The results of the experiments are given Table 7

For all the experimental runs, S/N ratios and sensitivities (βs) are calculated and they are as provided in Table 8.7.

For all the factors, average S/N ratio and sensitivity analysis are calculated and these average effects are as shown in Fig. 8.6.

8.3.2.5 Two-Step Optimization

Based on these averages and sensitivities, the two-step optimization is performed as follows:

Step 1 In this step the combination that maximizes SN ratio is identified. From the S/N ratio graph, it is clear that except for factors B and F, other factors exert significant effects on the output. Therefore, A1 C3 D3 E1 G1 H3 are considered as best levels for S/N ratio analysis.

Step 2 The factors B and F have no impact on S/N ratios and therefore using them to adjust sensitivities will not affect overall S/n ratio. Interestingly these factors have impact on the sensitivities. For example B2 and F3 have

Table 8.6 Experimental results

									M_1: 0.008				M_2: 0.016				M_3: 0.032				M_4: 0.064			
	A	B	C	D	E	F	G	H	II_1		II_2		II_1		II_2		II_1		II_2		II_1		II_2	
	1	2	3	4	5	6	7	8	Q_1	Q_2	Q_1	Q_2	Q_1	Q_2	Q_1	Q_2	Q_1	Q_2	Q_1	Q_2	Q_1	Q_2	Q_1	Q_2
1	1	1	1	1	1	1	1	1	4.8	0.9	5.8	0.8	8.5	6.5	11.5	6.8	20.4	13.2	25.0	16.2	36.9	32.7	43.5	34.5
2	1	1	2	2	2	2	2	2	4.5	2.5	5.7	3.2	12.5	9.6	13.0	10.0	23.5	20.3	25.1	21.4	42.0	36.0	43.2	36.1
3	1	1	3	3	3	3	3	3	5.9	5.2	6.8	5.9	10.6	9.3	11.4	10.2	23.5	22.0	24.3	22.5	42.9	40.3	43.8	40.6
4	1	2	1	1	2	2	3	3	4.5	2.1	5.7	3.0	12.1	8.9	14.3	10.5	22.1	16.9	24.2	20.0	41.0	34.0	42.4	37.6
5	1	2	2	2	3	3	1	1	6.5	2.1	7.8	3.2	12.3	6.9	13.2	8.6	23.3	17.2	24.3	18.3	44.3	36.9	48.9	37.2
6	1	2	3	3	1	1	2	2	5.0	4.2	5.8	4.3	11.5	9.4	12.3	9.9	20.8	16.8	21.0	18.5	43.0	40.2	43.1	41.0
7	1	3	1	2	1	3	2	3	5.2	4.0	5.6	4.5	11.8	9.1	12.3	10.1	21.2	17.5	20.0	18.3	40.3	36.2	42.2	38.2
8	1	3	2	3	2	1	3	1	2.4	0.0	4.3	2.8	6.7	4.0	7.2	3.6	16.3	11.1	18.3	12.3	30.1	27.8	34.3	30.6
9	1	3	3	1	3	2	1	2	6.3	4.8	7.8	6.1	12.1	9.3	13.5	11.9	24.4	19.6	26.3	22.3	48.5	40.3	50.2	44.0
10	2	1	1	3	3	2	2	1	2.1	0.0	2.9	0.0	4.9	0.0	7.4	4.2	18.3	9.5	17.7	10.8	32.0	26.3	35.3	28.1
11	2	1	2	1	1	3	3	2	4.9	1.2	7.6	1.8	11.3	6.5	15.3	6.8	23.4	15.0	25.1	17.2	40.1	33.2	50.5	35.5
12	2	1	3	2	2	1	1	3	5.1	4.4	6.4	4.4	10.1	7.8	11.2	8.5	21.7	18.7	22.1	20.1	43.1	41.2	44.4	41.5
13	2	2	1	2	3	1	3	2	2.1	0.0	5.4	0.6	6.7	1.2	7.3	2.3	13.4	9.4	16.4	11.1	38.9	27.9	43.3	31.1
14	2	2	2	3	1	2	1	3	5.9	5.0	6.8	5.2	13.3	12.0	14.2	13.3	24.9	23.1	26.3	25.4	47.9	46.3	49.7	47.2
15	2	2	3	1	2	3	2	1	3.2	0.0	3.9	1.8	8.7	3.2	9.6	5.1	13.2	7.9	19.5	11.1	38.2	32.1	42.5	33.0
16	2	3	1	3	2	3	1	2	4.1	2.7	5.9	4.4	12.3	8.7	13.7	9.2	24.3	18.9	25.5	20.2	44.3	39.0	47.7	42.4
17	2	3	2	1	3	1	2	3	2.3	0.8	3.2	2.1	10.2	8.0	12.5	8.8	21.6	16.5	23.6	20.4	38.8	32.4	41.1	36.6
18	2	3	3	2	1	2	3	1	1.2	0.0	5.1	1.2	7.8	2.3	13.0	5.0	20.3	11.1	21.2	12.4	40.1	31.6	45.1	32.0

Table 8.7 S/N Ratios and sensitivities (βs)

	A	B	C	D	E	F	G	H		
	1	2	3	4	5	6	7	8	S/N	β
1	1	1	1	1	1	1	1	1	44.0	573.0
2	1	1	2	2	2	2	2	2	47.4	634.3
3	1	1	3	3	3	3	3	3	53.2	667.9
4	1	2	1	1	2	2	3	3	46.9	617.4
5	1	2	2	2	3	3	1	1	45.3	651.6
6	1	2	3	3	1	1	2	2	52.5	644.2
7	1	3	1	2	1	3	2	3	51.4	613.6
8	1	3	2	3	2	1	3	1	45.0	465.6
9	1	3	3	1	3	2	1	2	48.9	718.0
10	2	1	1	3	3	2	2	1	41.6	454.3
11	2	1	2	1	1	3	3	2	42.4	622.1
12	2	1	3	2	2	1	1	3	53.3	657.1
13	2	2	1	2	3	1	3	2	40.0	504.5
14	2	2	2	3	1	2	1	3	55.3	755.9
15	2	2	3	1	2	3	2	1	41.5	527.1
16	2	3	1	3	2	3	1	2	48.5	679.2
17	2	3	2	1	3	1	2	3	46.3	590.7
18	2	3	3	2	1	2	3	1	41.2	556.3

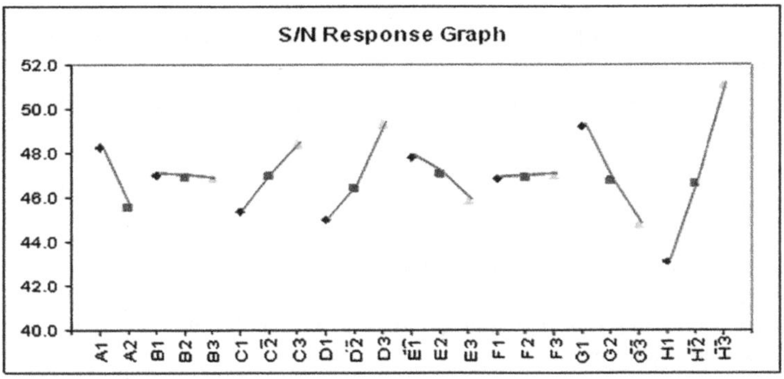

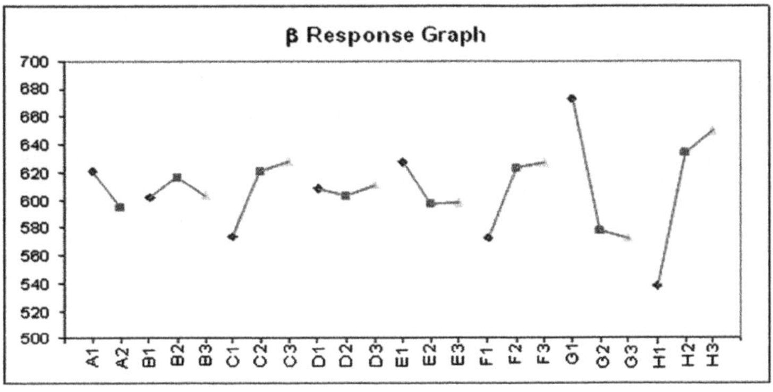

Fig. 8.6 Average response analysis for S/N Ratios and Sensitivities (βs)

Table 8.8 Summary of predictions

Predictions	S/N	b
Initial design	47.51	634.2
Optimum design	59.52	789.3
Gain	12.01 dB	24.9%

Table 8.9 Results of confirmation run for initial design and the optimal design

		M_1 0.008	M_2 0.016	M_3 0.032	M_4 0.064	S/N	β
Initial design	N_1Q_1	4.8	11.1	23.1	42.0	47.56	63.47
	N_1Q_2	1.2	8.6	18.1	36.0		
	N_2Q_1	5.7	13.0	25.1	43.2		
	N_2Q_2	4.4	11.8	21.4	37.6		
Optimum design	N_1Q_1	5.3	12.2	24.6	49.3	57.37	757.9
	N_1Q_2	4.6	10.1	23.1	47.1		
	N_2Q_1	5.8	13.2	25.0	50.1		
	N_2Q_2	5.4	11.9	24.3	48.2		
						9.81	19.0%

higher sensitivities. Therefore B2 and F3 are considered to important levels for the factors B and F.

Based on two-step optimization, the optimal combination is: A1-B2-C3-D3-E1-F3-G1-H3.

The summary of predictions for this combination is given in Table 8.8.

A confirmation run is performed for this combination and they closely match with predictions. Table 8.9 gives the results of the confirmation run for initial design and the optimal design.

Acknowledgements The author is grateful to Dr. Genichi Taguchi for his tireless efforts in promoting the robust engineering concepts and for allowing to use circuit stability design case example in this paper. The author is also grateful to Mr. Shin Taguchi for allowing to use design of brake system case study in this paper.

References

Jugulum, R. and D. D. Frey (2007). Toward a taxonomy of concept designs for improved robustness. Journal of Engineering Design 2:139–156.

Phadke, M. S. (1989). Quality Engineering Using Robust Design, Prentice Hall, Englewood Cliffs, NJ, USA.

Taguchi, G. (1987). System of Experimental Design, Vol. 1 and 2. ASI press, Dearborn, MI, USA.

Taguchi, G., Jugulum, R., and Taguchi, S. (2004). Computer Based Robust Engineering: Essentials for DFSS. ASQ Quality Press, Milwaukee, WI, USA.

Chapter 9
Distributed Collaborative Designs: Challenges and Opportunities

Abhijit Deshmukh, Timothy Middelkoop and Chandrasekar Sundaram

Abstract This chapter focuses on distributed product development, and highlights issues that can arise when distributed designer's attempt to reach a consensus on a good design alternative. Specifically, we address the issue of quality of the overall design as a function of negotiations between different designers, as well as the correctness of the solution. First, we present an iterative agreement protocol for resolving conflict, and examine the solution characteristics and conditions for convergence. Second, we present a diagnostic tool to determine if the design team's ordering of design choices, after aggregating individual designer's preferences, is rational. We identify the characteristics of individual designer's utility functions that lead to irrational outcomes, and present parametric approaches for manipulating utility curves to avoid irrational outcomes.

9.1 Collaborative Product Development

Product development is typically a collaborative activity involving multiple individuals. Different disciplinary experts and domain specific tools are needed to bring specialized knowledge and analysis to bear on different aspects of new product design. Moreover, for large scale design problems, decomposing the overall design problem and assigning different parts of it to separate individuals or design teams is the only way to manage the problem's complexity. Since design decisions made by the different individuals are not independent, the team must coordinate and integrate their decisions in order to reach a final design agreeable to all participants. Even when all the participants can be brought together in one place,

---

A. Deshmukh (✉) and C. Sundaram
235G Zachry Engineering Research Center, Industrial and Systems Engineering,
Texas A&M University, 3131 TAMU, College Station, TX 77843-3131
e-mail: deshmukh@tamu.edu

T. Middelkoop
University of Florida, Department of Industrial and Systems Engineering,
303 Weil Hall, PO Box 116595, Gainesville, FL 32611-6595
e-mail: middelk@ise.ufl.edu

N.R.S. Raghavan and J.A. Cafeo (eds.), *Product Research: The Art and Science Behind Successful Product Launches*, DOI 10.1007/978-90-481-2860-0_9,

reaching the best compromise to a constrained multi-criteria decision problem can be challenging. When the participants work for different organizations located at different geographic locations, the problem takes on a new dimension of difficulty.

One way to view a product development task is as a distributed optimization problem involving a search over a high-dimensional *design space* for the product configuration that gives performance that is *best* in some sense. What the different dimensions of the design space represent depends on the specific design problem. For the design of an airplane they might, for example, represent factors such as the number and size of the engines, the passenger capacity, and physical dimensions such as length, width and weight. The choices made along each of these dimensions contributes in different ways to the resulting performance of the final design, which for a commercial aircraft would probably center around the expected income of the resulting design. Since performance is generally not a separable function of the different design parameters, the product configuration needs to consider all the parameters in a coordinated manner. For example, the number of passengers impacts physical dimensions, which impacts weight, which impacts wingspan and engine type, which limits which airports the plane can fly to and determines fuel efficiency. All of this impacts manufacturing and other life-cycle costs, which impacts selling price, which impacts customer demand, which impacts expected income of the final design. It is impossible to look at decision on any one factor in isolation.

Being a bit loose mathematically, let us define a design process as an optimization problem of the form,

$$\min_{x \in X}(y = f(x)) \tag{9.1}$$

subject to,

$$g(x) \leq 0 \tag{9.2}$$

where x is a vector of numerical design parameters, X is a space of feasible values for the design parameters, y is a scalar performance measure. The challenge of the design problem is that while the space of possible designs might be huge, the space of designs meeting all of our performance requirements might be quite small, particularly when all of the constraints are taken into account. Moreover, it is often the case in complex design problems that the function f is not known explicitly. In this case, designers must build prototypes or use numerical tools such as finite element analysis or simulation in order to evaluate different design alternatives. Similarly, the constraint function g may also not be known in explicit form, requiring complicated computations to check design feasibility. These considerations generally preclude a centralized solution to a design problem, and we must resort to *distributed decision making* methods.

9.1.1 Issues in Distributed Collaborative Design

All applications of distributed collaborative designs require rational solutions, of a desired quality, in a specified amount of time. However, decisions made by a collective of designers suffer from the same perils as group decision making, namely the difficulty of satisfying the rationality, convergence and optimality conditions. Arrow's theorem asserts the non-existence of a *democratic* preference aggregation method which guarantees rational outcomes for a group composed of rational individual preferences (Arrow 1951). Moreover, distributed design systems, especially the ones that use iterative procedures, do not always guarantee convergence to a solution in a specific time period. Furthermore, the optimality or robustness of the solutions are not assured. It is important to understand these limitations and plan appropriate diagnostic tests to guarantee quality of the results generated by distributed design systems.

Moreover, dealing with errors, misinformation and incorrect information is an important issue in distributed design system since there may not be a single entity with a centralized world view to verify or authenticate the information. Product designers need to pay close attention to the authentication and data consistency issues, and incorporate evaluation mechanisms in the design architecture. In a resource constrained environment, it may not always possible nor desirable to validate all information accurately. Even though design team members may not maliciously present false data, errors can occur in data transmission or translation. In such cases, apportioning the blame becomes an interesting legal issue.

As distribute design systems move out of the controlled laboratory and simulated environments into real world product development applications, serious thought needs to be given to answering fundamental questions about the validity and quality of the decisions made by these distributed, autonomous design team members. This paper focuses on two issues related to distributed design teams: (1) quality of the overall product as a result of negotiations between different designers, and (2) correctness of the solution.

9.2 Negotiation Among Designers

When using a collaborative approach for design, designers representing different subproblems have to negotiate in order to reach agreement on design parameters of common interest. Negotiation provides a conceptually appealing way to coordinate distributed decision making in large-scale design problems. In this section we develop a modeling framework for a study of several important issues related to distributed negotiations, including: What collective agreement will the different decision makers converge to? Will they converge at all? How good are the resulting decisions and consequently the resulting design?

Negotiation is a rich and diverse field of study which draws from may areas. It is important to be able to distinguish *formal* negotiation from ad-hoc methods. Ad-hoc methods provide important context and understanding of negotiation in a human context but often do not provide the rigorously required for formal analysis. The work by Raiffa (1985) contains a good introduction and background of the issues and complexities of negotiation.

Table 9.1 Weights assigned to different issues by individual designers

	Designer A	Designer B	Designer C
1	0.8/0.0/0.2	0.0/0.4/0.6	0.4/0.1/0.5
2	0.2/0.0/0.8	0.0/0.6/0.4	0.3/0.4/0.3
3	0.0/1.0/0.0	1.0/0.0/0.0	0.3/0.5/0.2

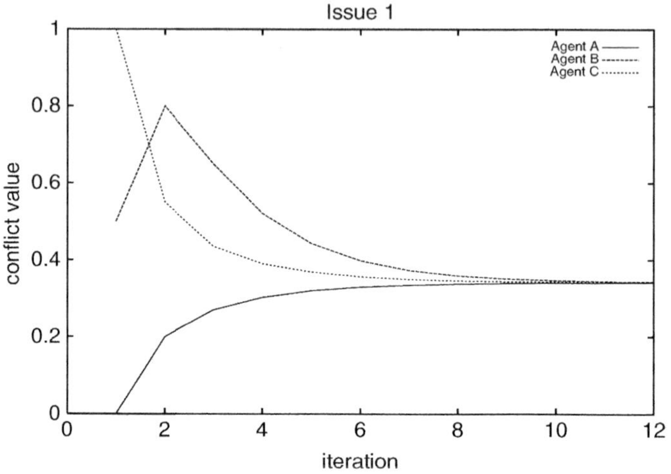

Fig. 9.1 Agreement dynamics for Issue 1

We present a motivating example to illustrate the problems that could arise when designers negotiate on different issues. Table 9.1 illustrates the use of the weights used in a three party and three issue agreement negotiation. The rows indicate issues and the columns indicate individual designers. The cells are divided into weights assigned to other participants (A/B/C respectively) by the individual in the column. The designers iterate over the values of the weights until an equilibrium solution is reached.

For instance on issue 1 party A wishes 0.8 of his *agreement allocation* to be used in setting his own value on issue 1 with the remainder being distributed to party C. From the constraints on the decision makers allocation the cells must sum to one. In order to prevent the decision makers from only choosing their own interests the allocation must also sum to one over the allocation of issues (columns).

The agreement allocation determines how much of a conflict variable is used in the determination of the next value (a measure how much the individual agrees on with another on a specific issue). Again looking at the example we see in Fig. 9.1 decision maker A starts with an initial value of zero and having the dominate allocation (**0.8**/0.4/0.5) therefore has the most influence on the outcome. This domination on issue 1 necessarily causes a weak standing in issue two as seen in Fig. 9.2 which has a large amount of concession (movement away from the initial value). We also

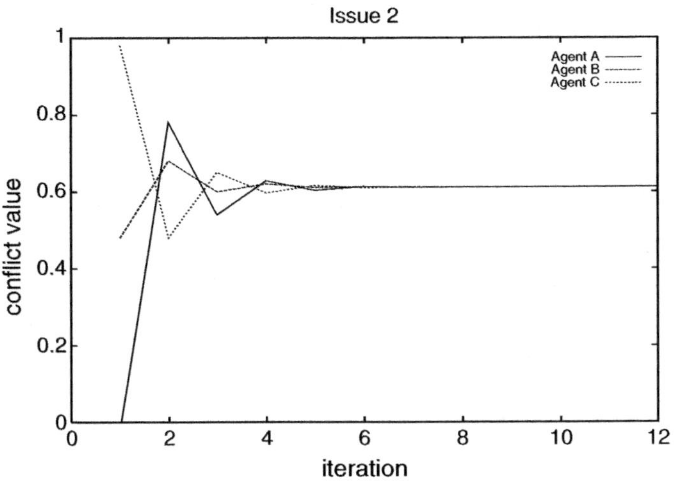

Fig. 9.2 Agreement dynamics for Issue 2

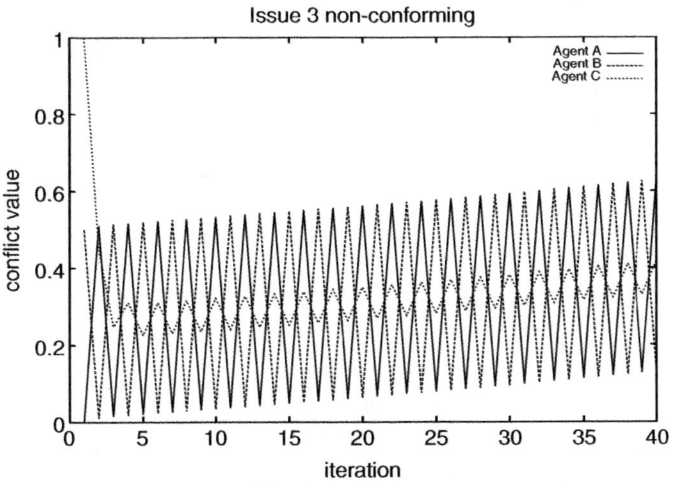

Fig. 9.3 Cyclic dynamics for Issue 2

note the importance of the constraints, even a small deviation can violate the convergence properties as seen in Fig. 9.3.

Now that we have a basic understanding of the protocol we look in Fig. 9.4 and see that the negotiation has entered a cycle. The weights used in this example were carefully chosen to produce this effect, however, this shows that the analytical analysis of the protocol is important otherwise these cases might be missed. Fortunately this problem is easily corrected by introducing a relaxation and the results can be seen in Fig. 9.5.

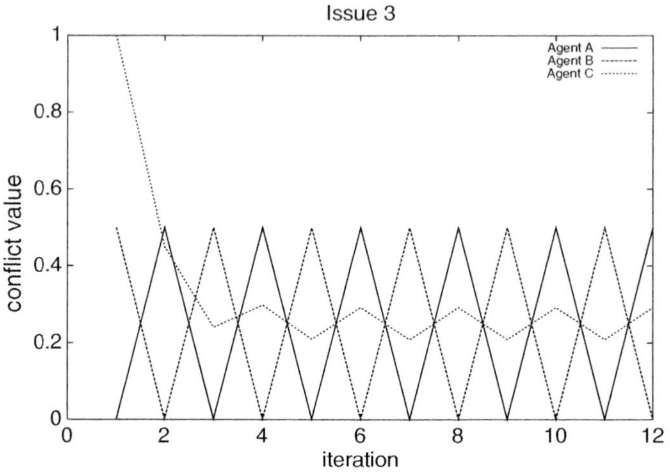

Fig. 9.4 Cyclic dynamics for Issue 3

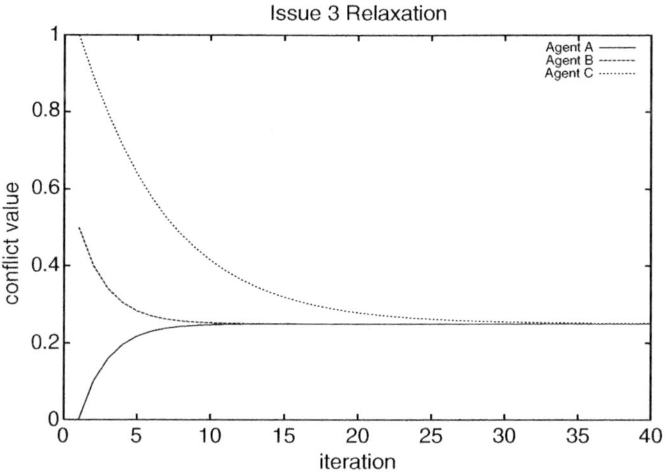

Fig. 9.5 Agreement dynamics for Issue 3 after relaxation

This discussion demonstrates the importance for individuals to have an expectation of the outcome, or performance, of the negotiation. The term *performance* is used instead of *optimality* because it is difficult to answer the question of *what is a good solution*. This is where system designers and participants goals can differ significantly.

9.2.1 Negotiation Framework

In the following section we develop the analysis framework by first defining the negotiation problem in terms of issues and conflict. We then break down the negotiation procedure into various parts that separate the designer (decision maker) from the negotiation and the negotiation from the underlying implementation. This framework will give us a basis for the analysis of the properties.

Negotiation Variables In order for a negotiation to take place, there must be issues to negotiate over. These issues are represented by *negotiation variables* and represent the outcome of the negotiation. We define a negotiation variable as a variable ($x \in \Re$) in which two or more individuals are interested in setting to a single value.

We will use the notation x_i where i is the *issue* being negotiated and x_i^j which is the portion appropriated to individual j over issue i. The term *issue* is used to denote a single individual item of contention between parties. In negotiation over an allocation the resource (i) would be the issue and the portions that each individual (j) receives is given by the value of the negotiation variable x_i^j. This leads to the notion of *divisible* vs. *indivisible* issues. An issue is indivisible when the number of negotiation variables needed to resolve it is exactly one.

Conflict and Agreement A conflict variable represents the desired value an individual wishes the corresponding negotiation variable to be set to. We will use the notation c_{ij}^k where i is the *issue* being negotiated over and the optional j which is the individual that the value is associated to with as k indicating that the conflict variable is controlled by individual k. For example individual k wishes that individual j portion (negotiation variable) to be η, then they wish that $x_i^j = \eta$ and is indicated by their conflict value $c_{ij}^k = \eta$. For the purpose of this paper we will consider the conflict variables as the sole mechanism used to communicate during the negotiation process.

The previous discussion allows us to represent the issues involved in the negotiation. We now characterize the exact nature of the conflict between the participants and how it relates to the set of conflict variables in the negotiation. We define an *agreement conflict* as a negotiation variable that must be set to a single value by two or more individuals. In this case the negotiation variable is again x_i with the participants, indicated by k, conflict variable being c_i^k.

Negotiation terminates when an agreement is reached. This occurs when all the conflict values are the same value for a negotiation variable as in Eq. 9.3. Additionally ϵ agreement is defined as when the values are with in some ϵ region.

$$c_i^k = x_i \ \forall k \tag{9.3}$$

$$\max_k |c_i^k - x_i| < \epsilon \quad \forall \, i \tag{9.4}$$

Similarly we can represent the amount of disagreement on an issue by δ-disagreement which we define as follows.

$$\delta_i = \sum_i |c_i^k - c_i^k| \tag{9.5}$$

Iteration Update Function A negotiation session is defined as a series of mappings of the conflict variables onto it self. Given the set of all possible conflict values C we have the following mapping which defines the iteration update function or negotiation iteration.

$$T : C \mapsto C, \quad C \in \Re^m$$

We can also represent it in the form $c(t+1) = T(c(t))$ where $c \in C$ and $c(t)$ is the conflict vector at time t. This representation clearly shows that the updates form a series $\{c(t)\}$ in which these updates happens is determined by the interaction protocol.

In order to prescribe convergence properties to the negotiation implementation we must ensure that the iteration update function is either contraction mapping or a non expansive mapping. These functions use a norm to ensure that all valid update mappings reduces the distance between two points when the mapping is applied in all cases. A contraction mapping is defined as follows:

$$\|T(x) - T(y)\| \leq \alpha \|x - y\|, \quad \forall x, y \in C \tag{9.6}$$

where $\|\cdot\|$ is some norm (a metric space) and $\alpha \in [0, 1)$ is called the contraction modulus. When the *contraction* property holds for an update function ($T : C \mapsto C$) the sequence of points will converge to a fixed point c^* where $c^* \in C$.

Negotiation Protocol The iteration update function is the manifestation of both the negotiation protocol and how the participating decision makers interact. In a purely closed cooperative design scenario the negotiation protocol solely determines the actions of the decision makers. However, this does not mean that the negotiation is centralized, it means that iteration update function extends into the decision making process of the participants. The closed cooperative nature means that this process can be designed optimally, and hence is deterministic from the protocol perspective. In this case it is not useful to decompose the update function into protocol and policy. In this case policy is protocol and protocol is policy.

Negotiation is not very interesting from the pure closed cooperative approach as more novel approaches could be applied from this centralized perspective. As we move away from this extreme and the designers gain more autonomy the need to separate the individual from the protocol becomes apparent. This separation, which we call *policy*, allows the analysis models to express the ability of the designers to make decisions by restricting the *policy*, not the decision maker. This is the public portion of the negotiation protocol and is what binds the actions of the participants. The degree of freedom that the individuals can exhibit in this framework is discussed in the next section.

Negotiation Policy The negotiation policy is the externally observable implementation of the decision makers in the decision making process. The negotiation policy is linked to the negotiation protocol through a vector which optionally has public constraints. This approach eliminates the need to fully characterize the decision makers' utility function because their exact nature may not be known at the time of analysis or even change during the course of the negotiation.

Fogerman-Soulie et al. (1988) use a similar methodology of removing the requirement that the decision makers strategies are known *a priori*. This is accomplished by allowing the decision makers to adjust their utility formulation in the negotiation progresses. The result of this adjustment is that the reasoning process is abstracted to a policy that changes through out the negotiation process.

Given an update interaction function $T : C^m \mapsto C^m$ we separate it into the policy and protocol portions. Formally we get the following:

$$T(c(t), \phi(u, c(t))) = c(t+1) \tag{9.7}$$

where we now have the iteration update function with the policy function ϕ and the designer's utility function u explicitly represented.

Interaction Protocol The interaction protocol determines the manner in which the updates on the iteration function interact. The most loosely coupled interaction protocols are fully asynchronous. In these protocols the decision makers recalculate their conflict values as new information is received, transmitting them without restriction. The only assumption made in the analysis is that the time between conflict value updates is finite, and thus the information will eventually propagate through the system.

Often in systems where arbitrary long delays destroy the convergence of the system we use a partially asynchronous interaction protocol. This is similar to the fully asynchronous protocol with the added restriction that the decision makers will update their conflict variables within a fixed number of iterations. By adding this assumption it is often possible to considerably reduce the maximum number of negotiation iterations needed to find a solution. It also restores convergence in cases where long delays in the propagation of information can cause oscillations in the negotiation by producing a dampening effect.

Additionally more sophisticated interaction protocols can be used to increase the speed of convergence and better prioritize the communication traffic. In general, these interaction protocols are independent from the convergence and performance parts of the analysis framework and can freely implement domain and communication optimizations with out negatively impacting performance.

9.2.2 Analyzing Negotiation-Based Product Development

We now introduce a new negotiation protocol called the multi-agreement protocol which can resolve conflict between multiple designers over multiple issues. The protocol is based on the single agreement protocol by Bertsekas (1989) which is extended to handle multiple decision makers in negotiation setting. The analysis of the distributed multi-agreement protocol is meant to be used as an example of the more general approach.

The basic principle of the distributed multi-agreement protocol is to move conflict variables in the best concession or compromise direction. Istratescu (1981) uses a similar method of searching the solution space in the compromise direction and uses a Gauss–Newton linearization to find the solution. The distributed agreement protocol uses a similar method but one that can be applied in a distributed manner.

We now define the distributed multi-agreement protocol as a negotiation protocol that considers n individuals trying to agree on the value of m continuous variables. Without loss of generality we will consider for the duration of the discussion that negotiation variables are $x_i \in [0, 1]$. The formulation of conflict variables is as follows:

$$c_i^k = x_i \tag{9.8}$$

where i is the variable to be agreed upon (issue) and c_i^k is the individual k's current proposal of the agreement issue value which is set according to the protocol.

The iteration update function is defined as:

$$c_i^k := \sum_{j=1}^{n} a_{k,j}^i c_i^j \quad \forall i, k \tag{9.9}$$

where the equation comes from the perspective of participant k on issue i. The matrix A^i is a table of weights for issue i relating how much of participant k agrees with participant j in issue i. In order to ensure convergence and other desirable properties we must constrain A^i.

$$\sum_{j=1}^{n} a_{k,j}^i = 1 \quad \forall i, k \tag{9.10}$$

$$\sum_{j=1}^{m} a_{k,j}^i = 1 \quad \forall k, j \tag{9.11}$$

$$a_{j,k}^i \in (0, 1) \quad \forall i, j, k \tag{9.12}$$

Eq. 9.10 distributes the agreement among the participants over an issue and Eq. 9.11 distributes an agreement weight over issues.

In addition to these constraints we must also relax the iteration update function for two reasons: the first is to eliminate potential oscillations, the second is to allow the decision makers to react to the progress of the negotiation. The designers react by adjust their polices are changing their associated $a_{k,j}^i$ values. The following shows the iteration update function in matrix form with the relaxed version.

$$c_i := A^i c_i \quad \forall i \tag{9.13}$$

$$c_i := (1-\gamma)c_i + \gamma A^i c_i \quad \forall i \tag{9.14}$$

where γ is the relaxation parameter. Finally we split the update function apart for a single decision maker k which yields the update function for the conflict variable c_i^k.

$$c_i^k := (1-\gamma)c_i^k + \gamma \sum_{j=1}^{n} a_{k,j}^i c_i^k \tag{9.15}$$

Now that the negotiation protocol is defined we explore its properties in the following subsections.

9.2.2.1 Convergence

The convergence property of negotiation is important to ensure that the negotiation will terminate in a reasonable period of time. Additionally, the convergence analysis can also characterize the uniqueness of the solution. These two properties are fundamental for the success of any negotiation protocol.

The convergence analysis consists of showing that the iteration update function always produce points that are closer when the update is applied than the non transformed points. This is done by proving that the iteration update function in conjunction with the interaction protocol has the property of being a *non-expansive mapping* or a *contraction mapping*.

An important part of this process is the selection of an appropriate norm. Convergence under one norm may fail under another norm making it crucial for the success of the analysis.

The multi-agreement protocol at its core uses a convex combination of the other conflict variables which, by its nature, guarantees that the δ-disagreement will always decrease ensuring convergence. This property is strong enough to hold even when the assumption that A is static is removed.

Many convergence proofs use a mapping function to reduce a metric over the entire space. For the multi-agreement protocol we show that the update function is a non-expansive mapping which is defined as the following:

$$\|f(c) - c^*\|_\infty \leq \|c - c^*\|_\infty, \quad \forall c \in \Re^n, \forall c^* \in C^* \tag{9.16}$$

where $f(c)$ is the update function over all issues and participants and $c^* \in C^*$ is the final agreement point. The norm used is the $\|\cdot\|_\infty$ or matrix norm which yields the maximum row sum. Given a matrix of dimension $\Re^{n\times n}$ the norm is defined as:

$$\|A\|_\infty = \max_i \sum_{j=1}^{n} |a_{ij}| \tag{9.17}$$

The previous material gives a brief outline of how the convergence works. We now show that the multi-agreement protocol converges by showing that it is of the following form which has been proven to converge by Bertsekas (1989). Given the partially asynchronous update function in the form:

$$c := f(c) = c - \gamma(Ac - b) \quad A \in \Re^{n \times n} \; c, b \in \Re^n \tag{9.18}$$

with the following properties:

1. For every k, we have
$$|1 - a_{kk}| + \sum_{j \neq k} |a_{kj}| \leq 1 \tag{9.19}$$
2. The set of solutions C^* is non empty.
3. A is irreducible.

is proven to converge. Given these properties it is easy to show that Eq. 9.14 can be rewritten as $c - \gamma(I - A^i)x$ which is in the form of Eq. 9.18. Here it is also easy to show that $I - A^i$ satisfies the first two condition due to the constraints imposed by Eq. 9.10. Furthermore, strictly positive condition of Eq. 9.12 satisfy the irreducible condition. Therefore the multi-agreement protocol satisfies all conditions which are necessary for convergence.

9.2.2.2 Solution Quality

Quality of solution as indicated earlier is a difficult problem when trying to prescribe the global performance characteristics on a system that is comprised of local maximizers. There are two approaches to solve this problem: the first is to try to develop a metric from properties of the negotiation protocol and the decision makers that interact with them; the second approach that we use is to develop the metric first and show that outcome the negotiation solution satisfies the optimization the of metric.

The generating performance metrics used to determine the solution quality using the second approach is done by generating global maximization problems on a *social utility function*. Any point that maximizes this problem is deemed to belong to the socially optimal set X^*.

Given the set of *social optimal* solutions

$$X^* = \{x \mid x = \arg\max_x S(u, x)\} \tag{9.20}$$

and the set of negotiation solutions

$$N^* = \{g(c^*) \mid T(\phi(u, c^*), c^*) = c^*\} \tag{9.21}$$

where $T : C \mapsto C$ is the protocol mapping function, $g(c)$ is the function that generates a negotiation variable solution from the conflict variables, and ϕ is the policy generating function from the utility functions. Given the previous definitions we can now define the quality of the negotiation in terms of the social function S as $N^* \subseteq X^*$, then we can say that the quality is of S. This framework allows multiple performance characteristics to be applied to a negotiation protocol by showing that the negotiation solution solves the social maximization problems.

When discussing the quality of solution the issues surrounding the negotiation *policy* become apparent. The issue of the balance between modeling or controlling the behavior of the decision maker with the ability to say anything useful about the protocol. In the following material we define a relationship between the decision maker's policy and its expressed form of utility. What we do not do is tell how the decision maker should transform this utility into a policy which will obtain its goals.

In this case we must distinguish between utility and goal. We define utility as an internal valuation of an independent state. How these utilities are combined to make a decision is the basis of what the goal formation is. In the following material we will show that the solution to the negotiation is the same as the one to the centralized maximization of the social function.

In order to show the quality of the solution obtained by the negotiation we show that the solution to the negotiation is the same as the solution to the maximization of the social utility. We define the social utility in terms of Eq. 9.20 and the solution to the negotiation problem by Eq. 9.21.

Once the social function is defined, we apply it to the maximization problem. In the case of the multi-agreement protocol the social function is a cost function so the we can equivalently formulate it as a minimization problem. We define the designer's exogenous utility as a cost factor (v) per deviation from the idea point $\hat{c} = c(0)$. Combining this with the social function and simplifying we get the following:

$$S(u(v), x) = \sum_i \sum_k v_i^k (\hat{c}_i^k - x_i)^2 \tag{9.22}$$

Expanding we get:

$$S(v, x) = \sum_i x_i^2 m_{2i} - 2 \sum_i x_i m_{1i} + m_0 \tag{9.23}$$

with the following constraints:

$$m_0 = \sum_i \sum_k v_i^k [\hat{c}_i^k]^2 \tag{9.24}$$

$$m_{1i} = \sum_k v_i^k \hat{c}_i^k \tag{9.25}$$

$$m_{2i} = \sum_k v_i^k \tag{9.26}$$

Solving this by setting $\nabla S(u, x) = 0$ we get the following set of equations:

$$2x_i m_{2i} - 2m_{1i} = 0$$

without loss of generality, we set $\sum_k v_i^k = 1$ giving us a member of the social optimal set X^* by solving

$$x_i^* = \sum_k v_i^k \hat{c}_i^k \tag{9.27}$$

Given now that we have a way of generating a solutions to the social optimal set, we can see how the results of the negotiation are characterized in this manner.

Given a set of designers with their characteristic polices we can estimate the result of the negotiation. Given that the A^i matrix is stochastic the result of the negotiation is the same as the limiting probabilities. Taking the limit of Eq. 9.14 we can find the result of A^i by the following set of equations for each issue

$$\pi_i^k = \sum_j a^i_{jk} \pi_i^j \tag{9.28}$$

$$\sum_j \pi_i^j = 1 \tag{9.29}$$

where π_i^k is the stationary point for issue i for decision maker k. Using this we get the estimated negotiation solution c^* as

$$c_i^* \approx \pi_i' \cdot \hat{c}_i \tag{9.30}$$

If the assumption is made that the agents goal is the maximization of the social utility, then the agents through negotiation, wish to find the solution of the maximization problem. Using the results of the social utility maximization and the estimation of the negotiation solution we can find for what conditions that this can be achieved.

Given this assumption we find the conditions that are true by finding the points that are common between the social and negotiation problems, in this case when $x^* = c^*$.

$$x^* = \min_x S(v, x) \tag{9.31}$$

$$c_i^* = \hat{c}_i \cdot \pi_i \tag{9.32}$$

By combining and expanding we get

$$\sum_k \hat{c}_i^k \pi_i^k - \sum_k \hat{c}_i^k v_i^k = \sum_k (\pi_i^k - v_i^k)\hat{c}_i^k = 0 \quad \forall i \tag{9.33}$$

which indicates that for the designers to obtain their goal they should adjust their policy so that $\pi_i^k = v_i^k$ to have $c^* \in X^*$.

9.2.2.3 Communication

The communication requirements imposed by a negotiation can be measured by analyzing the interaction protocol. The protocol yields information about the destination, frequency, and volume of information transmitted during the negotiation. Performance measures such as the value of information can be calculated by combining the communication volume with convergence rate of the δ-disagreement value. This information is obtained by analyzing the message traffic required by the protocol. As the δ-disagreement reduces so does the value of the messages as their effect on the negotiation outcome decreases as time passes.

The properties of the communication traffic is important in restricted environments by providing bandwidth requirements. The value of information can be used to prioritize the communication traffic, or in high loss conditions determine what error correction codes should applied. Decision makers can also prioritize computation based the value of information by focusing attention on more important decisions. This demonstrates that information about communication is valuable in directing efforts in the environment both at the system and individual designer levels.

As the negotiation progresses the convergence property of the multi-agreement protocol gives us a system where the δ-disagreement (Eq. 9.5) converges at a similar rate as shown in Fig. 9.6. As the negotiation approaches the agreement point the individual decision makers have less ability to effect the outcome. This inability to change the relative result of the negotiation reduces the value of the information submitted as part of the iteration update function. Conversely we see that the early updates have the most effect on the outcome.

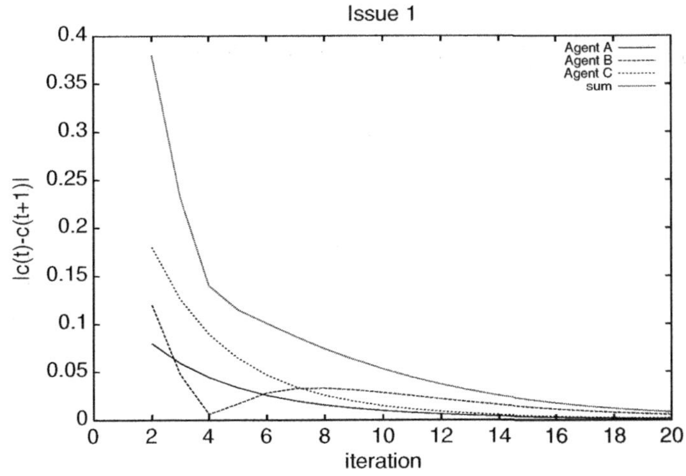

Fig. 9.6 δ-disagreement on Issue 1

The above discussion presents a framework for distributed decision makers involved in the design of complex systems to arrive at a *good* solution through an iterative negotiation process. Furthermore, we identify conditions under which this process converges to a fixed point, and characterize the quality of the consensus solution obtained. In order to ensure that distributed design teams generate good final products, we need to develop coordination protocols that have known convergence properties, under both perfect and imperfect information conditions.

9.3 Rationality of Collaborative Designs

Arrow's (1951) Impossibility Theorem raises serious questions about the *rationality* of the overall solution if a group decision process is used by design teams to select an alternative from a set of choices. Although each designer may assign utilities or preferences to different alternatives which are perfectly rational, the outcome of the selection process may not reflect majority opinions or transitive orderings. Several such paradoxes have been documented in literature (Arrow 1951; Fishburn 1974; Sen 1995). Collaborative product development teams need to be aware of the conditions under which their consensus results are valid and identify decision strategies which minimize the probability of irrational outcomes for a specific application scenario.

We motivate the issues related to rational outcomes using the following example. Consider that a group of designers use a collective ordering scheme to select a design alternative based on their individual preferences and the available choices. We assume that, designers form preference ordering over the choices based on their utility functions. Referring to Fig. 9.7, V_1, V_2 and V_3 are the utility functions of the three

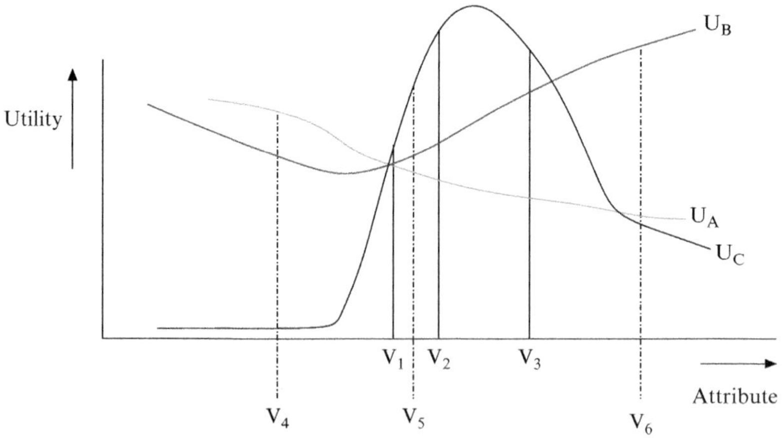

Fig. 9.7 Co-existence of rational and irrational outcomes for the same set of decision makers' utility functions

designers trying to arrive at a decision over three design choices. We assume that the decision makers use pairwise comparison of alternatives (Condorcet scheme) to rank order the different choices. For one set of choices, C_{11}, C_{12} and C_{13}, the design team arrive at a rational solution of $C_{12} > C_{11}, C_{12} > C_{13}, C_{13} > C_{11}$, where C_{12} is the outcome of the decision process. But for another set of choices, C_{21}, C_{22} and C_{23}, the same group of designers, with their utility functions unaltered, arrive at an irrational solution of $C_{21} > C_{22} > C_{23} > C_{21}$. This also called, a cyclic outcome, where there is no clear winner.

9.3.1 Rationality Tester

In this paper we describe a diagnostic tool that can be used by collaborative product development teams, the Rationality Tester (RaT) that allows users to ascertain if the system level ordering of choices, after aggregating preferences from all the agents involved, is *rational*. We emphasize that in the context of this discussion we refer to rational outcome as one which satisfies the transitive preference ordering relationship. The RaT tool has three major thrust areas. First, we focus on relating agent utilities to the irrational outcome space in a given preference ordering scheme. Second, we identify the characteristic structures of agent utility functions which lead to irrational outcomes. And, third, we develop parametric approaches for manipulating the utility curves in order to avoid irrational outcomes.

Referring to the example discussed earlier, we can observe that for arbitrary utility functions, same set of decision maker's can produce rational and irrational outcomes for different sets of choices. The primary motivation for RaT is to give the designers a tool which will guide them in selecting appropriate utility functions for agents involved in a group decision process, and identify regions of the parameter space where the group is likely to produce irrational outcomes. Although the discussion in this paper is in the context of the Condorcet preference aggregation scheme, we can address the issues of rank reversal and different outcomes under other preference aggregation schemes, such as plurality, Borda count or other aggregation schemes, using similar approaches. Readers are referred to Saari (1995) for details of the geometric representation of preference aggregation schemes, which forms the basis of the RaT diagnostic tools.

Consider a set of decision makers involved in selecting a product configuration from three choices, for the sake of simplicity. The number of choices can be generalized to any arbitrary number without significant difficulty. Each designer arrives at their choice ordering based on their utility function. We assume that decision makers taking part in the voting processes, make their decision on the choices based on a strict linear ordering of the choices themselves, that is, each voter compares each pair of choices in a transitive manner without registering indifference between any two choices. For a given set of three choices, we have six different strict linear ordering over the choices themselves. The decision maker's profile is a 5-simplex residing in 6-dimensions, with each dimension representing one persons profile type only. Extending this argument, for n contesting agents, we have n! agent profile types, and so the space of normalized agent profiles to a n! – 1 dimensional object in a host space of dimension n!

The first step in establishing a relationship between designer's utility functions and irrational outcomes is to find the pre-image of the irrational region of the outcome space in agent voter profile. We can find the pre-image by, transforming the irrational region in $\mathcal{R}^3$ to another region in $\mathcal{R}^3$, drawing the transformed object out along the three suppressed dimensions, and finally tweaking and rotating the object in $\mathcal{R}^6$. The first result to note is that the convexity of election outcomes is preserved. Moreover, the pre-image of the irrational region is given by: $x_1 - x_2 + x_3 - x_4 + x_5 - x_6 \leq 1$, $x_1 + x_2 + x_3 - x_4 - x_5 - x_6 \geq 1$, $x_1 - x_2 - x_3 - x_4 + x_5 + x_6 \leq 1$, $-x_1 - x_2 + x_3 + x_4 + x_5 - x_6 \leq 1$, $x_1 + x_2 + x_3 + x_4 + x_5 + x_6 = 1$ and $x_i \geq 0, \forall i$.

Applying $x_1 + x_2 + x_3 + x_4 + x_5 + x_6 = 1$ and $x_i \geq 0, \forall i$, we get the 5-simplex in $\mathcal{R}^6$ (x_i are the agent profile types in each of the six dimensions) (Sundaram 2001). Using this information we can say that, in a three person case, for irrational outcomes to occur, one of the agents' utility functions should be concave, one should be convex, and the other should be either monotonically increasing or decreasing function, as shown in Fig. 9.8 (where C_i are the choices and 1, 2 and 3 represent each agent's utility function). Note that this result is similar to the Black Single Peakedness result (Black 1958). However, we can extend the approach used to determine the pre-image to higher dimensional problems and also to other preference aggregation methods.

Now we find out the ranges of the choices that result in irrational outcomes. We find the ranges of choices C_1 and C_2 for a given value of C_3, and we move C_3 to its left and ascertain new ranges of C_1 and C_2 that lead to irrational outcomes, and so on. Hence, we arrive at a collection of sets of C_1, C_2 and C_3, that will always lead to irrational outcomes, as shown in Fig. 9.9. Note that there exists a unique transition point for C_3 beyond which irrationality ceases to exist. This unique point of C_3 results in a singleton set for C_2 for irrational outcomes to occur, and beyond

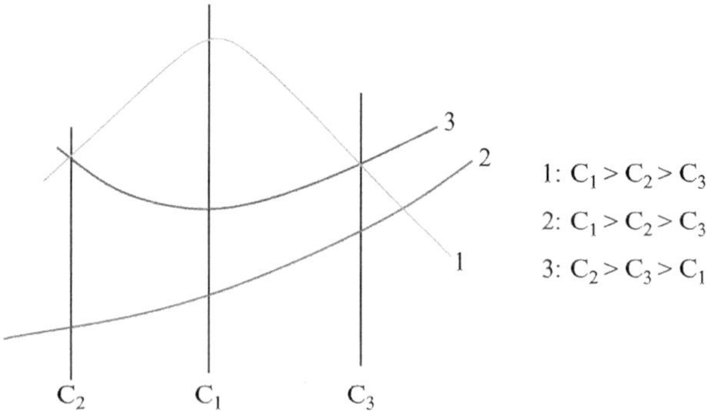

Fig. 9.8 The combination of agent's utility functions that lead to irrational outcomes

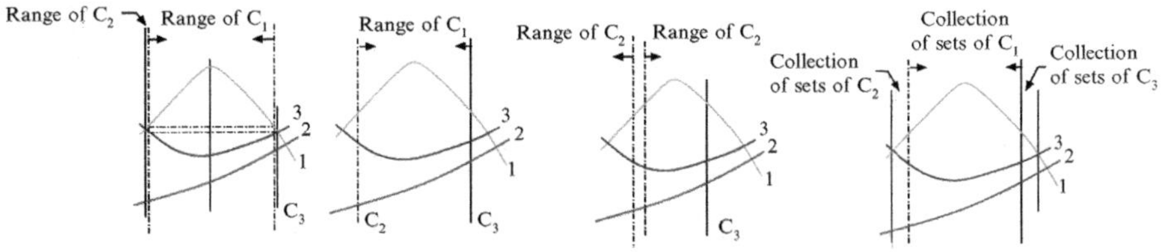

Fig. 9.9 Ranges of irrationality and collection of sets

this point of C_3 range of C_2 is a null set. In the first figure, we show the ranges of C_1 and C_2, that lead to irrational outcomes, for a given value of C_3. The second figure shown that there exists a unique point for C_3 beyond which irrationality ceases to exist. At this point, C_2 and C_3 have equal utilities with respect to two agents. The third figure illustrates this result, where one can notice that there are two ranges of C_2 with no intersection, for a certain C_3 beyond the unique transition point of irrationality. Finally, the fourth figure shows the collection of sets of C_1, C_2 and C_3 that lead to irrational outcomes.

We can use the above observations to manipulate the utility curves in order to avoid irrational or undesirable outcomes. There are various methods of changing utility functions, where an extreme case may be to completely replace the given utility function with a different one. However, in this paper we only focus on simple modification of utility functions. Note that simple translations will not have any effect on the overall ordering of the outcomes. We consider manipulating a curve by *shearing* a profile to get the desired shape. Let us say a positive shear is one in which a portion of the object is removed, which is, with respect to the curves equivalent to reduction in the height of the curve, and a negative shear is one in which a portion of the object is added to it, which is equivalent to increase in the height of the curve. So, one can see that curve 1 can only undergo positive shear and curve 2 only negative shear while curve 3 can undergo either of the two to manipulate the result. In our utility choice space, we assume that choices are lined along the x axis and their utility is represented along y axis. This manipulation of utility curves performed by simple shear operations can be represented mathematically as a transformation matrix $\begin{pmatrix} 1 & b \\ 0 & 1 \end{pmatrix}$, where b is the shear parameter. One can see that a complex problem of manipulation is now reduced to altering a single parameter to achieve the desired outcomes. In our problem, value of b either increases or decreases, and there exists a limiting value or critical effort to just manipulate the result. So, one needs to estimate this critical value of b that is need to manipulate the outcomes. For more than one issues in a given problem, the utility choice-space is more than two-dimensional, and so the size of the transformation matrix will be more than two, and so more than one parameter (similar to b) will determine the critical effort needed to manipulate the result. So it is necessary to identify the parameters and their limits to have an idea of what happens when utility of a choice changes. RaT allows design teams to investigate the process of manipulating utility functions of decision makers where the level of manipulation, b, is associated with a cost. Hence, in a multi-dimensional case, we can formulate the optimization problem of modifying utility functions at the lowest cost while guaranteeing rational outcomes for a desired range of choice parameters. RaT can also be extended to include order reversals in Borda count and other weighted preference aggregation methods.

The methods developed for RaT allow us to relate designer's utility functions to the group rationality outcomes in different preference aggregation schemes. Moreover, we can use this information to manipulate the utility functions in order to obtain rational outcomes.

9.4 Summary

It is important to note that the key intellectual questions in realizing truly distributed collaborative product development revolve around distributed decision making. As of today, the selection and operation of design teams that are ideally suited for complex systems design remains more of an art than science. This paper attempts to highlight some of the key issues involved in understanding distributed teams and developing a scientific basis for constructing collaborative product teams that are *functional*. The ability to guarantee rationality of the final outcome, solution quality and time it takes to reach a good solution is essential in transforming collaborative design into a predictable process.

Acknowledgments The research presented in this chapter was supported in part by NSF Grant # DMI-9978923, NASA Ames Research Center Grants # NAG 2-1114, NCC 2-1180, NCC 2-1265, NCC 2-1348, and DARPA/AFRL Contract #F30602-99-2-0525.

References

Arrow, K.J. (1951), Social choice and individual values. Wiley, New York

Bertsekas, D., Tsitsiklis, J. (1989), Parallel and distributed computation: Numerical methods. Prentice–Hall, NJ

Black, D. (1958), The Theory of Committees and Elections. Cambridge University Press, London

Fishburn, P.C. (1974), Paradoxes of voting. The American Political Science Review **68**, 537–546

Fogelman-Soulie, F., Munier, B.R., Shakun, M.F. (1988), Bivariate Negotiation as a Problem of Stochastic Terminal Control, chap. 7. Holden-Day

Istratescu, V.I. (1981), Fixed Point Theory: An Introduction. Kluwer

Raiffa, H. (1985), The Art and Science of Negotiation. Harvard University Press

Saari, D.G. (1995), Basic geometry of voting. Springer

Sen, A. (1995), Rationality and social choice. The American Economic Review **85**, 1–24

Sundaram, C. (2001) Multi–agent rationality. Master's thesis, University of Massachusetts, Amherst, MA

Zeleny, M. (ed.) (1984), Multiple Criteria Decision Making: Past Decade and Future Trends, Vol. 1. JAI Press

Part III
Customer Driven Product Definition

Chapter 10
Challenges in Integrating Voice of the Customer in Advanced Vehicle Development Process – A Practitioner's Perspective

Srinivasan Rajagopalan

Abstract Traditional automotive product development can be segmented into the Advanced and Execution phases. This paper focuses on three specific aspects of the Advanced Vehicle Development Process. Voice of the Customer is an important and integral part of the Advanced Vehicle Development Process. We highlight the different issues involved in understanding and incorporating the Voice of the Customer in the product development process. A catalog of questions is provided to help the decision maker understand better the information provided and make the right decisions.

Keywords New product development · Voice of the Customer · Market research · Survey methods · Conjoint analysis

10.1 Introduction

Traditional automotive product development process can be thought of in terms of a series of steps that start from the identification of the Voice of the Customer to the final design through a series of gates and checks (Cooper 1990). These gates and checks evaluate the status of the current design with respect to meeting the Voice of the Customer based requirements. The automotive product development process can be segmented in two phases – Advanced/Preliminary and Execution. The Advanced or Preliminary Vehicle Development Phase (A-VDP) is described by Donndelinger (2006) as a "structured phase-gate product development process in which a fully cross-functional product development team rigorously evaluates the viability of a new vehicle concept". This paper will not focus on the stage-gate process (Cooper 1990) itself, but on a specific aspect of the Advanced/Preliminary VDP and the challenges faced in this process.

S. Rajagopalan (✉)
General Motors Advanced Vehicle Development Center, Mail Code 480-111-P56, 30500 Mound Road, Warren, MI 48090, USA
e-mail: srinivasan.rajagopalan@gm.com

N.R.S. Raghavan and J.A. Cafeo (eds.), *Product Research: The Art and Science Behind Successful Product Launches*, DOI 10.1007/978-90-481-2860-0_10,

10.2 Voice of the Customer

The A-VDP consists of a variety of conflicting voices (Engineering, Best Practice, Manufacturing, Design, Finance, Brand, etc.). Exploring the design space in the A-VDP stage in order to identify feasible solutions involves balancing these various voices. It is critical for the decision maker to evaluate these conflicting voices in an objective manner and have the right reference while trading-off and balancing these voices. An effective way to balance these conflicting voices is to use the Voice of the Customer as a guide. In the absence of customer related information, the decision maker is likely to use internal metrics such as cost or complexity to evaluate trade-offs. A cost or complexity driven balance may help a company satisfy internal targets and position itself as a cost leader, but it does not guarantee that the product will be one that has an inherent demand in the marketplace. Knowing the customers' wants and needs is critical for creating new and innovative products that can provide a sustainable competitive advantage. Studies (Katz 2006) have shown that Product Development companies that practice Voice of the Customer are more successful than the ones that don't. The importance of understanding the Voice of the Customer and incorporating it is well documented in literature.

Co-creation can be considered as the ultimate example of listening to the customer needs and wants. Prahalad and Ramaswamy (2004) focus on the next stage of customer–corporate interaction – moving from products and services to experiences. In the case of co-creation, customer value is created at the points of interaction between the customer and the company. They highlight various examples of co-creation in their book, such as Sumerset Houseboats, Lego (Mindstorms), Starbucks and Walt Disney. Sumerset Houseboats go well beyond doing market research and understanding what the customers want. Their customers are part of their boat design process. Customers can interact with the company and co-operatively design their desired boat by making the right trade-offs and also work with the factory and be an active participant in the boat's construction. Sumerset also allows their customers to view the manufacturing process of the boat and monitor the progress of their boat's construction. This is a win-win situation because customers can be specific about the details of their boats and can provide their desires and wants (at the point of production) and the company engineers and decision makers do not have to make decisions in a vacuum. They have precise information about what attributes the customers are willing to pay for, which attribute they think as a must-have and which attributes they may be willing to sacrifice. Sumerset Houseboats has changed the whole boat purchasing experience and made it a focal point of interaction with their customers. Companies like Starbucks and Disney also focus actively on the customer experience and pay detailed attention to the desires, likes and behavior of their customers. These companies have recognized that customer driven innovation is important for their growth.

All the aspects under discussion in this paper focus on the customer. The phrase Voice of the Customer in this paper refers to the general process of understanding the customers' views, their needs, desires and wants. It does not specifically focus on any single approach of estimating and obtaining the customers' input.

We will consider three main aspects of the A-VDP in this paper and examine in detail the current approaches and the questions faced in a proper implementation of these approaches:

(a) Understanding the Voice of the Customer
(b) Incorporating the Voice of the Customer
(c) Global customer voices

Within each of these three aspects, the paper will attempt to provide a brief overview of the various techniques and tools that are currently available for use and then follow it up with a catalog of questions. The answers to these questions will help the decision maker understand the capability and limitations of the Voice of the Customer activity used within his/her company.

10.3 Understanding and Interpreting the Voice of the Customer

Automotive companies have been listening to the Voice of the Customer for a very long time. JD Power has been instituting surveys to measure customer satisfaction since 1968 (http://www.jdpower.com/about). Consumer Reports has been evaluating vehicles and acting as a consumer advocate (and representing the Voice of the Customer). They have been surveying people for opinions on products since 1940 and their first automobile evaluation began with the Toyota Corona in 1965 (http://www.consumerreports.org/cro/aboutus/history/printable/index.htm). Automotive companies have been conducting market research in various forms since the 40s and 50s – be it through surveys or customer reaction, to concepts and products at automotive shows and autoramas.

Market research surveys have traditionally asked customers for their reaction to the features and attributes of a product. These surveys provide the customer a 5, 7 or 10 point scale that rates their impression on the various attributes. The scale typically ranges from "Extremely satisfied" to "Extremely dissatisfied" with a few intermediate anchors. The scales typically tend to be symmetrical though it is not necessary. These market research surveys have helped understand customers' reaction to a product (post-mortem), thereby providing insight into what features/attributes may need to be tweaked for the next generation. They have also been a great feedback mechanism for things gone wrong (or problems) with the current design.

10.3.1 Conjoint Analysis

Conjoint analysis started in the mid-1960s and has been heavily used by automotive companies (Orme, 2006) to understand which combination of attributes/features makes a vehicle more attractive from a customer preference standpoint. Over the years, both the methods of conducting conjoint clinics (going from cards to computer graphic based explanation of attributes to web based clinics) and the techniques for analyzing the data (adaptive conjoint analysis, discrete choice based conjoint models) have evolved and progressed significantly over the past 30 years. The improvements in Information Technology have increased significantly, both the effectiveness and usage of conjoint analysis techniques.

10.3.2 S-Model

Donndelinger and Cook (1997) proposed an approach to understand the impact of an attribute change on the pricing and profitability of a product. As part of this approach, they used value curves to estimate the customer preference for a given attribute. They use a continuous function to represent the customer value for an attribute (as long as it is a continuous attribute). They use three specific instances of an attribute (ideal point or point of maximum customer value, baseline point or reference point, and critical point or point of no value) and estimate the value function based on the customers' reaction to the attribute value at these points. The value curve will vary by individual and will depend on both the demographic and psychographic aspect of the individual. Depending on the shape of the value curve, it can also be related to the Kano model.

10.3.3 Quantitative vs. Qualitative Market Research

Corporations have been increasingly emphasizing the quantitative aspects of market research in the past 20–30 years. The increased sophistication of various choice modeling techniques has translated into this being one of the primary focus areas in market research – especially those supporting new product development. While traditional quantitative market research techniques use ratings and rankings to understand customer behavior and extract product/attribute preference information, choice models use trade-offs to extricate the inherent value of a product/attribute from a customers' standpoint. The use of choice models helps the decision maker understand the value of a product/attribute in terms of price and thereby make the right choices while balancing various attributes. However, quantitative market research has a couple of drawbacks when it comes to innovative or new product development:

(a) Responses provided by the customers are dependent on their experience with the product or attribute. Customers' responses to attributes/products that they are unaware of may drive the decision maker towards the wrong choice. For example, would customer research have predicted the success of the ipod in the early stages of product development or would market research have indicated that the inability to replace the battery as a deal breaker for the customers and thereby have killed the product?

(b) Quantitative research also requires that the right question be asked. Customers will only respond to the question they are asked. The research literature and the experience of market research practitioners provide significant information on the proper way to frame a question and the appropriate verbiage for a survey. However, it is imperative for the company conducting the research to ensure that all the right questions (and attributes to be traded-off) are included in the survey. Missing out on a key attribute or question may result in the wrong conclusions being drawn from the survey/choice model results and thereby resulting in the wrong attribute balance for the product.

Qualitative research on the other hand is extremely useful when considering unmet needs of the customer. Lifestyle research or ethnography focuses on shadowing the customer and observing their day to day activities and extracting information about their needs and desires through their actions. The qualitative methods go beyond the traditional focus groups and interviews. The goal of the ethnography or lifestyle research is to get at how the product fits within the customers' life and try to obtain deep insights and understanding of the individual customers' decision making process. Qualitative research is very useful in getting this deep insight, however it is predicated on knowing the right customer to shadow. Eric Von Hippel advocated the use of lead users as the primary focus for this qualitative research (Von Hippel 1986). Lead users for a given product are defined as those that have a need which is not satisfied by the current technology but have identified some means of working around the limitations. Qualitative research, with the help of lead users will help a company understand the effectiveness of a unique and innovative product within the early adopter community. However, it may not provide answers required by the decision maker for balancing the product attributes that will help a quick spread from the lead users to the mass market. Extrapolating the results from qualitative research (which tends to focus on a few individuals) to the mainstream market can result in erroneous trade-offs being made.

10.3.4 Kano Model

The Kano model (developed by Professor Noriaki Kano in the 80s) is another approach of classifying attributes into categories. The categories specified by Kano can be paraphrased as Exciters/Delighters, Criticals (Must-have's) and Performance as shown in Fig. 10.1. This model focuses on the product attributes and not on customer needs (http://en.wikipedia.org/wiki/Kano_model). However, Kano also produced a methodology for mapping consumer responses to questionnaires onto this model.

The above paragraphs describe briefly the various techniques and methods of understanding the customers' needs and wants. These techniques when used properly can provide the decision maker with an incredible set of tools and will help him understand the impact of each trade-off/balance decision that is made. Listed below is a catalog of questions that the decision maker must be aware of before taking as gospel the results provided to him/her from any Voice of the Customer study.

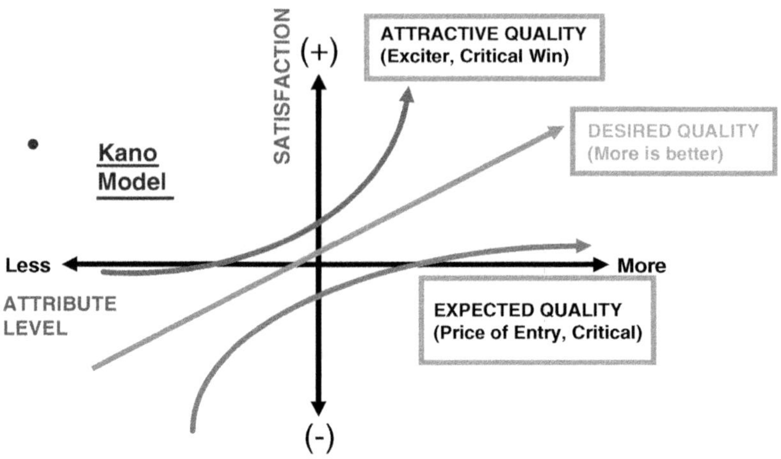

Fig. 10.1 A schematic of the Kano model

Understanding the answer to these questions will strengthen the decision maker's knowledge of the strength and limitations of various customer research techniques and thereby help him/her make more informed decisions.

10.3.5 Questions

- How much do we trust customers' responses to surveys? (Stated vs. Derived vs. Revealed.)
- How do we discriminate the "real" answers from the ones that customers give us because they perceive that it is what we want to hear? Also how do we account for customers who tend to answer based not on their circumstances, but what they perceive as others customers might think? (Interviewer bias/Social bias.)
- How can we be sure that their behavior on the surveys is indicative of their behavior when they shop for vehicles? (External validity.)
- How do we capture the intangibles? Conjoint based analysis is useful in capturing attributes/features that can be easily explained or quantified, but may not help us in understanding the softer aspects of a vehicle. (e.g., irrespective of the actual dimensions of the vehicle, the perception of being too big or too small can influence a customers' choice.)
- How do you capture preference for styling? (Styling can be highly polarizing – how do we know that the right customers are attracted to the styling?) (Heterogeneity.)
- How do we project for changing preferences over time? Even if we do accurately capture a customers' desired preferences, how can we estimate a customers' preferences for attributes over time? Customers' preferences for Fuel Economy can be highly dependent on the prevailing gas prices and hence vary over time.

- The attributes/features in an automobile can be classified as primary attributes (Fuel economy, Reliability, Price, Acceleration, etc.) and secondary attributes (sun roof, CD changer, etc.) Primary attributes are those that are fundamental to the vehicle from a product development standpoint. The secondary attributes are those that can be added or deleted from the automobile without a significant change to the product development process. Conjoint models traditionally limit the number of attributes that can be considered. The more the attributes, the larger the conjoint experiment. Which attributes should we consider in our analysis? Should only primary attributes be considered (Size, Price, Quality, FE, Acceleration, etc.)? What about the secondary attributes and features? Does an improvement in a secondary attribute have inherent value if the primary attributes are not set at the right value?
- New vehicles being developed have increased in complexity significantly. This is partly because of the increased functionality being provided to the end customer (NAV systems, real time traffic information, infotainment systems, etc.) Increased number of attributes may necessitate web based/handheld based conjoint tools. How do we ensure that the customers' perception of the attribute and understanding of the attribute's description is accurate when the survey is conducted over the web?
- How do we understand a customers' unmet needs (can we ask them about things they haven't seen yet or do not know about yet). Ethnographic research (garage visits/home visits) help us understand the customers' lifestyle and needs better but they require that we have a fair idea of who our customer is.

10.4 Incorporating the Voice of the Customer

The process of estimating the Voice of the Customer can be a significant challenge. However, incorporating the Voice of the Customer within the product development process is important for effectively delivering new and innovative products in the marketplace. This section will focus on some of the challenges involved in the A-VDP process, especially with the challenges of incorporating and realizing the Voice of the Customer. Once the voice of customer has been identified along with the associated priorities and importance, the AVD factory has the responsibility of figuring out the detailed physical solution set and its associated impact in terms of cost and manufacturing footprint. As described in Fig. 10.2, the decision maker needs to comprehend the various conflicting voices and be able to make the right decisions and balance. This work is traditionally done by a cross-functional workgroup that usually translate and interpret the Voice of the Customer. The translation from the customer voice into engineering typically occurs in companies within the QFD framework (www.wikipedia.com/QFD). The second and third steps in the QFD process involve identifying the key product attributes that affect the Voice of the Customer and set appropriate targets for these metrics. Design Integration tools such as iSight (Dassault systems), Modelcenter (Phoenix Integration)

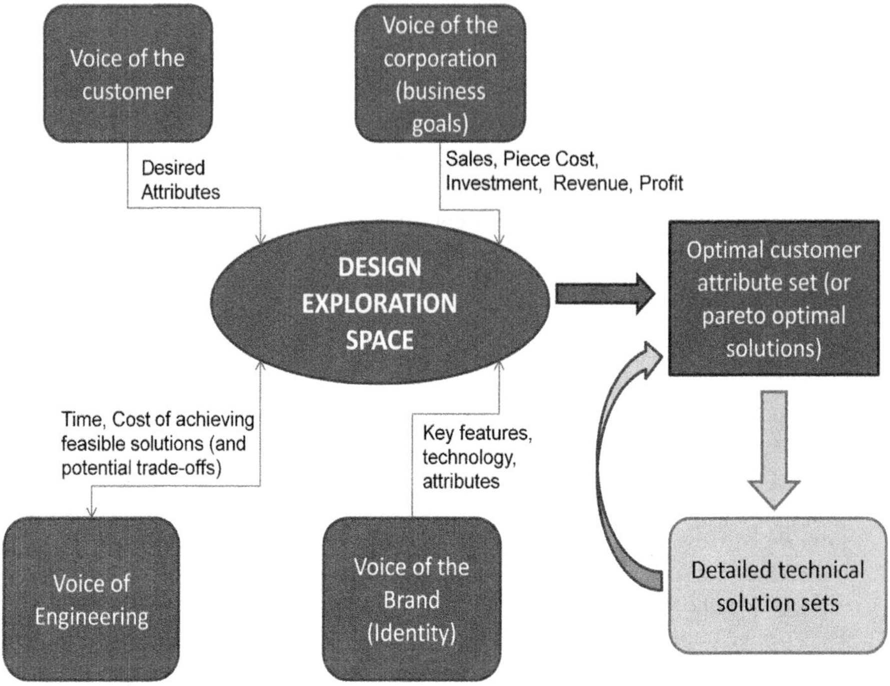

Fig. 10.2 Schematic showing conflicting voices that need to be balanced in the AVD process

are useful visual process integration tools. These tools provide a framework for the engineers within which they can integrate the different software and tools they use in their design exploration phase of A-VDP.

10.4.1 Questions

- How do we translate from customer speak to the associated engineering metrics that the AVD factory can work towards? Does the traditional QFD process also provide us with the right sensitivity for the engineering attributes, so we can balance between the engineering metrics and targets? Is this approach heavily dependent on existing in-house expertise and knowledge?
- How do we develop transfer functions between customer preferences and product attributes for new products and complex systems at the A-VDP stage without building expensive prototypes?
- How do we manage trade-offs? – Can we use optimization to set the right mix of attributes? How do we add the right constraints and bound the optimization to ensure "feasible" solutions?

- How do we know that the Voice of the Customer is balanced without going into detailed engineering solutions? Is it worth having the engineering factory spending 2–3 months working on trying to achieve a given solution only to find out that the solution set is either infeasible or requires extremely large investments?
- When multiple products are built off the same platform, how do we differentiate between these products and still maintain the tractability of the problem?
- How do we enable the cross-functional workgroup to become active participants in the process of identifying the optimal product attributes and features that will provide the best balance across all the conflicting voices (Voice of Customer, Engineering, Brand, Corporation, Manufacturing, etc.)?
- Given the complexity of the problem, the multi-dimensional nature of identifying the optimal set of vehicle attributes in a vehicle, and the diverse cross-functional nature of the workgroup, do tools exist that visually represent the impact of various trade-offs (across multiple dimensions)?

10.5 Global Voice of the Customer

GM Ex-CEO highlighted the need to be aware of global competition in 2003 in his remarks to the Nikkei Global Management Forum in Tokyo Japan (Wagoner 2003). Since then, the automotive industry has become even more global with vehicles being engineering in one country, manufactured in another and sold in a completely different country. The opening of markets and the increase in consumerism in India, China, Brazil and Russia have made them significant automotive markets. The rapid increase in IT and the easy access to all kinds of data have helped consumers in these emerging markets become very aware of product quality as well. These consumers are no longer willing to settle for 1 or 2 generation old technologies, but instead demand on having the latest gadgets and equipment and also desire the products to be tailored to their needs and wants. In the same speech in 2003, Wagoner also highlighted the need to find the optimal balance between global and local and termed it the "race to the middle". As GM and other companies continue on this battle to find the right balance between global and local, it has become even more important to understand the Voice of the Customer across different regions and identify the areas that can be "globalized" and the ones that need to be "localized".

10.5.1 Questions

According to Whirlpool, a company successful in adjusting its products to local markets, there are eight questions related to successful globalization (Shelton 2003).

- Is it a homogenous global market or heterogeneous collection of local markets?
- What are the commonalities that exist across local markets?
- What challenges do we face in trying to serve local markets?

- How does globalization impact our business in ways that are non-obvious?
- What are some of the non-obvious challenges for global companies?
- Do products have to be culturally appropriate to succeed?
- Are there drivers of innovation that can be found in the local markets?
- What is the future of global companies trying to succeed in local markets?

Questions 1, 2, 3 and 6 from the list above all require a company to understand the customers' needs and wants in the local market in order to be able to answer it. In addition to the questions above, three additional questions that need to be added to the catalog of the decision maker's list of queries are:

- Is there a systematic framework that will help in aggregating and balancing the global voice of customer and help decide when voices can be merged globally and when local voices should be maintained?
- Can we use global migration patterns and regional trends to predict trends for the future?
- How do we incorporate the current customer interaction/experience trends such as co-creation and crowd-sourcing (especially on a global scale)?

10.6 Conclusions

The A-VDP phase of a vehicle development process is critical in setting the right requirements for the vehicle. It is imperative that the decision-maker makes the right balance between conflicting voices and make the right trade-offs between the various engineering attributes. In order to ensure this, it is important to use the Voice of the Customer (i.e., the customers' needs and wants) as the guiding principle for making the balancing and trade-off decisions. Although there are many different approaches to estimating the Voice of the Customer, the decision-maker must understand the capability and limitations of the customer information that he/she is using for the trade-offs. Incorporating the Voice of the Customer as part of the advanced vehicle development process poses additional challenges that may get complicated further by the global nature of business. The paper highlights a series of questions that are designed to help the decision-maker probe further into the type of information being used and the capability of the organization in obtaining the information, incorporating it as part of the standard process and propagating it globally within the organization. In order to provide winning products globally and do it in an efficient manner, understanding and overcoming the challenges posed in integrating the Voice of Customer in A-VDP is critical.

References

Cooper, R.G. 1990, "Stage-Gate systems: a new tool for managing new products – conceptual and operational model", *Business Network* (www.bnet.com)

Donndelinger, J. 2006, "A Decision-Based Perspective on the Vehicle Development Process," *Decision Making in Engineering Design*, Ch. 19, ASME Press, New York, NY.

Donndelinger J.A. and Cook H.E. 1997, "Methods for Analyzing the Value of Automobiles", *SAE International Congress and Exposition*, Detroit, MI.

http://www.jdpower.com/about

http://www.consumerreports.org/cro/aboutus/history/printable/index.htm

http://en.wikipedia.org/wiki/Kano_model

Katz G.M. 2006, "Hijacking the Voice of the Customer", *Applied Marketing Science Inc.*, http://www.asm-inc.com/publications/reprints/voc_hijack.pdf

Nelson B. and Ross D., "Research Shows Understanding Customer Needs is Critical for Effective Innovation", www.innovare-inc.com/downloads/Customer_Needs_Innovation_Effectiveness.pdf

Orme, B. 2006, "A short history of conjoint analysis", *Getting Started with Conjoint Analysis: Strategies for Product Design and Pricing Research*, Ch. 4, Research Publishers LLC, Madison, WI.

Prahalad C.K. and Ramaswamy V. 2004, "The Future of Competition – Co-Creating Unique Value with Customers", *Harvard Business Press*.

Shelton E. 2003, "Globalization, Technology and Culture Initiative Takes Product Development Down to the Local Level", *The Journal of the International Institute*, Vol. 11, No. 1.

Von Hippel E. et al., 1986, "Lead Users: A source of novel product concepts", *Management Science 32*, no. 7, pp 791–805.

Wagoner G.R. 2003, "The Race to the Middle: Managing in Today's Global Automotive Industry", *Speech at the Nikkei Global Management Forum*, Tokyo, Japan.

www.wikipedia.com/QFD

Chapter 11
A Statistical Framework for Obtaining Weights in Multiple Criteria Evaluation of Voices of Customer

R.P. Suresh and Anil K. Maddulapalli

Abstract In this work, we formulate the problem of prioritizing voices of customer as a multiple criteria decision analysis problem and propose a statistical framework for obtaining a key input to such an analysis, i.e., importance (or weights) of data sources used in the analysis. Engineers usually resort to surveys from different sources for gathering data required to analyze the voices of customer. However, different surveys use different scales, statements, and criteria for gathering and analyzing the data. To account for these issues with survey data and to aggregate the data from different surveys for prioritizing voices, we apply a popular multiple criteria decision analysis technique called Evidential Reasoning (ER). Like any other multiple criteria decision analysis approach, ER approach handles the conflicting criteria by querying the decision maker for the weights associated with the criteria. However providing weights for different surveys is difficult because of the external factors involved in conducting a survey. Here, we investigate a statistical approach for obtaining the weights of the surveys instead of directly querying the decision maker. We identify several factors that could affect the weights of surveys and propose a framework for statistically handling these factors in calculating the weights.

Keywords Survey importance · Survey design · Coverage probability · Credibility of survey

11.1 Introduction

Quality Function Deployment (QFD) has now become a standard practice by many leading firms and has been successfully implemented worldwide (Chan and Wu 2002). QFD is a cross-functional planning methodology, commonly used to ensure that customer needs (referred to as the voices of customer) are deployed through product planning, part development, process planning and production planning. The successful implementation of QFD depends on the identification and prioritization

R.P. Suresh (✉) and A.K. Maddulapalli
GM R&D, India Science Lab, Bangalore-560066, India
e-mail: suresh.rajagopalan@gm.com

N.R.S. Raghavan and J.A. Cafeo (eds.), *Product Research: The Art and Science Behind Successful Product Launches*, DOI 10.1007/978-90-481-2860-0_11,

of the voices of customer. A customer voice is a description, stated in the customers' own words, of the benefit to be fulfilled by a product or service. Identifying the voices of customer is primarily a qualitative research process.

The prioritization of the voices of customer has attracted much attention in the research community and a number of prioritization methods have been suggested in the literature. Examples include the analytic hierarchy process (Saaty 1994; Kwong and Bai 2002), the entropy method (Chan et al. 1999; Chan and Wu 2005) and several other methods as mentioned in (Olson 1996). In these studies, it is assumed that customer needs can always be expressed using the specific frameworks as designed in these methods. However, this may not be the case if multiple data sources are used for prioritizing the voices of customer, because the analysts may not always be able to control the design of data formats, in particular if data are generated from external sources.

Product engineers usually resort to surveys from different sources for gathering data required to analyze the voices of customer. Different surveys concentrate on different aspects of the product and target different parts of the population. So it is important for the engineers to consider data from all available sources before prioritizing the voices of customer. However, different surveys use different scales, statements, and criteria for gathering and analyzing the data. In a previous work, it has been identified that Evidential Reasoning (ER) algorithm (Yang and Singh 1994; Yang 2001) is a good approach for handling the above mentioned issues with different surveys and aggregating the information from the surveys. Refer to (Xie et al. 2008; Yang et al. 2008) for more details.

ER approach is based on Dempster-Shafer theory of evidence and is applied to a variety of problems (Yang and Xu 1998, 2005; Chin et al. 2007). Like any other multiple criteria decision analysis approach, ER algorithm handles the conflicting criteria by querying the decision maker on the trade-offs associated with the criteria. These trade-offs are captured in the form of importance or weights between the criteria. Usually the decision maker is queried directly to obtain the weights of the criteria. However providing weights for different surveys is difficult because of the external factors involved in conducting a survey. In this work we investigate a statistical approach for obtaining the weights of the surveys instead of directly querying the decision maker. To the best of our knowledge, this work is the first attempt in the literature for statistically estimating the weight of a survey to use in a multiple criteria decision making technique.

The importance weight of a survey is affected by many factors. In this work, after discussions with heavy users of survey data, we have identified four key factors that influence the weight of a survey. These factors are: Design for selecting respondents of a survey, Source for identifying the respondents (i.e., coverage of a survey), Credibility of agency conducting the survey, and Domain experience of respondents in a survey. We have decided not to include the scale used in a survey as a factor because ER algorithm can be used to transform different scales on to a common scale. Also, survey design and questionnaire structure are explicitly not included as factors and are assumed to be part of the credibility of the agency conducting the survey. Sample size of a survey is implicitly taken into account in the design factor.

We believe that each of the above factors induce a bias/error in the data collected from a survey and propose that the weight of a survey be inversely proportional to the bias/error of that survey. We discuss some possible approaches for estimating the error induced into a survey from each of the influencing factors and estimate the weight of a survey as an inverse function of the cumulative error. Our intent in this paper is to highlight a new (and important from industrial practitioners' view) area of research and suggest a framework for further investigations. In this work we will propose some approaches and suggest using some other approaches without any validation studies. The validation of the framework and approaches discussed in the paper will be taken up as future work.

The remainder of this paper is organized as follows. In Section 11.2, we will formulate the voice of customer prioritizing problem as a multiple criteria decision making problem and discuss how ER algorithm can be used to solve the problem. Next in Section 11.3, we discuss in detail the rationale behind identifying the key factors affecting the weights of the survey and propose some approaches for handling these key factors in determining the weight of a survey. In Section 11.4, we demonstrate voice prioritizing using ER algorithm with the help of a simple example in which our statistical framework is used for determining the survey weights. Finally, we conclude the paper with a summary and discussion on future work in Section 11.5.

11.2 Voice of Customer Prioritization Using ER Algorithm

Figure 11.1 depicts a generic criteria hierarchy for evaluating voice of customer using data from different surveys. As shown in Fig. 11.1, data pertaining to a voice could be obtained from different surveys. Some of these surveys could be mass surveys (i.e., mail-in survey, web based surveys, and so on) internally conducted by an organization. Some of the surveys could be bought from third party external agencies that conduct independent mass surveys (e.g., J.D. Power, Consumer Reports). Some of the surveys could be focused workshops or clinics that are conducted internally by an organization or bought from an independent external agency.

More than one statement could be used in a survey for gathering data on a particular voice. This happens frequently in surveys conducted by external agencies as the organization of interest does not have control on the questionnaire used in those surveys. In an internal survey, usually the questionnaire is designed following the template of customer voices developed by the organization. In Fig. 11.1, for clarity, the branch with criteria is shown only for one statement of a survey and similar branches can be added for all the statements of a survey that are related to a voice of interest.

The data from a statement of survey is typically used to evaluate a voice of interest using different criteria. A typical criterion could be that mean rating of the organization's product is greater than the competitor product's mean rating. Another criterion could be that the cumulative frequency of the ticks in the top

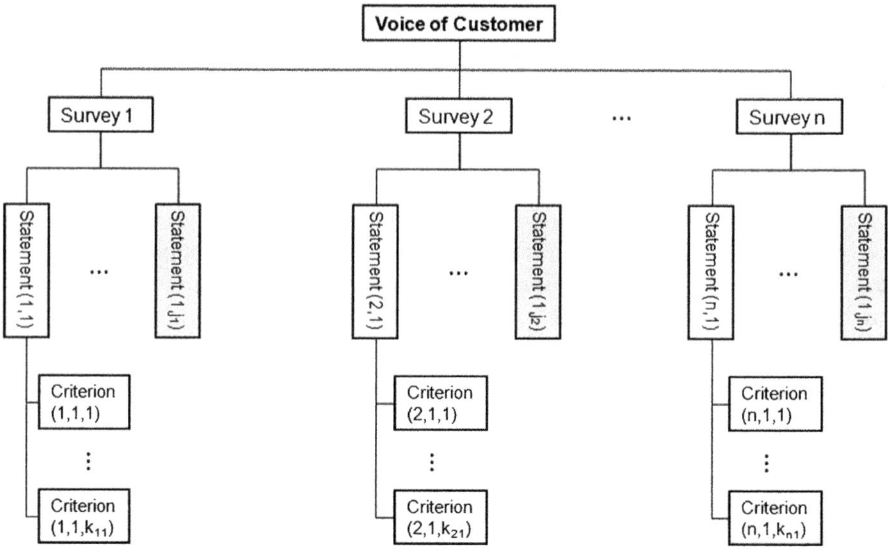

Fig. 11.1 Generic criteria hierarchy for evaluating voice of customer

two boxes corresponding to better satisfaction is more for the organization's product when compared to the competitor's product. A criterion in Fig. 11.1 could have sub-criteria and the hierarchy shown in Fig. 11.1 could be expanded vertically and horizontally.

So, it is possible to formulate the voice prioritization problem as a multiple criteria decision making problem as shown in Fig. 11.1. However, the unique feature of the hierarchy shown in Fig. 11.1 is that different surveys use different scales to collect the data and there is a lot of uncertainty in the data reported by the surveys. Also, the scales used in surveys are often qualitative. In addition, some voices might have data from some surveys only leading to more missing information. Taking into account all these issues, we have decided that the Evidential Reasoning (ER) algorithm would be a good approach for aggregating information from different surveys.

The main attraction of ER algorithm is that it allows the analyst to intuitively transform data from different surveys onto a common scale. ER algorithm captures the uncertainty in the data using the distributed belief structure of Dempster-Shafer theory (Yang and Singh 1994; Yang 2001). Also the aggregation using ER algorithm is shown to overcome rank-reversal problem that are seen in other multiple criteria decision making methods like Analytical Hierarchy process (Saaty 1994). Also, the aggregation process of ER algorithm is very transparent and allows an analyst to trace up and down the criteria hierarchy.

Next, we briefly discuss the details of ER algorithm.

11.2.1 Evidential Reasoning Algorithm

The details of ER algorithm are only discussed briefly here and further details can be found in (Yang and Singh 1994; Yang 2001). The first step in ER algorithm is to transform the data from the bottom nodes of criteria hierarchy into a common scale. The common scale allows for a common ground for aggregating and comparing data from different surveys. For a specific application, each grade of the common scale needs to be defined in detail and guidance should be provided about how to differentiate between these grades. This forms part of a process for gathering and structuring assessment knowledge. For voice prioritization one could propose to use a five-point common scale as follows: Top Priority (H_1), High Priority (H_2), Medium Priority (H_3), Low Priority (H_4) and No Priority (H_5). Some rules for transforming data from the original scales of a survey to the common scale are discussed in Xie et al. (2008) and Yang et al. (2008).

In general the ER algorithm involves the following steps:

Step 1: Generate distributed assessments (belief degrees) on the common scale. This involves transforming data from original surveys to the common scale using mapping functions (Xie et al. 2008; Yang et al. 2008).

Step 2: Get normalized weights at different levels of criteria hierarchy. For standard criteria that are understood quantitatively or qualitatively, many methods are available for eliciting weight information (Olson 1996; Saaty 1994). However, using those methods it is not easy to provide weights for surveys because surveys are not single attribute and involve many external factors (see Section 11.3 for details).

Step 3: Assign basic probability mass, which is defined as the product of belief degree and weight.

Step 4: Generate combined probability mass. The details of the aggregation can be found in (Yang and Singh 1994; Yang 2001).

Step 5: Normalize combined belief degrees and get overall assessment.

Step 6: Convert the combined belief degree into utility score. This is done using a utility function for transforming each grade of common scale into a cardinal utility value that lies between zero and one. This conversion is done for rank ordering the voices after aggregating the data.

Next we discuss the impact of weight of survey on the final ranking of voices.

11.2.2 Impact of Weight of Survey

We have implemented the above discussed ER algorithm on a simple case study using data from four different surveys. The first survey is an internally conducted survey and is done on focused groups of respondents. The second survey is a mass mail-in survey that is internally conducted. The third survey is mass mail-in external survey and the fourth survey is also an external survey. We have used the available data from these four surveys on four voices of interest and computed the aggregated belief degree and the overall utility of each voice. We have used a linear utility function with H_1 mapped to one and H_5 mapped to zero. The details of the voices and the corresponding data are not produced here due to confidentiality reasons. We then conducted a sensitivity study to understand the impact of the weight of survey in determining the rank order of the voices. Figure 11.2 below shows the results of the sensitivity study.

The vertical line (i.e., "Given weight") shown in Fig. 11.2 depicts the actual weight given to Survey 1 by the analysts. The analysts arrived at this weight based on their intuition and were not sure if they are certain about the weight. So we conducted a sensitivity analysis by varying the weight of Survey 1. While varying the weight of Survey 1, we have proportionally changed the weights of other surveys in order to maintain the sum of weights to one. From Fig. 11.2, it is clear that the overall utility of all voices is affected when the weight of Survey 1 is changed. The relative rank order of Voice B and Voice C is also affected by changing the weight of Survey 1. So, it is clear that the weight associated with a survey has a clear impact on the overall utility and rank order of the voices. Hence, it is very important to arrive at the appropriate weights while using multiple criteria decision making techniques like ER algorithm for prioritizing voices using data from multiple surveys. Our experience has shown that the analysts are comfortable providing weights for the quantitative criteria used in prioritizing voices but are very anxious while providing weights for the surveys. So, in the next section we discuss the different factors that could affect the weight of a survey and propose a framework for arriving at the weight of the survey.

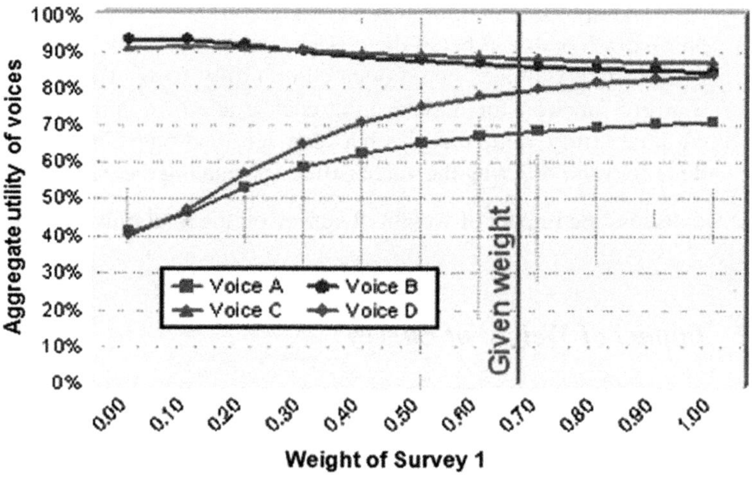

Fig. 11.2 Sensitivity of overall utility to weight of Survey 1

11.3 Factors Influencing the Weight of a Survey

After some preliminary discussions with the stakeholders involved in voice prioritization within our organization, we have arrived at four key factors that influence the weight of a survey. These four factors are not exhaustive and more factors could be added depending on the product and application. The key factors we identified are:

- Design for selecting respondents of a survey
- Source for identifying the respondents of a survey
- Credibility of agency conducting the survey
- Domain experience of respondents

We provide the rationale for choosing each of these factors in the following subsections.

11.3.1 Design for Selecting Respondents

Many techniques exist in the literature for sampling the respondents appropriate for a survey. Some of these techniques include stratified sampling, clustered sampling and so on (Churchill and Iacobucci 2005). Loosely classified, there are two types of surveys that are usually conducted by an organization. The first type is a survey on a focused group of respondents, which is often referred to as clinic. The objective of such a survey is to get opinion among people who are knowledgeable about the product or service, and their opinion will reflect that of the whole market. Examples of such surveys include capturing preferences on

- Voices about features like GPS based navigation system which are usually expensive and are seen in luxury vehicles.
- Voice related to purchase decision of mobile phones which are based on cultural aspects of a country. In Japan the main drivers of mobile purchase are cost of service package, design and color. In china the main drivers are brand, design, and screen size and in US the main driver is the service plan (see Hirsh 2007).

The other type of surveys is mass surveys involving people from the whole market, viz., all stakeholders in the market. These mass surveys could be web based, mail based, or telephone based.

Usually the sampler resorts to random sampling techniques in order to reach the length and breadth of the population. Depending on the characteristic of the population, e.g., homogeneity, geographical vastness of the population and so on, one may use Simple Random Sampling, Stratified Sampling, Cluster Sampling, Multi-stage Sampling etc. The error or variance associated with Simple Random Sampling (SRS) and Stratified Random Sampling (STS) schemes are different and are given in Eq. 11.1 (see Churchill and Iacobucci 2005, for further details.). The variances corresponding to other sampling schemes can be derived similarly.

$$V_{SRS}(SD) = \frac{\sigma^2}{n}$$
$$V_{STS}(SD) = \sum_h (W_h)^2 \, \sigma_h^2 \, / \, n_h \tag{11.1}$$

In Eq. 11.1, $V_\tau(SD)$ stands for mean of the variance from the survey data using a sampling scheme "τ" (i.e., SRS, STS and so on). σ^2, σ_h^2 are the variances in the data from the whole sample and from the hth stratum respectively. n, n_h are the sample sizes of the whole sample and of a hth stratum respectively. W_h is the weight (i.e., percentage population in the hth stratum when compared to the whole population) associated with the hth stratum.

The variance given in Eq. 11.1 are applicable to the second type of surveys discussed above as the potential population is infinite. However, since the first type of survey considers only a certain class of respondents satisfying certain criteria, which restricts the population, (for example, a survey among mobile users may be restricted to only those who have been using mobiles since five years) the sampling variances are different for this type of surveys and are given in Eq. 11.2

$$V_{SRS}(SD) = \frac{\left[\frac{N-n}{N-1}\right]\sigma^2}{n}$$
$$V_{STS}(SD) = \sum_h (W_h)^2 \left[\frac{N_h - n_h}{N_h - 1}\right] (\sigma_h^2)/n_h \tag{11.2}$$

In Eq. 11.2 N and N_h are the population size of the whole population and that of the hth stratum respectively. The other symbols in Eq. 11.2 are similar to the ones in Eq. 11.1. It may be noted here that, in general, the sample sizes in the first type of surveys tend to be much smaller than the second type of Survey, which will be reflected in the variance or error associated with the survey. *Also, Eqs. 11.1 and 11.2 account for the sample size of surveys and this is the reason why we have not explicitly included sample size as one of the factors influencing the survey weight.*

The error due to sampling design can be calculated using the following steps:

Step 1: Identify the ideal sampling scheme for the voice and survey of interest. For example, in an application, if the population is known to be heterogeneous, then stratified sampling may be the appropriate scheme.

Step 2: Calculate the variance from the survey data using the formula appropriate to the sampling scheme used in the survey, $V_s(SD)$.

Step 3: Re-group the sample data to suit ideal sampling scheme identified in Step 1 and calculate the variance using the formula appropriate to ideal sample scheme, $V_{Ideal}(SD)$. For example, use post-stratification to identify samples in each stratum, and compute $V_{Ideal}(SD) = V_{STS}(SD)$ using the post-stratified sample.

Step 4: Error induced into a survey due to sampling design, $\in_{SD}$, can then be obtained using Eq. 11.3. In using Eq. 11.3, care should be taken that $V_{Ideal}(SD)$ is greater than V_{SD}.

$$\in_{SD} \propto \left(1 - \frac{V_s(SD)}{V_{Ideal}(SD)}\right) \tag{11.3}$$

11.3.2 Source for Identifying the Respondents

The agencies conducting surveys adopt different means for identifying the respondents for a survey. Some of these means could be email addresses if the survey is an e-survey, IP addresses if survey is web based, telephone directory if the survey is telephonic, and mail addresses from registered organizations if the survey is mail-in or in-person. Some agencies set up special kiosks at popular places like malls, cinema halls, airports, etc. for conducting in-person interviews.

Each source for identifying respondents has its own advantages and disadvantages. For example if a telephone directory is used to identify respondents for telephonic surveys, it is possible that many people are not listed in the telephone directory and also many people might not even have a telephone (especially in developing countries). This leads to errors in surveys due to non-coverage and some work has been reported in literature for handling such issues (Raghunathan et al. 2007). An estimation of non-coverage probability of a telephonic survey can be obtained by estimating the proportion of people not having telephone as reported in the national census. For example, 5.2% of the households in U.S.A did not have telephones as per the 1990 census. The error estimation associated with this factor, namely source for identifying respondents ($\in_1$ SR), is directly proportional to the non-coverage probability, which can be estimated using the procedure of Raghunathan et al. (2007).

$$\in_{SR} \propto \Pr(non\text{-}coverage) \tag{11.4}$$

11.3.3 Credibility of Agency Conducting the Survey

Perhaps the most important factor influencing the weight of a survey is the credibility of a survey. Credibility by definition is very subjective and depends on how the analysts perceive a survey. To our knowledge, there is no reported work on estimating the credibility of agencies conducting surveys. However there is literature reported on credibility estimation in advertising industry which could be taken as a starting point for conducting research on estimating survey agencies' credibility.

In advertising industry, credibility estimation literature is divided into two groups. The first group focuses on the credibility of the person (e.g., a celebrity) appearing in an advertisement (Ohanian 1990). The second group focuses on the credibility of the corporation depicted in the advertisement (Mackenzie and

Lutz 1989; Newell and Goldsmith 2001). Of the two groups the research on corporate credibility seems relevant to our problem of estimating the credibility of the agency conducting the survey.

Newell and Goldsmith (2001) have proposed and validated a scale for measuring the corporate credibility. Through their research they have identified that the two main factors to consider in estimating corporate credibility are "expertise" and "trustworthiness". They have devised their scale around the two factors and validated the scale through some experiments conducted using survey responses from undergraduate students. The truthfulness and honesty factors of a corporate are included in the "trustworthiness" factor of their scale.

In general organizations and consumers look for "expertise" and "trustworthiness" of agencies while buying survey data and the scale proposed by Newell and Goldsmith (2001) takes these factors into account. So their scale could be readily used for conducting an internal survey inside an organization for estimating the perception of analysts on the survey they use. The results of the survey could be used to estimate the belief error induced into the weight of a survey from lack of credibility on a survey. Our future research would focus on designing an internal survey following the guidelines of (Newell and Goldsmith 2001) for estimating the survey agencies' credibility.

For the purposes of this paper, we propose to use the number of users (both analysts in the organization and consumers, if available) of the survey data as a surrogate for survey agencies' credibility. It may be noted that some standard external survey agencies do report an estimate of the number of customers using their surveys. Eq. 11.5 gives the error from lack of credibility on agency conducting the survey, i.e.,$\in_{cs}$.

$$\in_{cs} \propto 1/n_u : n_u \textit{ is the number of users of a survey} \tag{11.5}$$

The presumption in the proposed approach is that, analysts and consumers use the survey data from an agency only if they think that the agency is credible. However, sometimes users are forced to use a survey data even if the agency is not credible because of lack of data from any other source and Eq. 11.5 ignores those aspects.

11.3.4 Domain Experience of Respondents

Domain experience refers to the experience of respondents with the product for which the survey is conducted. For surveys conducted before launching a new product, domain experience would refer to experience with similar products in the market. Domain experience is an important factor influencing the weight of a survey because the quality of responses obtained in a survey depends on the domain experience of the respondents.

The timing of a survey related to a product launch varies across different industries. In automobile industry, clinics are usually conducted before launching a new vehicle. Some surveys are conducted on buyers with around three months of ownership experience. Some other surveys are conducted on buyers with more than a year of ownership experience. Obviously the domain experience of the respondents varies significantly in the above mentioned three kinds of surveys. Usually, the domain experience of the respondents attending clinics before a product launch is minimal unless they have used a very similar product before. The domain experience of respondents with three months of ownership is slightly better and the domain experience of respondents with more than a year ownership is considered the best.

The usage of a product within an ownership period would provide a good measure for estimating the domain experience of respondents. Many surveys query the respondents on the typical usage and the frequency of usage for a product. The response to these queries could be used to arrive at a metric for domain experience as follows.

For a voice of interest, determine the appropriate usage types queried in a survey and identify the number of respondents that could be classified as heavy users ($\mu_1 H$), medium users ($\mu_1 M$), and light users (μ_L) (one could use more classifications for frequency of usage). For each frequency of usage, based on ownership of respondents, assign a value that corresponds to domain experience, i.e., α_H for μ_H; α_M for μ_M, and α_L for μ_L, where $0 \leq \mu_H, \mu_M, \mu_L \leq 1$. As an example, for respondents with more than a year's ownership, α_H could be 1 because these users are the most experienced in terms of ownership α_H and usage. The domain experience of the respondents (f_{1DE}) can then be calculated as a sum product of the frequency of users and value of usage as shown in Eq. 11.6.

$$f_{DE} = \alpha_H \cdot \mu_H + \alpha_M \cdot \mu_M + \alpha_L \cdot \mu_L : \\ 0 \leq \mu_H \cdot \alpha_H \cdot \mu_M \cdot \alpha_M \cdot \mu_M \cdot \alpha_2 \leq 1 \tag{11.6}$$

The error induced into the survey from lack of domain experience, $\in_{DE}$, is inversely proportional to f_{DE} as shown in Eq. 11.7.

$$\epsilon_{DE} \propto 1 - f_{DE} \tag{11.7}$$

If a survey does not query the respondents about the usage of the product then alternate approaches need to be developed for estimating the domain experience. We are currently pursuing research in that direction.

11.3.5 Weight of a Survey

As mentioned above we model the weight of a survey as inversely proportional to the cumulative error induced from different factors, see Eq. 11.8.

$$W_j \propto \frac{1}{(\in_{SD} + \in_{SR} + \in_{CS} + \in_{DE})} \tag{11.8}$$

For simplicity, in Eq. 11.8, we have assumed that all factors influencing the survey weight are independent of each other. We will further investigate this independence assumption and modify Eq. 11.8 accordingly as part of our future research.

Next we will demonstrate the application of the approaches discussed above with a simple example.

11.4 Demonstrative Example

In this section, we demonstrate the approaches recommended for estimating the error (and hence the weight) induced in a survey with the help of a simple example. We suppress many details of the surveys used in our example due to confidentiality reasons.

In this example, we wish to assess two important customer voices using data from three surveys. The first voice, V_1, refers to an exterior feature of an automobile and the second voice, V_2, refers to an interior feature of an automobile. The first surve S_1 is an internally conducted clinic on focused group of respondents. The second survey S_2 is also an internal survey but is a mass mail-in survey. The third survey S_3 is conducted by a popular external agency and is also a mass mail-in survey. We apply our recommended approaches to determine the weights of the three surveys and then apply ER algorithm for aggregating data from the three surveys. Note that from here on in this paper, we use $\in_{\gamma S_j}^{V_k}$ to represent the error from the factor "γ", where S_j stands for the jth survey and V_k stands for the kth voice.

11.4.1 Influence of Sampling Design on Survey Weights

As mentioned before, S_1 is a clinic conducted on a focused group of respondents, where as S_2 and S_3 are mass mail-in surveys. All the three surveys use the simple random sampling (SRS) scheme for selecting the respondents. However, for the voices of interest in this example, we believe that a stratified sampling (STS) scheme, where in the strata are identified based on the age of respondents, would be ideal for all three surveys. In particular, we wish sampling for the surveys had been done with three strata, where respondents in the age group 15–29 years fall in the first stratum, respondents in the age group 30–49 years fall in the second stratum, and the respondents with age greater than 50 years fall into the third and last stratum.

For calculating the error induced into the survey data from the difference in the ideal and actual sampling schemes, we follow the steps identified in Section 11.3.1. We have identified STS as the ideal sampling scheme and then calculate the mean variance in the survey data using the appropriate formulae for the SRS and STS

Table 11.1 Variance in V_1 data for SRS and STS sampling schemes

V_1						
	S_1		S_2		S_3	
	Variance	Sample size	Variance	Sample size	Variance	Sample size
Stratum 1 (age between 15 and 29, 26.48%)	1.2500	80	0.7139	4970	3.3738	4763
Stratum 2 (age between 30 and 49, 38.77%)	0.9448	386	0.7998	18567	3.5141	19551
Stratum 3 (age greater than or equal to 50, 34.75%)	0.9273	168	0.7556	37206	3.7201	32375
Whole sample	0.9801	634	0.7685	60743	3.6332	56689
Mean variance from post-stratification	2.03E-03		1.90E-05		9.06E-05	
Mean variance from original simple random sampling	1.45E-03		1.27E-05		6.41E.05	
Error from difference in sampling methods	0.2862		0.03341		0.2923	

schemes. Table 11.1 shows the variance in the data corresponding to V_1 using different schemes for all three surveys in the example and Table 11.2 shows the same for V_2 .

In Tables 11.1 and 11.2, the variance, σ_h^2, and sample size, n_h, for each of the stratum after post-stratification of the data is given in the first three rows. The variance, σ^2, and sample size, n, of the whole sample is given in the fourth row. The mean variance using STS scheme, $V_{STS}(SD)$, is shown in the fifth row and the mean variance using SRS scheme, $V_{SRS}(SD)$, is shown in the sixth row. The error induced by using a sampling scheme that is different from the ideal sampling scheme is calculated using Eq. 11.3 and is shown in the last row of Table 11.1.

For calculating the mean variance using STS sampling scheme in Eqs. 11.1 and 11.2, we need to identify the appropriate W_h values for each stratum. To set these W_h values, we have used the 2000 U.S.A. census data from http://www.census.gov/, to identify the percentage of population in the age groups 15–29 years, 30–49 years, and greater than 50 years. We calculated the percentages taking only the population that is greater than 15 years. Using the 2000 census data we obtain $W_1 = 0.2648$ (i.e., 26.48%) for stratum 1 (i.e., age group 15–29 years), $W_2 = 0.3877$ for stratum 2 (i.e., age group 30–49 years), and $W_3 = 0.3475$ for stratum 3 (i.e., age greater than 50 years).

Since S_1 is a clinic conducted on focused group of respondents, we have used Eq. 11.2 for calculating the mean variance using the SRS and STS schemes. For

Table 11.2 Variance in V_2 data for SRS and STS sampling schemes

V_2	S_1		S_2		S_3	
	Variance	Sample size	Variance	Sample size	Variance	Sample size
Stratum 1 (age between 15 and 29, 26.48%)	1.0112	80	0.7038	4991	2.4038	4852
Stratum 2 (age between 30 and 49, 38.77%)	0.7970	386	0.7489	18626	2.6641	20010
Stratum 3 (age greater than or equal to 50, 34.75%)	0.6621	168	0.7550	37118	3.0771	33861
Whole sample	0.7948	634	0.7534	60735	2.9034	58723
Mean variance from post-stratification	1.59E-03		1.84E-05		6.57E-05	
Mean variance from original simple random sampling	1.17E-03		1.24E-05		4.94E-05	
Error from difference in sampling methods	0.2626		0.3253		0.2477	

$V_{SRS}(SD)$, we have set $N = 10{,}000$ and for $V_{STS}(SD)$ we have set $N_1 = 2648$, $N_2 = 3877$, and $N_3 = 3475$, following the percent population obtained from 2000 census data. For S_2 and S_3, we have used Eq. 11.1 as these surveys are mass surveys.

11.4.2 Influence of Source of Respondents on Survey Weights

For S_1 the current vehicle that a respondent owns is used as a qualifying metric for inviting the respondents to the clinic. The data on the qualifying vehicle is obtained from a well known registration company. S_2 and S_3 are mail-in surveys conducted on new car buyers with an ownership period of around three months. The details of the owners of the vehicles are obtained from the same registration company as S_1. This registration company is very reliable and many organizations use the data from this company. So, in this example, there is no error induced due to the source of respondents. Hence, we assign the error due to non-coverage resulting from the source of selecting the respondents as $\in_{SR_{s_j}}^{V_k} = 0$ for $j = 1, 2, 3$ and $k = 1, 2$.

11.4.3 Influence of Agency Credibility on Survey Weights

In Section 11.3.3, for estimating the credibility of a survey agency we recommended conducting an internal survey with scale proposed by Newell and Goldsmith (2001) as the ideal approach. For the purpose of this paper, we suggested to use the number of users of a survey as proxy for estimating the agencies' credibility. Two of the surveys in our example are internal surveys and are used by people within the organization. The third survey, S_3, is external and people within and outside the organization use this survey data. But we don't have a measure of the number of people using S_3 data outside the organization. So we only use the number of users within the organization for estimating the agencies' credibility.

Due to confidentiality reasons we cannot disclose the actual number of users for each of the three surveys within the organization. However the users are such that for every 100 users of S_1, there are 90 users of S_2, and around 65 users of S_2. Using these numbers we can arrive at the error induced due to lack of agencies' credibility for the three surveys as $\in_{CS_{S_1}^{V_1}} = \in_{CS_{S_1}^{V_2}} = 0.010$; $\in_{CS_{S_2}^{V_1}} = \in_{CS_{S_2}^{V_2}} = 0.011$; and $\in_{CS_{S_3}^{V_1}} = \in_{CS_{S_3}^{V_2}} = 0.015$.

11.4.4 Influence of Domain Experience on Survey Weights

In the first survey S_1, the respondents are asked to give details on two types of usages for their qualifying vehicles. Both these usages are related to the voices of interest in this example and hence are considered for estimating the domain experience of respondents using Eq. 11.6. There are 159 respondents in the survey which translates to 318 combined responses for the two usages combined. Of the 318 responses, there are 185 responses that correspond to heavy users which translates to a $\mu_H = 0.5818$, 115 responses that correspond to medium users which translates to a $\mu_M = 0.3616$ and 16 responses that correspond to light users which translates to a $\mu_L = 0.0503$ (there are two missing responses). Based on the ownership of the qualifying vehicle for the respondents in the clinic, we have assigned the value of frequency of usage in determining the domain experience as $\alpha_H = 1.0$, $\alpha_M = 0.7$ and $\alpha_L = 0.4$. Using Eq. 11.6, the domain experience of the respondents of S_1 can then be calculated as $f_{DE}^{S_1} = 0.8550$. The error in S_1 survey data due to lack of domain experience can then be calculated using Eq. 11.7 as $\in_{DE_{S_1}^{V_1}} = \in_{DE_{S_1}^{V_2}} = 0.1450$.

In the second survey S_2, the respondents are asked to give details on fourteen types of usages for the vehicles they bought. From the responses to these fourteen usage types, we considered the top four usage types for each respondent. Since the ownership period of the respondents in S_2 is at most three months, we assign the value of the frequency of their usage in determining the domain experience as $\alpha_H = 0.9$, $\alpha_M = 0.7$, and $\alpha_L = 0.4$. A total of 261,408 responses for the four usage types are used to obtain $\mu_H = 0.6770$, $\mu_M = 0.2052$, and $\mu_L = 0.0385$ (there are some missing responses also). Using Eq. 11.6, the domain experience of the respondents of S_2 can then be calculated as $f_{DE}^{S_2} = 0.7683$. The error in S_2 survey data due to lack of domain experience can then be calculated using Eq. 11.7 as $\in_{DE}{}_{S_2}^{V_1} = \in_{DE}{}_{S_2}^{V_2} = 0.2317$.

The usage data from the third survey is S_3 not available with us and domain experience could not be calculated. But based on other information available with us we estimate the error in S_3 survey data due to lack of domain experience as $\in_{DE}{}_{S_3}^{V_1} = \in_{DE}{}_{S_3}^{V_2} = 0.3333$.

11.4.5 Estimating Survey Weights

Using Eq. 11.8 of Section 11.3.5, we estimate the weight of the three surveys in our example as inversely proportional to the cumulative error from the four factors. Table 11.3 shows the errors induced into the three surveys from the four factors (see rows 1–4). The cumulative error in each survey is given in the fifth row, followed by the non-normalized weight in the sixth row. Since ER algorithm requires the weights to add up to one, we normalize the weights in the sixth row to obtain the normalized survey weights for V_1 and V_2 as given in the last row of Table 11.3. Note that there is some difference in the survey weights for V_1 and V_2, and this difference is due to the difference in error from the first factor, sampling design. Many of the multiple criteria decision making methods assume that the weight structure for criteria is same across all alternatives. However, as we have shown in this example it is possible to have different criteria weight structure for different alternatives, especially while using data from surveys. The algorithm we use from aggregating data, i.e., the ER algorithm, is flexible enough to accommodate different criteria weight structure.

Table 11.3 Survey weight estimate using errors from different factors

Errors from different factors	V_1			V_2		
	S_1	S_2	S_3	S_1	S_2	S_3
Error from sampling design ($\in_{SD}$)	0.2862	0.3341	0.2923	0.2626	0.3253	0.2477
Error from source of respondents($\in_{SR}$)	0.0000	0.0000	0.0000	0.0000	0.0000	0.0000
Error from credibility of source ($\in_{CS}$)	0.0100	0.0110	0.0150	0.0100	0.0110	0.0150
Error from domain experience ($\in_{DE}$)	0.1450	0.2317	0.3333	0.1450	0.2317	0.3333
Cumulative Error	0.4412	0.5768	0.6406	0.4176	0.5680	0.5961
Weights	2.2665	1.7336	1.5609	2.3946	1.7605	1.6777
Normalized Weights	0.4076	0.3117	0.2807	0.4105	0.3018	0.2876

11.4.6 Application of ER Algorithm for Voice Prioritization

For assessing the two voices of interest, V_1 and V_2, we use a criteria hierarchy similar to the one shown in Fig. 11.1. In all three surveys, there is only one statement corresponding to each voice. Under S_1 there are three criteria and there are eight criteria each under S_2 and S_3. The assessments on common scale for voices V_1 and V_2 under all criteria in the hierarchy are given in Tables 11.4 and 11.5 respectively. The assessment on common scale for a criterion is given in the order $\{H_1, H_2, H_3, H_4, H_5\}$ (recall Section 11.2.1).

We used ER algorithm for aggregating the assessments shown in Tables 11.4 and 11.5. We have used equal weights for all the criteria. For the surveys we have used equal weights in one run and the weights calculated using our framework (refer last row of Table 11.3) in the other run. The overall assessment of the voices and the corresponding utility (using a linear utility function as discussed in Section 11.2.2) for both the cases are shown in Table 11.6.

From Table 11.6, it can be clearly seen that the difference in the utility between V_1 and V_2 has increased when survey weights from our framework are used (see last column of Table 11.6). In fact, the overall utility of both voices has increased with weights from our framework. This means that when compared to other voices, V_1 and V_2 could be ranked higher using the weights from our framework. Clearly, the

Table 11.4 Assessment on common scale for V_1

S_1		S_2		S_3	
Criterion(1,1,1)	{1,0,0,0,0}	Criterion(2,1,1)	{1,0,0,0,0}	Criterion(3,1,1)	{0,0,0,0,1}
Criterion(1,1,2)	{0,0,0, 0.854, 0.146}	Criterion(2,1,2)	{1,0,0,0,0}	Criterion(3,1,2)	{0,0,0,0,1}
Criterion(1,1,3)	{1,0,0,0,0}	Criterion(2,1,3)	{1,0,0,0,0}	Criterion(3,1,3)	{0,0,0,0,1}
		Criterion(2,1,4)	{0,0,0,0,1}	Criterion(3,1,4)	{0,0,0,0,1}
		Criterion(2,1,5)	{0,0,0,0,1}	Criterion(3,1,5)	{1,0,0,0,0}
		Criterion(2,1,6)	{0,0,0.372, 0.628,0}	Criterion(3,1,6)	{0,0.081, 0.919,0,0}
		Criterion(2,1,7)	{0,0,0,0,1}	Criterion(3,1,7)	{0,0,0,0,1}
		Criterion(2,1,8)	{0,0,0,0,1}	Criterion(3,1,8)	{0,0,0,0,1}

Table 11.5 Assessment on common for V_2

S_1		S_2		S_3	
Criterion(1,1,1)	{1,0,0,0,0}	Criterion(2,1,1)	{1,0,0,0,0}	Criterion(3,1,1)	{0,0,0,0,1}
Criterion(1,1,2)	{0.064, 0.936,0,0,0}	Criterion(2,1,2)	{1,0,0,0,0}	Criterion(3,1,2)	{0,0,0,0,1}
Criterion(1,1,3)	{1,0,0,0,0}	Criterion(2,1,3)	{1,0,0,0,0}	Criterion(3,1,3)	{0,0,0,0,1}
		Criterion(2,1,4)	{0,0,0,0,1}	Criterion(3,1,4)	{0,0,0,0,1}
		Criterion(2,1,5)	{0,0,0,0,1}	Criterion(3,1,5)	{1,0,0,0,0}
		Criterion(2,1,6)	{0,0,0.359, 0.641,0}	Criterion(3,1,6)	{0,0.829, 0.171,0}
		Criterion(2,1,7)	{0,0,0,0,1}	Criterion(3,1,7)	{0,0,0,0,1}
		Criterion(2,1,8)	{0,0,0,0,1}	Criterion(3,1,8)	{0,0,0,0,1}

Table 11.6 Overall assessment and utility of voices V_1 and V_2

Voice	Equal weight from surveys		Weights from our proposed approach	
	Assessment on common scale	Utility using linear utility function	Assessment on common scale	Utility using linear utility function
V_1	{0.396, 0.002, 0.037, 0.093, 0.472}	0.44	{0.462, 0.002, 0.030, 0.117, 0.389}	0.508
V_2	{0.409, 0.077, 0.035, 0.025, 0.455}	0.49	{0.478,0.106, 0.028, 0.021, 0.366}	0.577

rank order of the voices is influenced by the survey weights and it is very important to arrive at the correct survey weights while prioritizing voices of customer. The framework we discussed in this paper would provide the decision maker a very good guidance value for the survey weights.

11.5 Summary

In this paper, we have formulated the problem of prioritizing voice of customer as a multiple criteria decision making problem. The surveys conducted for gathering data on voice of customer are used as criteria in the proposed formulation. Our experience has indicated that it is not trivial for the analysts to assign weights for the surveys as used in the formulation. Hence, in this work we proposed a statistical framework for estimating the weights of the surveys.

We have identified four key factors that influence the weight of the surveys. These factors are not exhaustive and more factors could be added depending on the application domain. We proposed to model the weight of a survey as inversely proportional to the cumulative error/bias induced into survey data from the influencing factors. We have proposed some approaches to estimate the error in survey data from individual factors. We have applied our approaches for calculating survey weights in a simple example.

To the best of our knowledge, this is first attempt in which a statistical framework is proposed for obtaining the weights of surveys that are in turn used in a multiple criteria decision making problem. In this paper, we concentrated on highlighting the need for such work and provided a framework without formal validation studies. In future, we will work to make the proposed framework more rigorous and conduct more validation studies.

Acknowledgements The authors would like to sincerely thank George Phillips and Srinivasan Rajagopalan of GM for all the support.

References

Chan LK, Wu ML (2002) Quality function deployment: A literature review. European Journal of Operational Research 143(3): 463–497.
Chan LK, Wu ML (2005) A systematic approach to quality function deployment with a full illustrative example. Omega 33(2): 119–139.
Chan LK, Kao HP, Wu ML (1999) Rating the importance of customer needs in quality function deployment by fuzzy and entropy methods. Intl J of Production Research 37(11): 2499–2518.
Chin KS, Yang JB, Lam L, Guo M (2007) An evidential reasoning-interval based method for new product design assessment. IEEE transactions on Engineering Management (in press).
Churchill GA, Iacobucci D (2005) Marketing research: methodological foundations. Thomson, South-western.
Hirsh S (2007) http://iresearch.wordpress.com/2007/05/11/understanding-cultural-differences-in-the-use-of-mobile-phones/.
Kwong CK, Bai H (2002) A fuzzy AHP approach to the determination of importance weights of customer requirements in quality function deployment. J of Intelligent Manufacturing 13(5): 367–377.
Mackenzie SB, Lutz RJ (1989) An empirical examination of the structural antecedents of attitude towards the ad in an advertising pretesting context. J Mark 53: 48–65.
Newell SJ, Goldsmith RE (2001) The development of a scale to measure perceived corporate credibility. J of Business Research 52: 235–247.
Ohanian R (1990) Construction and validation of a scale to measure celebrity endorsers' perceived expertise, trustworthiness, and attractiveness. J Advertising (19): 39–52.
Olson DL (1996) Decision aid for selection problems. Springer, New York.
Raghunathan TE, Xie D, Schenker N, Parsons VL, Davis WW, Dodd KW, Feuer EJ (2007) Combining information from two surveys to estimate county-level prevalence rates of cancer risk factors and screening. J of American Statistical Association 102(478): 474–486.
Saaty TL (1994) Fundamentals of decision making and priority theory with the Analytic Hierarchy Process. RWS Publications, Pittsburgh.
Xie X, Yang JB, Xu DL, Maddulapalli AK (2008) An investigation into multiple criteria vehicle evaluation under uncertainty. (under review).
Yang JB (2001) Rule and utility based evidential reasoning approach for multiple attribute decision analysis under uncertainty. European Journal of Operational Research 131(1): 31–61.
Yang JB, Singh MG (1994) An evidential reasoning approach for multiple attribute decision making with uncertainty. IEEE Transactions on Systems, Man, and Cybernetics 24(1): 1–18.
Yang JB, Xu DL (1998) Knowledge based executive car evaluation using the evidential reasoning approach. Advances in Manufacturing Technology XII (Baines, Taleb-Bendiab and Zhao eds), Professional Engineering Publishing, London.
Yang JB, Xu DL (2005) The IDS multi-criteria assessor software. Intelligent Decision System Ltd, Cheshire.
Yang JB, Xu DL, Xie X, Maddulapalli AK (2008) Evidence driven decision modelling and reasoning based decision support for prioritising voices of customer. (under review).

Chapter 12
Text Mining of Internet Content: The Bridge Connecting Product Research with Customers in the Digital Era

S. Shivashankar, B. Ravindran, and N.R. Srinivasa Raghavan

Abstract Primary and secondary market research usually deal with analysis of available data on existing products and customers' preferences for features in possible new products. This analysis helps a manufacturer to identify nuggets of opportunity in defining and positioning of new products in global markets. Considering the fact that the number of Internet users and quantum of textual data available on the Internet are increasing exponentially, we can say that Internet is probably the largest data repository that manufacturer's cannot ignore, in order to better understand customers' opinions about products. This emphasizes the importance of web mining to locate and process relevant information from billions of documents available online. Its nature of being unstructured and dynamic, an online document adds more challenges to web mining. This paper focuses on application of web content analysis, a type of web mining in business intelligence for product review. We provide an overview of techniques used to solve the problem and challenges involved in the same.

Keywords Market research · Opinion mining · Sentiment analysis · Buzz analysis · Web content mining · Natural language processing

12.1 Introduction

The role of information gathering and analysis as regards customers' needs, both expressed and latent, needs no exaggeration, especially during the early stages of new product development. It is in every innovative firm's interest that it lays its

---

S. Shivashankar and B. Ravindran (✉)
Department of Computer Science and Engineering, Indian Institute of Technology, Chennai-600036, India
e-mail: shivas@cse.iitm.ac.in,ravi@cse.iitm.ac.in

N.R.S. Raghavan
General Motors, R&D, India Science Lab, Bangalore-560066, India
e-mail: srinivasa.raghavan@gm.com

N.R.S. Raghavan and J.A. Cafeo (eds.), *Product Research: The Art and Science Behind Successful Product Launches*, DOI 10.1007/978-90-481-2860-0_12,

foundations for product development strong, by spending its resources in this important endeavor. Some of the typical tasks in such information gathering are usually in the purview of the market research function of an organization. Conducting consumer surveys, both for identifying unmet needs, as well as feedback on existing products of the company, are vital for the product development function. It is the inputs from such studies that usually shape the nature and portfolio of new products that the company can prioritize given limited budget for innovation. Take the case of an automotive company. Some of the popular studies include the surveys done by J.D. Power and Consumer Reports. Many prospective automotive customers resort to the opinion that is held out by such reports. It may be noted that survey reports usually do a numerical aggregation using statistical methods, on the various attributes of the product in question. On the other hand, there are other types of online surveys that are today typically packaged along with recommender engines. Such surveys seek the preferences of customers as regards to new features that could be offered in new products. To illustrate the significance, it may be noted that decisions regarding, say, new interior features like DVD players, audio control on steering wheel, roominess inside a vehicle, trunk capacity, drive type to be offered, etc., are prioritized based on such online survey data.

It may be noted that most surveys tend to capture views of customers in a stereotypical format, e.g., 'What is your rating for your overall experience regarding the interiors of vehicle ABC', 'What is your importance for fuel economy in the vehicle type you selected', etc. In today's digital world, with social networks becoming increasingly popular on the Internet, it is important to consider the impact of views and opinions of existing customers and experts in the industry, on the purchasing consideration of new customers. Put differently, there are hidden treasures in terms of unmet needs that are often ignored and unobtainable through traditional market surveys. Such information is usually contained in blogs and online product reviews by experts. Given the thousands of such sources available, and their dynamic nature, it becomes a tedious task to summarize such Internet content to make sense for product definition. Thus, it becomes extremely important to be able to automatically analyze the online content and make inferences as appropriate. The purpose of this paper is to appraise the reader on possibilities in that direction.

Web mining is an application of data mining that aids in discovering patterns from the web. According to the targets analyzed, web mining can be divided into three different types. They are web usage mining, web content mining and web structure mining (Cooley et al. 1997; Madria et al. 1999). This paper focuses on application of web content analysis/text mining for product review, which aids in identifying the popularity of a product, extract unbiased feedback from blogs and other Internet sources. The techniques that can be used are simple buzz analysis and opinion mining which involves sentiment classification of text. In market research it fits into strategic intelligence that deals with searching for opinions in the Internet (http://en.wikipedia.org/). The pattern/knowledge gained after processing the information can enrich customer satisfaction research and product positioning research. Section 12.2 provides an overview of types of web mining. Section 12.3 deals with text mining techniques for product review. We conclude this paper in Section 12.4.

12.2 Overview of Web Mining Types

As mentioned in introduction, depending on targets analyzed and patterns identified web mining can be classified into three types – web usage mining, web content mining and web structure mining (Cooley et al. 1997; Madria et al. 1999; Han and Kamber 2000; http://en.wikipedia.org/). Web usage mining is a data mining technique primarily used for business intelligence of a company. If a click is made on any part of the web page it gets recorded in access logs of the company servers. Discovery of such navigation patterns could serve as an effective knowledge base. This information can be used to develop marketing strategies. By accessing the server logs we could get information on the user tastes and interests. The effectiveness of a web campaign could also be estimated by knowing how many customers click on the area of the web page. Identifying the navigation patterns could also help the company to reorganize the web page. Many tools have been developed for web usage mining and they could be broadly classified as pattern discovery tools and pattern analysis tools. Typical sources of data are automatically generated data stored in server access logs, referrer logs, agent logs and client-side cookies, user profiles and metadata.

The World Wide Web (WWW) is a network of innumerable web sites linked together. Web structure mining tries to unravel web documents and retrieves hyperlinks present in the web documents. A web graph could be drawn to show the relationship of web pages. The nodes of the graph would contain the web pages and the links would indicate the relationship. This helps in categorizing and establishing relationship between the web pages. Similar web pages could be clustered based on structural similarities. When relationships are established between the web pages, indexing of these web pages become easier for the web crawlers of the search engine. Web structure is a useful source for extracting information such as finding quality of web page, analyzing interesting web structure, pages to crawl, mirror sites detection etc. Google's PageRank, HITS algorithm are a few famous examples.

Web content mining is the process of extracting useful information from the contents of web documents. It is also called as text mining. Few typical applications of text mining include text categorization, text clustering, concept mining, opinion mining and document summarization. Research activities in this field also involve using techniques from other disciplines such as Information Retrieval (IR) and Natural Language Processing (NLP).

12.2.1 Information Retrieval

IR deals with querying of unstructured text data. It differs from DBMS in the following ways

1. IR deals with unstructured text whereas DBMS queries with structured data.
2. It allows querying with partial keyword.

3. It does not deal with transaction and concurrency control, update techniques which is important in DBMS.

There are billions of documents on the web, so it is necessary to locate and return documents that are relevant to the domain of interest. As the number of documents returned could be in thousands it is necessary to rank them according to the relevance. Each domain must have an ontology that has concept hierarchy with relevant terms to be looked upon. Once the ontology is created, the problem of identifying relevant documents, narrows down to finding relevance of terms within the ontology, in the input documents.

TF-IDF: One simple way to find the relevance of a term ‘t’ to a document ‘d’ is using frequency of the term occurring in the document. This leads to following problems

1. If the length of the document is more, the frequency might be high. This does not mean that the document is more relevant.
2. We cannot say that if a term appears 10 times in a document, it is 10 times more relevant than the document where it appears once.

So, it makes it necessary to consider the relative frequency of the term in document. Hence a formula as shown below can help:

$$\mathrm{TF}\,(\mathrm{d},\mathrm{t}) = \log\left(1 + \mathrm{n}\,(\mathrm{d},\mathrm{t})\,/\mathrm{n}\,(\mathrm{d})\right) \tag{12.1}$$

Where, n (d, t) – number of occurrences of the term ‘t’ in document ‘d’

n (d) – number of terms in the document ‘d’.

Concepts in a hierarchy can contain multiple keywords, so the relevance measure will be estimated by combining the relevance measures of each keyword. A simple way is to add the relevance measures, but the problem here is not all terms used as keywords are equal. In fact certain terms have little or no discriminating power in determining relevance. For instance, a collection of documents on the auto industry is likely to have the term auto or vehicle in almost every document. It needs a mechanism for attenuating the effect of terms that occur too often in the collection to be meaningful for relevance determination. An approach here is to scale down the term weights of terms with high collection frequency, defined to be the total number of occurrences of a term in the collection.

$$\mathrm{IDF}\,(\mathrm{t}) = 1/\mathrm{n}\,(\mathrm{t}) \tag{12.2}$$

n (t) – denotes number of documents that contain the term ‘t’.

One then computes the TF-IDF (term frequency-inverse document frequency) value for all terms, which is the product of IDF(t) and TF(t). The above approach is called the TF-IDF approach. It can be used to identify the relevance of a document to the domain of the product analyzed. For instance it can be used to identify whether the blog post (or any other Internet content) talks about Chevrolet brand or not. If the document is identified to be talking about Chevrolet (or any topic of interest), it can be then be used for further processing. In a way it is a simple statistical technique to locate relevant information.

12.2.2 Natural Language Processing

The ultimate goal of research on Natural Language Processing (NLP) is to parse and understand language. Before processing through the entire text, certain pre-processing methodologies are used to extract only the selective portions of the sentence – identifying entities spoken about, extract words that relate to the feature of the product or words that describe attributes of those features, and extracting text that convey the opinion about the entity in a sentence. Named Entity Recognition (NER), tagging and chunking are common pre-processing methodologies used in web content analysis after the text is extracted from the html document (Manning and Schutze 1999). NER is a popular information extraction technique that relates to identifying certain entities from unstructured text. For, e.g., a simple news NER for English would recognize Barack Obama as a person and United States as a location in the text "Barack Obama takes the oath of office as the 44th president of the United States". In tagging, the words of a sentence are tagged with their respective part-of-speech (POS). POS tagging focuses on semantic understanding of the text. Instead of parsing through the entire text, tagging helps us to partially parse it. Word sense disambiguation could also arise in tagging. For example consider the sentence "The bully used to cow the students". The word cow in the sentence is a verb, but for most cases it is considered the bovine animal. Though there are possibilities of such ambiguity to arrive, current POS tagging techniques have an accuracy of about 97%. Part of speech tagged data is used for deriving higher order structure between words. Once the text is tagged we could extract the noun groups, adjectives and the verb groups by means of chunking. These phrases represent relationships between entities and can be used for feature extraction and opinion phrase extraction.

12.3 Product Review

The following section deals with product analysis (based on what customers talk about on the Internet) in two parts – buzz analysis and opinion mining. Buzz gives us a quantitative measure of what is popular while opinion mining allows us to find what is liked and not liked.

12.3.1 Buzz Analysis

What is hot and what is not.

According to Merriam-Webster buzz is speculative or excited talk or attention relating especially to a new or forthcoming product or event; an instance of such talk or attention.

Buzz is the somewhat indescribable and often evanescent quantity that measures how much a particular event or product has impinged on the social consciousness. Buzz has always been a valued commodity, and 'buzz' marketing techniques have been adopted frequently by companies while launching new products. With an increasing number of people, especially in the valued 18–30 demographic, turning to the Internet as their primary source of information and opinions, online buzz analytics is becoming a valued marketing tool (http://www.blogpulse.com/; http://www.nielsen-online.com/products.jsp?section = pro_buzz ; Leskovec and Horvitz 2007;). The primary approach to measuring buzz about a particular entity (e.g., Malibu) in a given community (e.g., auto bloggers) is to estimate the baseline average frequency of mentions of the entity in the community. One would expect several automobile models to be talked about routinely in an online community of automotive enthusiasts. Hence a high frequency of mention alone is not a strong indicator of buzz. Only if the frequency increases by a significant amount above a base line, then the term is being 'buzzed' about.

There are several issues that need to be addressed here, some of which are common to all NLP tasks, some peculiar to analyzing blogs, and some unique to buzz analytics.

12.3.1.1 Named Entity Recognition

Named entity recognition (NER) is the process of finding mentions of predefined things in input text. The implication for buzz analysis is, to identify what is the entity that is being buzzed about – is it identical to the entity of interest to us? For instance to analyze the opinion of customers about Chevrolet Tahoe, it is necessary to identify the model spoken about in the input text. The following are the types of learning while training NER (Nadeau and Sekine 2007).

1. Supervised: The idea of supervised learning is to study the features of positive and negative examples of named entities over a large collection of annotated documents and design rules that capture instances of a given type. Supervised learning techniques include Hidden Markov Models (HMM), Decision Trees, Maximum Entropy Models, Support Vector Machines and Conditional Random Fields.
2. Semi-supervised Learning: The term semi-supervised is relatively recent. The main technique for semi supervised learning is called bootstrapping and involves a small degree of supervision, such as a set of seeds, for starting the learning process. For example, a system aimed at identifying car models might ask the user to provide a small number of example names. Then the system searches for sentences that contain these names and tries to identify some contextual clues common to those examples. Then, the system tries to find other instances of car models that appear in similar contexts. The learning process is then re-applied to

the newly found examples, so as to discover new relevant contexts. By repeating this process, a large number of car models and a large number of contexts will eventually be gathered.
3. Unsupervised: Basically, the techniques rely on lexical resources (e.g., WordNet, POS Tagger), on lexical patterns and on statistics computed on a large unannotated corpus. For example, one can try to gather named entities from clustered groups based on the similarity of context. Noun/Noun phrases are commonly used patterns for identifying named entities in text.

NER has reached a level of maturity that standard open source tools are available (http://nlp.stanford.edu/ner/index.shtml; http://alias-i.com/lingpipe/).

12.3.1.2 Establishing a Baseline

One of the important issues to be addressed includes what constitutes a baseline buzz for a particular entity. In automotive websites in general, the baseline frequency of mentions might be low for, say, Saturn Vue, but in a site devoted exclusively for GM vehicle owners, it might be very high. But this should not be counted as new buzz about the model in such a site. One way to address this issue is to estimate the baseline by observing each and every site for a period of time, but this leads to several issues like how long the observation period should be? Can you boot strap off earlier learning, etc.

12.3.1.3 Cleaning the Data

While noisy text – ungrammatical, badly formatted, slangy is something all approaches dealing with online sources have to deal with, one of the main problems with blog mining is the enormous number of mirror sites that are present in the blogosphere. Some very popular bloggers columns are duplicated verbatim by many sites, hence inflating the buzz. It could be argued that this does constitute additional buzz, but if this is an exact duplicate of the original posting, it should not be considered as a new mention.

12.3.1.4 Weighing the Opinions

One way of handling the problem of multiple copies is to use some methods similar to Page ranking schemes (Brin and Page 1998), with mentions by highly popular bloggers generating higher buzz counts. Popularity could encompass, number of page views, number of comments, number of duplicate feeds, frequency of updates, etc. Such ranking schemes will also mitigate attempts by users to artificially increase buzz by a large number of mentions in a closed set of pages.

12.3.2 Opinion Mining

There are two types of information available over Internet – Facts and opinion. Search engines retrieve facts as they are represented using keywords. But they do not search for opinions.

Wikipedia's definition of opinion is it is a person's ideas and thoughts towards something. It is an assessment, judgment or evaluation of something. An opinion is not a fact, because opinions are either not falsifiable, or the opinion has not been proven or verified. If it later becomes proven or verified, it is no longer an opinion, but a fact.

It deals with the question "What is liked and what is not". While buzz is more likely to surround a new entity (design, model, etc.), blogs are also rich source of detailed information about entities already in existence, as well as the newer ones. While buzz is pretty much transient, sentiments about entities, whether they are liked or not, build up over time, and determine the success or failure of a product. Analysis of blogs can allow us to gauge the sentiment about a product prevailing in a particular community or the blogosphere at large. While it makes sense to look at a wider audience in case of popular topics like movies, when we want more focused information about certain products, analyzing more knowledgeable communities, like automotive blogs, is more preferable.

The advent of the Internet has seen everyone become a publisher. Blogs are seen as a means of influencing public opinion, with everyone from Google, to Political Parties, to Scott Adams, to Joe, using this to push their opinions and agenda. Often bloggers are seen as trendsetters in the online community. Several manufacturers are now looking at automated mechanisms to analyze the information available in blogs. This is called as opinion mining (Liu 2006). It is a recent discipline at the crossroads of information retrieval, text mining and computational linguistics which tries to detect the opinions expressed in the natural language texts. These opinions have many applications such as business intelligence. Companies can use it as feedback to improve business, customer satisfaction research, product positioning research, ad placements. Prospective customers also could use it to find the right product.

If the product review articles have pros and cons part separately then it is not necessary to do a sentiment analysis on text: it is enough to identify the entities spoken about. But most of the blog reviews have free format, where the bloggers type their detailed reviews. Here positive and negative statements can be in same paragraph and even a single sentence can reflect both positive and negative sentiments. So this makes it necessary to adapt techniques for opinion mining relative to the complexity of the text.

Opinion mining can be done by using the following techniques:

(i) Sentiment analysis of the document (document level)
(ii) Feature based opinion extraction and summarization
(iii) Comparative sentence and relation extraction

Sentiment classification at document level is useful, but it does not find what the reviewer liked and disliked. A negative sentiment on an object does not mean that the reviewer dislikes everything. A positive sentiment on an object does not mean that the reviewer likes everything. It is similar to topic-based text classification where sentiment words replace topic words.

Receiving more attention lately is a more 'fine-grained' sentiment analysis. So the question is no longer 'is the sentiment in this page favorable about a product or not', but which aspects of the product have generated what kind of sentiment and to what degree. Thus one could analyze a page discussing a new model camera to find out that the optics is very good, the controls are ok, the size is very bad (too bulky), and the over all sentiment is mildly positive. Such a fine-grained analysis would necessarily mean that several of the earlier issues discussed need proper treatment. As any automated system working with Internet content, it is also important to have the ability to handle noisy text. Apart from this it would require sophisticated NER, tagging and chunking to identify and map sentences to features. As mentioned above even a single sentence can reflect both positive and negative sentiments, for e.g., "Camera has very good clarity but very less battery stand time". So this makes it necessary to do opinion phrase extraction that will associate the feature term with the corresponding opinion term mentioned in the sentence. It should look at finer nuances of language to determine which aspects of the product a particular adjective attaches to, etc.

Also there is another issue to be handled for domain specific implementation, for instance the review statement "The basic layout of the Buick Enclave's drivetrain is very ambitious" talks about engine/transmission of the car (drivetrain is a domain specific term that must be captured). So it is necessary to bootstrap domain specific knowledge to the system apart from general lexicon used. Both implicit and explicit features must be identified, and sentiment orientation of the feature must be estimated. Sentiment orientation depends on the opinion word associated to the feature word (Hu and Liu 2004).

Opinions can either be direct or comparisons. If it is a comparative opinion then we should go for the third type. Steps that are generally followed would be – identify comparative sentences, classify them into non-equal gradable, equative, and superlative and then extract the relations (Jindal and Liu 2006a, b).

There are two main approaches to sentiment analysis from blogs: a purely statistical approach, and an approach that is driven by natural language processing ideas. The simplest statistical approach looks for repetitions of words representing positive or negative sentiments in a given piece of text, and assigns a score based on relative frequency. More complicated approaches look to build training data sets of documents representing positive and negative sentiments. They then train classifiers that can identify a new document as belonging to one or the other class. While these methods do go beyond simple counting, they can be mislead by modifiers ('I do not like this product'), compound phrases ('I like this product, not.'), and sarcasm. NLP inspired methods take a more traditional approach of understanding the structure of the text, and then determining the sentiment espoused there in. The main problem with adopting this approach is the poor quality of the text, which is not amenable to syntactic analysis that is usually available from online sources. One compromise solution is to enrich statistical methods with more semantic information, such as word net (http://wordnet.princeton.edu/). Sentiment classification of text could depend on a set of words called the appraisal group. Appraisal groups are a set of words that indicate the emotional reactions of a person (the appraiser) about a particular object or event (appraised) based on Appraisal theory. Appraisal usually comprises many of the following components:

Attitudes: These are the linguistic resource by which the speaker expresses her judgment and emotional responses about the appraised.

- Affect: It refers to personal emotional state of the appraiser. For example, *I feel "happy" to have bought this pen, I feel "miserable" to have failed in my exams.* Here the words happy, miserable indicate the attitudes of the appraiser.
- Judgment: It refers to the social judgment of the other people by the appraiser. For example consider the statement – *He was so "fortunate" to inherit the wealth.* Here the appraiser makes a judgment that the other person would be fortunate.
- Appreciation: Appreciations refers to the words that describe the properties of the object. Words such as beautiful, thin would describe appreciation.

Graduation: Graduation words could be classified under two categories.

- Force: Words that indicate the intensity of the description. For example, *I liked the book "very" much.* The word "very" indicates the intensity of how much the appraiser likes the object in picture.
- Focus: Consider the sentence – *I have a "slight" inclination towards the color.* The word "slight" would determine the amount of focus the object gets from the appraiser.

Orientation: Orientation could be categorized as either positive or negative. This simply indicates the sentiments of the appraiser.

Polarity: Certain words in a sentence could entirely change the meaning of the sentence with its inclusion. An appraisal could be marked or unmarked considering the presence of these words. For example the word 'not' in the sentence *I do not like the book very much*, acts highly upon the words following it changing the entire meaning of the sentence. Such an appraisal is considered marked whose exclusion makes the sentence unmarked.

Appraisal theory was combined with bag of words approach to classify movie reviews (Whitelaw et al. 2005). The given sentiments could be explored for appraisals and each appraisal could be brought into a framework such as Appraiser: Author, Appraised: the object, Attitude: appreciation/reaction, Orientation: positive or negative.

Every appraisal group contains a head adjective that defines the attitude and additional options called the appraisal modifier. We could construct a lexicon that contains all the appraisal words which could be used to classify text. Every appraisal attribute is defined in the lexicon with an appraisal adjective. In order to build the lexicon we would have to consider various seed terms from various sources, such as word net (http://wordnet.princeton.edu/). Candidate lists are generated from the seed terms based upon the various categories. The candidate lists have to be manually inspected to produce the final set of appraisal words. The lexicon thus obtained may be used to label sentences accordingly.

12.4 Conclusions

Product research in the digital age is incomplete if manufacturers don't pay attention to the voice of customers that are often available from blogs on the Internet. In this paper, we provided an overview of how web content analysis can help in identifying popularity of the product, customers' feedback about various features of the product compared to the competitors, etc., that can in turn help the manufacturer in better positioning of the product global markets. With respect to the information processed from the Internet data, they can be loosely associated to a few types of market research – positioning research and customer satisfaction research. There are two dominant approaches, one based purely on statistics, and the other on NLP to build such a system. Both have pros and cons: depending on the requirement, automated systems can be built using methodologies from both the approaches. Also it has to be noted that the general dictionary like word net will not be enough to build domain specific opinion mining systems. Enough domain knowledge needs to be bootstrapped in such cases.

References

Brin, S. and Page, L, "The Anatomy of a Large-Scale Hypertextual Web Search Engine", 1998.

Cooley, R. Mobasher, B. and Srivastave, J., "Web Mining: Information and Pattern Discovery on the World Wide Web", In Proceedings of the 9th IEEE International Conference on Tool with Artificial Intelligence, 1997.

Jiawei Han and Micheline Kamber, "Data Mining: Concepts and Techniques", The Morgan Kaufmann Series in Data Management Systems, 2000.

http://www.blogpulse.com/

http://en.wikipedia.org/

http://wordnet.princeton.edu/

http://www.nielsen-online.com/products.jsp?section = pro_buzz

http://nlp.stanford.edu/ner/index.shtml

http://alias-i.com/lingpipe/

Minqing Hu and Bing Liu, "Mining and summarizing customer reviews", Proceedings of the ACM SIGKDD International Conference on Knowledge Discovery & Data Mining (KDD-2004, full paper), Seattle, Washington, USA, Aug 22–25, 2004.

Nitin Jindal and Bing Liu, "Identifying Comparative Sentences in Text Documents", Proceedings of the 29th Annual International ACM SIGIR Conference on Research & Development on Information Retrieval (SIGIR-06), Seattle 2006a.

Nitin Jindal and Bing Liu, "Mining Comprative Sentences and Relations", Proceedings of 21st National Conference on Artificial Intelligence (AAAI-2006), July 16.20, 2006b, Boston, Massachusetts, USA.

Jure Leskovec and Eric Horvitz, "Worldwide Buzz: Planetary-Scale Views on an Instant-Messaging Network", Microsoft Research Technical Report MSR-TR-2006–186, June 2007.

Bing Liu, "Web Data Mining Exploring Hyperlinks, Contents and Usage Data", Springer, December 2006.

S.K. Madria, S.S. Rhowmich, W.K. Ng, and F.P. Lim, "Research issues in Web data mining", In Proceedings of Data Warehousing and Knowledge Discovery, 1999.

Chris Manning and Hinrich Schutze, "Foundations of Statistical Natural Language Processing", MIT Press, Cambridge, MA: May 1999.

David Nadeau and Satoshi Sekine, "A survey of named entity recognition and classification", Journal of Linguisticae Investigationes, 2007.

Whitelaw, C., Garg, N., and Argamon, S, "Using appraisal groups for sentiment analysis", In Proceedings of the 14th ACM international Conference on information and Knowledge Management (Bremen, Germany, October 31 – November 05, 2005). CIKM '05. ACM, New York, NY, 625–631. DOI = http://doi.acm.org/10.1145/1099554.1099714.

Part IV
Quantitative Methods for Product Planning

Chapter 13
A Combined QFD and Fuzzy Integer Programming Framework to Determine Attribute Levels for Conjoint Study

Malay Bhattacharyya and Atanu Chaudhuri

Abstract In a recent paper, Chaudhuri and Bhattacharyya propose a methodology combing Quality Function Deployment (QFD) and Integer Programming framework to determine the attribute levels for a Conjoint Analysis (CA). The product planning decisions, however, are typically taken one to two years before the actual launch of the products. The design team needs some flexibility in improving the Technical Characteristics (TCs) based on minimum performance improvements in Customer Requirements (CRs) and the imposed budgetary constraints. Thus there is a need to treat the budget and the minimum performance improvements in CRs as flexible rather than rigid. In this paper, we represent them as fuzzy numbers instead of crisp numbers. Then a fuzzy integer programming (FIP) model is used to determine the appropriate TCs and hence the right attribute levels for a conjoint study. The proposed method is applied to a commercial vehicle design problem with hypothetical data.

Keywords Quality function deployment · Conjoint analysis · Fuzzy integer programming

13.1 Introduction

Wasserman (1993) considers cost of resources that go into QFD planning and proposes a linear decision model for attribute prioritization. Bode and Fung (1998) incorporate product design budget into QFD planning and put forward an improved prioritization approach to effectively allocate design resources to the more important TCs. Park and Kim (1998) present a 0–1 integer programming model for prioritizing TCs. They also incorporate a cost constraint and calculate customer satisfaction. But

M. Bhattacharyya (✉)
Indian Institute of Management, Bannerghatta Road, Bangalore 560076, India
e-mail: malayb@iimb.ernet.in

A. Chaudhuri
Deloitte Support Services India Pvt. Ltd., Hyderabad 500081, India
e-mail: atanuch1@yahoo.com

N.R.S. Raghavan and J.A. Cafeo (eds.), *Product Research: The Art and Science Behind Successful Product Launches*, DOI 10.1007/978-90-481-2860-0_13,

they measure customer satisfaction in terms of TCs that are addressed in the final product. Dawson and Askin (1999) suggest a non-linear programming model to determine optimum TCs considering constraints on costs and development time. They point out that dependence among TCs also needs to be considered. Fung et al. (2002) include financial issues in attaining individual targets of TCs. They represent the correlation between TCs as the incremental change in one TC to change in another by one unit. The costs of improving the degree of attainment of a TC are formulated as a non-linear function of its degree. They introduce the concepts of actual and planned attainment, and primary, actual and planned costs for the attainment of TCs. Franceschini and Rossetto (1998) present a method to determine the existence of dependence among TCs and formulate a 'set covering' problem to choose the minimum set of TCs to cover all CRs. They found that the set of TCs obtained by the traditional prioritization method is not necessarily the same as that obtained by their 'set covering' approach.

Karsak et al. (2003) and Chen and Weng (2006) use goal programming for product planning using QFD while Raharjo et al. (2006) use quality loss function and 0–1 goal programming to prioritize quality characteristics in a dynamic QFD. They also consider budgetary constraints and minimum customer satisfaction level in their model. Lai et al. (2007) use Kano's model and goal programming in optimizing product design for personal computers.

Vanegas and Labib (2001) incorporate constraints on time, cost and technical difficulty. They define fuzzy membership functions for each constraint and an aggregate 'type 2 fuzzy set' is calculated for each TC. The fuzzy set represents the desirability with respect to meeting customer satisfaction and the optimum value of TC is the one with the maximum degree of membership in the aggregate fuzzy set. Karsak (2004) presents a fuzzy multi-objective programming approach to determine the level of fulfilment of design requirements. The author incorporates the relationships between CRs and TCs, importance of customer needs, sales point data and technical difficulty of design requirements in his model by using linguistic variables. Uncertain cost data are represented by triangular fuzzy numbers. By using a multi-objective approach, the author is able to incorporate the objectives of maximizing the extendibility of design and minimizing the design difficulty, apart from the objective of maximizing the fulfilment of design requirements. Khoo and Ho (1996) provide a framework for fuzzy QFD. Kim et al. (2000), Chen et al. (2005), Chen and Weng (2006), Liu (2005), Fung et al. (2002), Kahraman et al. (2006) are the others who use different forms of fuzzy modeling to capture the fuzziness of parameters used in QFD.

In a recent paper, Chaudhuri and Bhattacharyya (2008) propose a methodology combing Quality Function Deployment (QFD) and Integer Programming framework to determine the attribute levels for a Conjoint Analysis (CA). The product planning decisions, however, are typically taken one to two years before the actual launch of the products. The design team needs some flexibility in improving the Technical Characteristics (TCs) based on minimum performance improvements in Customer Requirements (CRs) and the imposed budgetary constraints. Thus there is a need to treat the budget and the minimum performance improvements in CRs as flexible rather than rigid. In this paper, we represent them as fuzzy numbers instead of crisp numbers. Then a fuzzy integer programming (FIP) model is used to determine the appropriate TCs and hence the right attribute levels for a conjoint study. The proposed method is applied to a commercial vehicle design problem with hypothetical data.

13.2 Solving Fuzzy Integer Linear Programs

Fabian et al. (1984) study the integer programming problem with fuzzy constraints and showed how the problem can be transformed into an auxiliary ILP problem with linear constraints, a modified objective function and some supplementary constraints and variables. Herrera and Verdegay (1995) provides three models of fuzzy integer linear programming with fuzzy constraints, fuzzy numbers in the objective function and fuzzy numbers defining the set of constraints. The authors show how a fuzzy integer linear programming (FILP) problems can be transformed into a parametric ILP using the representation theorem of fuzzy sets. Solving a parametric integer program (PIP) requires finding an optimal solution for every programme in the family.

Marsten and Morin (1976) shows how a generalization of the branch and bound approach for solving integer programming problems can be used to solve a parametric integer programme.

Bailey and Gillett's (1980) develop a contraction approach for solving families of integer linear programming problems in which the problems have identical objective coefficients and constraint matrix coefficients but the right hand side of the constraints have the form $b+\theta d$ where θ varies from 0 to 1. The authors provide two contracting algorithms one for the problem where $d \geq 1$ and in the other problem where d can take negative values.

13.3 Converting a Fuzzy Integer Linear Programming (FILP) Problem to Parametric Integer Linear Programming (PILP) Problem

Herrera and Verdegay (1995) consider the following ILP with fuzzy numbers defining the set of constraints:

$$\max\ z = cx = \sum_{j=1}^{n} c_j x_j$$

$$s.t. \sum_{j \in N} \tilde{a}_{ij} x_j \lesssim \tilde{b}_i,\ i \in M = \{1, \ldots, m\}. \tag{13.1}$$

$$x_j \geq 0, j \in N = \{1, \ldots, n\}, xj \in \mathbf{N}, j \in N.$$
$$\tilde{a}_{ij}, \tilde{b}_i j \in F(\mathbb{R})$$

where $\mathbb{N}$ is the set of integer numbers, $c \in \mathbb{R}^n$. $F(\mathbb{R})$ is the set of real fuzzy numbers. The symbol $\lesssim$ means that the decision-maker permits certain flexible accomplishment for the constraints. Thus the following membership functions are considered:

For each row (constraint),

$$\exists \mu_i \in F(\mathbb{R}) \; such\ that\, \mu_i : F(\mathbb{R}) \rightarrow [0, 1], i \in M$$

which defines the fuzzy number on the right-hand side.

For each $i \in M$ and $j \in N$,

$$\exists \mu_{ij} \in F(\mathbb{R}) \; such\ that\, \mu_{ij} : F(\mathbb{R}) \rightarrow [0, 1],$$

defining the fuzzy numbers in the technological matrix.

For each row (constraint),

$$\exists \mu^i \in F(\mathbb{R}) \; such\ that\, \mu^i : F(\mathbb{R}) \rightarrow [0, 1], i \in M$$

giving, for every $x \in \mathbb{R}^n$, the accomplishment degree of the fuzzy number, that is the adequacy between the left-hand side and the corresponding right-hand side for each constraint. They show that the above problem can be reduced to a parametric integer linear programming problem.

Let $\tilde{t}_i$ be a fuzzy number, fixed by the decision maker, giving his allowed maximum violation in the accomplishment of the i-th constraint. Then one can propose the following auxiliary problem to solve the above problem (13.1)

$$\max\ z = cx = \sum_{j=1}^{n} c_j x_j$$
$$s.t. \sum{}_{j \in N} \tilde{a}_{ij} x_j \preceq \tilde{b}_i + \tilde{t}_i\,(1-\alpha)\,, i \in M = \{1, \ldots, m\}, \tag{13.2}$$
$$x_j \geq 0, j \in N = \{1, \ldots, n\}, x_j \in \mathbf{N}, j \in N, \alpha \in (0, 1]$$

where $\preceq$ represents a relation between fuzzy numbers. Treating this relation as a ranking function and using the first index of Yager and assuming that the fuzzy numbers are LR type, the auxiliary problem can be reduced to a parametric integer linear programming problem of the following form

$$\max\ z = cx = \sum_{j=1}^{n} c_j x_j$$

$$s.t. \sum_{j \in N} a_{ij}^* x_j \leq b_i^* + t_i^* (1 - \alpha), i \in M = \{1, \ldots, m\}, \quad (13.3)$$
$$x_j \geq 0, j \in N = \{1, \ldots, n\}, x_j \in \mathbf{N}, j \in N, \alpha \in (0, 1]$$

where a_{ij}^*, b^*i, and t_i^* are obtained from $\tilde{a}_{ij}$, $\tilde{b}_i$, and $\tilde{t}_i$ respectively.

13.4 A Contraction Algorithm for Solving a PILP (Bailey and Gillett 1980)

For simplicity let us re-write (13.3) as below:

$$\max z = cx = \sum_{j=1}^{n} c_j x_j$$
$$s.t. \sum_{j \in N} a_{ij} x_j \leq b_i + \theta t_i, i \in M = \{1, \ldots, m\}, \quad (13.4)$$
$$x_j \geq 0, j \in N = \{1, \ldots, n\}, x_j \in N, j \in \mathbf{N}, \theta \in (0, 1]$$

In fact Bailey and Gillett's (1980) consider a problem with $\theta \in [0, 1]$. For the sake of completeness, we reproduce the Bailey and Gillett algorithm below. This algorithm assumes that the elements $t_i \geq 0$.

Step 1 With $\theta = 1$ set up the initial tableau using the lex-dual column simplex scheme. Let $\hat{\theta} = 1$.

Step 2 Solve the ILP with the current value of θ using Gomory's cutting plane algorithm.

Step 3 Determine which constraints, if any, are "active" constraints by substituting the optimal integer solution into each one.
Let I be the set of active constraints. If I is non-empty set $\theta^* = \hat{\theta}$ and go to step 4. Otherwise, determine the θ_i necessary to make constraint "I" active for i = 1,..., m. Let I be the set of constraints, k, such that $\theta_k = \max\{\theta_i\}$, and let $\theta^* = \max\{0, \theta_k\}$, $1 \leq i \leq m$.

Step 4 Record θ^*, the current optimal integer solution and the corresponding objective function value. This solution will be optimal on the interval $[\theta^*, \hat{\theta} + \in)$ if $\hat{\theta} < 1$ and on $[\theta^*, \hat{\theta}]$ if $\hat{\theta} = 1$. If ($\theta^* < \in$) go to step 9.

Step 5 Let $\hat{\theta} = \theta^* - \in$. Attach to the tableau the constraint(s)

$$\sum_{j \in N} a_{kj} x_j \leq b_k + \hat{\theta} t_k, k \in I.$$

Step 6 Perform one iteration and discard the newly added row(s).

Step 7 Use the lex-dual column simples to find the new current optimal integer solution.

Step 8 Test the new integer solution in each of the constraints

$$\sum_{j \in N} a_{ij} x_j \leq b_i + \hat{\theta} t_i, i = 1, \ldots, m.$$

If the solution is feasible in each constraint go to step 3. If the solution is infeasible in any constraint, say q, attach it to the tableau as a cut and go to step 6.

Step 9 Terminate.

13.5 The Model Description

In this subsection we present the mathematical formulation for determining the attribute levels of the selected features using the information in the HoQ of the QFD approach. Since our task is to determine the percentage changes in TCs from a set of finite alternatives and to choose specific features that need to be introduced (or not introduced) given budgetary and performance constraints with fuzzy RHS, integer programming (FILP) is used to solve the problem. To explain the model, we introduce some notations below.

Let us define the following:

i – attribute number, where $i = 1, \ldots, p$, j – TC number, where $j = 1, \ldots, q$, k_j – percentage improvement in TC$_j$, $j = 1, \ldots, q$, m – feature number (*present/absent* type feature), where $m = 1, \ldots, r$

R_{ijk_j} – percentage change in attribute i due to k_j percent improvement in TC$_j$

$R1_{im}$ – percentage change in attribute i due to introduction of feature m in the product

C_{jk_j} – cost of improving TC$_j$ by k_j percent

$C1_m$ – cost of providing feature m in the product

w_i – importance score of attribute i

B – the budget

$Y_{jk_j} = 1$, if TC$_j$ is improved by k_j percent

$= 0$, otherwise

$X_m = 1$, if feature m is introduced

$= 0$, otherwise

Then the model is

$$\text{Maximize F} = \sum_i \sum_j \sum_{k_j} \left(w_i R_{ijk_j}\right) Y_{jk_j} + \sum_i \sum_m \left(w_i R1_{im}\right) X_m \tag{13.5}$$

such that

$$\sum_{j}\sum_{k_j} Y_{jk_j} C_{jk_j} + \sum_{m} X_m C1_m \leq B \tag{13.6}$$

where B varies with ⊘ as 125 + 25⊘

$$\sum_{k_j} Y_{jk_j} \leq 1, \text{ for each } j, \text{ where} j = 1, \ldots, q \tag{13.7}$$

$$\sum_{j}\sum_{k_j} Y_{jk_j} R_{ijk_j} + \sum_{m} X_m R1_{im} \geq 0, \text{ for each } i,$$

$$\text{where} \, i = 1, \ldots, p \tag{13.8}$$

$$Y_{jk_j} \in \{0, 1\}, X_m \in \{0, 1\} \tag{13.9}$$

The objective function F in (1) depicts the total weighted change in the overall product for various changes in the TCs. The constraint (2) states that the total cost involved in changing the TCs by certain percentages and providing certain features, if any, should not exceed the budget B. Constraint (3) requires that TC_j can be improved by only one of the possible percentages k_j. Constraint (4) requires that TCs and features should be chosen such that improvement in each attribute i is greater than or equal to zero.

Furthermore, constraints can be added to ensure that minimum improvements are achieved in some or all attributes. The additional constraint will be

$$\sum_{j}\sum_{k_j} Y_{jk_j} R_{ijk_j} + \sum_{m} X_m R1_{im} \geq P_i, \text{for each } i \text{ where} \, i = 1, \ldots, p \tag{13.10}$$

where P_i denotes the minimum improvement threshold for attribute i.

Now by varying the budgetary and other limits as per the functions shown above, we get different solutions and hence different sets of attribute levels.

We can also apply the procedure for different segments to generate more product profiles. For example, one customer segment may have more preference for attributes 1 and 2, while another customer segment may have more preference for attributes 3, 4 and 5. In this way the entire range of attribute levels could be determined for generating product profiles to be used for conjoint analysis.

Herera and Verdegay (1995) have shown how a parametric solution of an FILP with fuzzy constraints can generate a fuzzy solution. We use Bailey and Gillett's (1980) contraction algorithm with positive coefficient of θ to generate the parametric solutions.

13.6 Application

For illustrating the proposed methodology, commercial vehicle is considered as an example product. In this problem, a product is already available in the market. The company wants to launch variants of the product, incorporating customer preferences. These products can also cater to different customer segments. The Voice of Customer (VoC) has been obtained to understand the CRs. The data used by Chaudhuri and Bhattacharyya (2008) are used in this paper. The relevant data are reproduced in Tables 13.1–13.4.

GAMS 21.0 with CPLEX solver is used to solve the family of problems shown in the FILP model given by (13.5)–(13.10). Bailey and Gillett's contraction algorithm for the case where d $\geq$ 1 has been used to obtain the optimal solutions for $0 \leq \theta \leq 1$.

We arbitrarily set triangular fuzzy numbers for the right hand sides of the constraints and then converted for ranking of fuzzy numbers

Budget: $(125, 150, 160) \rightarrow 150 + [(160–150) – (150–125)]/3 = 145$
Fuel Economy: $(4, 6, 7) \rightarrow 6 + [(7–6) – (6–4)]/3 = 5.666$
Payload: $(4, 5, 6) \rightarrow 5 + [(6–5) – (5–4)]/3 = 5.00$
Max cruising speed: $(3, 4, 7) \rightarrow 4 + [(7–4) – (4–3)]/3 = 4.6666$
Driver comfort: $(3, 5, 9) \rightarrow 5 + [(9–5) – (5–3)]/3 = 5.6666$

Similarly, we arbitrarily set triangular fuzzy numbers for the inequality relations of the constraints and then converted for ranking of fuzzy numbers

Budget constraint fuzzy relation: $(2, 3, 4) \rightarrow 3 + [(4–3) – (3–2)]/3 = 3$
Fuel Economy constraint fuzzy relation: $(2.0, 3, 3.5) \rightarrow 3 + [(3.5–3) – (3–2.0)]/3 = 2.834$
Payload constraint fuzzy relation: $(1, 3, 4) \rightarrow 3 + [(4–3) – (3–1)]/3 = 4.3333$

Table 13.1 Customer requirements and product attributes for commercial vehicles

Customer requirements (CRs)	Product attributes
Fuel efficiency	1. Fuel economy
Good load carrying capacity	2. Payload
Good pickup	3. Power to weight ratio
Easy drivability on highways	4. Max cruising speed
Easy to climb slopes and navigability in hilly terrain	5. Gradability
Comfortable driver's cabin	6. Driver's cabin

Table 13.2 Technical characteristics for commercial vehicles

Technical characteristics (TCs)	
Engine	1. Maximum torque
	2. Compression ratio
	3. Maximum engine RPM
	4. Turbo charger efficiency
	5. Combustion efficiency
Gear Box and axle	6. Maximum axle reaction
	7. Overdrive ratio
	8. Variable ratio power steering
Cabin	9. Maximum sleeper berth area
	10. NVH resistant body panel

Table 13.3 The QFD chart for the commercial vehicle design problem with proposed relationship matrix

Customer requirements	Attributes	Engine												Gear Box and axle						Cabin			Importance ratings	Current ratings	Current ratings of competitor	Target ratings	Improvement ratio	Scores	Percentage scores
		1. Maximum torque			2. Compression ratio			3. Max. engine RPM		4. Turbo charger efficiency		5. Combustion efficiency		6. Maximum axle reaction			7. Variable ratio PS	8. Overdrive ratio		9. Maximum sleeper berth		10. NVH resistant body							
		10%	15%	20%	3%	5%	7%	5%	10%	3%	5%	5%	10%	5%	10%	15%		2%	5%	2%	5%								
(1)	(2)	(3)	(4)	(5)	(6)	(7)	(8)	(9)	(10)	(11)	(12)	(13)	(14)	(15)	(16)	(17)	(18)	(19)	(20)	(21)	(22)	(23)	(24)	(25)	(26)	(27)	(28)	(29)	(30)
Fuel efficiency	1. Fuel economy									0.8	1.25	3.5	7.5									2	9	8.25	8.5	8.75	1.06	9.55	21.62
Good load carrying capacity	2. Payload													3.5	6	8							9	8	7.5	8.5	1.06	9.56	21.66
Good pickup	3. Power to weight ratio	6	8	12	1.5	2.25	3.5			0.2	0.25	2.5	4.5										7.5	7	7	7	1.00	7.50	16.99
Easy drivability on highways	4. Maximum cruising speed	−4	−6.5	−8				2	3.5									1.5	3.75				7	6.5	6.5	6.75	1.04	7.27	16.47
Easy to climb slopes and navigability in hilly terrain	5. Gradability	3	5		0.5							1	2							-0.5	-1.25	3	6	5	4.5	5	1.00	6.00	13.59
Comfortable driver's cabin	6. Driver's cabin																5			1	3	8	4	3.75	4	4	1.07	4.27	9.67
Weights		76.85	96.84	72.14	32.28	38.23	59.46	32.93	57.63	20.70	31.28	131.75	265.81	75.82	129.97	173.30	48.33	24.70	61.75	2.87	12.01	161.35	1606.00					44.14	100.00
Percentage weights		4.78	6.03	4.49	2.01	2.38	3.70	2.05	3.59	1.29	1.95	8.20	16.55	4.72	8.09	10.79	3.01	1.54	3.85	0.18	0.75	10.05	100.00						
Costs (in thousands of rupees)		30	40	75	45	60	75	20	30	12	20	40	50	30	40	75	15	15	25	15	30	75							

Table 13.4 Additional impacts on CRs due to correlations between TCs

Customer requirements	Attributes	Engine												Gear box and axle						Cabin		
		1. Maximum torque			2. Compression ratio			3. Max. engine RPM		4. Turbo charger efficiency		5. Combustion efficiency		6. Maximum axle reaction			7. Variable ratio PS	8. Overdrive ratio		9. Maximum sleeper berth		10. NVH resistant body
		10%	15%	20%	3%	5%	7%	5%	10%	3%	5%	5%	10%	5%	10%	15%		2%	5%	2%	5%	
(1)	(2)	(3)	(4)	(5)	(6)	(7)	(8)	(9)	(10)	(11)	(12)	(13)	(14)	(15)	(16)	(17)	(18)	(19)	(20)	(21)	02)	(23)
Fuel efficiency	1. Fuel economy									0.3	0.5	0.1	0.15									
Good load carrying capacity	2. Payload																					
Good pickup	3. Power to weight ratio	0.5	1.25	2	0.25	0.6	1			0.1	0.15	0.075	0.125									
Easy drivability on highways	4. Maximum cruising speed				−0.25	−0.4	−0.7															
Easy to climb slopes and navigability in hilly terrain	5. Gradability	0.3	0.5		0.1																	
Comfortable driver's cabin	6. Driver's cabin																					

Max cruising speed constraint fuzzy relation: $(0.5, 1, 2) \rightarrow 1 + [(2–1) – (1–0.5)]/3 = 1.1666$
Driver comfort constraint fuzzy relation: $(1.5, 2, 3.5) \rightarrow 2 + [(3.5–2) – (2–1.5)]/3 = 2.3333$
The final right-hand sides of the constraints of the PILP will be as follow:
Budget: $145 + 3*\theta$
Fuel Economy: $5.667 + 2.834*\theta$
Payload: $5 + 2.667*\theta$
Max cruising speed: $4.6666 + 1.1.666*\theta$
Driver comfort: $5.6666 + 2.3333*\theta$

13.7 Results

Constraints on fuel economy and driver comfort	Constraints on payload and maximum cruising speed
For $\theta = [0, 0.647]$	For $\theta = [0, 0.374]$
$Y_{combeff,10} = 1$	$Y_{mengrpm,10} = 1$
$X_{NVH} = 1$	$Y_{maxler,10} = 1$
$X_{VPS} = 1$	$Y_{combeff,10} = 1$
TC = 140	$Y_{odr,5} = 1$
Z = 432.265	TC = 145
	Z = 515.152
For $\theta = (0.647, 1]$	For $\theta = (0.374, 1]$
$Y_{combeff,10} = 1$	$Y_{mengrpm,10} = 1$
$Y_{tceff,5} = 1$	$Y_{maxler,15} = 1$
$X_{NVH} = 1$	$Y_{odr,5} = 1$
TC = 145	$X_{VPS} = 1$
Z = 415.187	TC = 145
	Z = 341.037

The above results show that with higher values of θ, i.e., with stringent constraints on minimum performance improvement and higher budget, the value of the objective function reduces. Thus with lower values of θ, the company can provide higher satisfaction to customers at lower cost.

13.8 Results with Symmetric Triangular Fuzzy Numbers

Constraints on fuel economy and driver comfort	Constraints on payload and maximum cruising speed
For $\theta = [0, 0.4]$	For $\theta = [0, 0.125]$
$Y_{combeff,10} = 1$	$Y_{combeff,10} = 1$
$Y_{mengrpm,5} = 1$	$Y_{maxler,10} = 1$
$Y_{maxlereac,10} = 1$	$Y_{odr,5} = 1$
$X_{VPS} = 1$	$TC = 115$
$TC = 125$	$Z = 457.507$
$Z = 477.035$	
For $\theta = (0.4, 0.5]$	For $\theta = (0.125, 0.4]$
$Y_{combeff,10} = 1$	$Y_{tceff,3} = 1$
$X_{NVH} = 1$	$Y_{combeff,10} = 1$
$TC = 125$	$Y_{maxler,10} = 1$
$Z = 383.915$	$Y_{odr,5} = 1$
	$TC = 127$
	$Z = 478.201$
For $\theta = (0.5, 0.6]$	For $\theta = (0.4, 1]$
$Y_{tceff,3} = 1$	$Y_{mengrpm,5} = 1$
$Y_{combeff,10} = 1$	$Y_{combeff,10} = 1$
$X_{NVH} = 1$	$Y_{maxler,10} = 1$
$TC = 137$	$Y_{odr,5} = 1$
$Z = 404.609$	$TC = 135$
	$Z = 490.447$
For $\theta = (0.6, 0.95]$	
$Y_{combeff,10} = 1$	
$X_{NVH} = 1$	
$X_{VPS} = 1$	
$TC = 140$	
$Z = 432.265$	
For $\theta = (0.95, 1]$	
$Y_{tceff,5} = 1$	
$Y_{combeff,10} = 1$	
$X_{NVH} = 1$	
$TC = 145$	
$Z = 415.187$	

TC = total cost and Z = objective function value.

The above results provide some interesting insights. When customers demand high levels of improvement in fuel economy (as denoted by the problem with $\theta = 1$), the value of the objective function, i.e., the overall weighted improvement across all CRs is lesser than the weighted improvement when the customer expects only minimum level of improvement in fuel economy (as denoted by the problem with $\theta = 0$). The results show that the manufacturer can provide the maximum customer satisfaction at the minimum cost if the customers are not too demanding on fuel economy.

Thus, as customers place more emphasis on fuel economy, the manufacturer has to trade in other attributes to meet its own budgetary constraints. This opens up two possibilities for the manufacturer. One is to educate customers about the overall performance of the vehicle and thus drive demand for products with moderate levels of fuel economy but better overall performance. The other option will be to strive for the highest fuel efficiency with moderate performance in other attributes. The above two approaches will lead to distinct products which can be targeted to specific customer segments. The relative attractiveness of the above two approaches can be ascertained from the utility and market share estimates from the Conjoint study.

When the customers are likely to attach more importance to performance attributes like payload and maximum cruising speed, the results are more intuitive. Higher is the requirement to improve the above attributes, higher is the cost incurred by the manufacturer with correspondingly higher overall satisfaction of the customer as captured by the objective function.

References

Akao, Y., 1990. *QFD: Integrating Customer Requirements into Product Design*. Cambridge, MA: Productivity Press.

Baier, D. and Brusch, M., 2005a. *Linking Quality Function Deployment and Conjoint Analysis for New Product Design. In*: Studies in Classification, Data Analysis, and Knowledge Organization, Vol. 29, 189–198.

Baier, D. and Brusch, M., 2005b. Improving the Predictive Validity of Quality Function Deployment by Conjoint Analysis: A Monte Carlo Comparison. *Operations Research Proceedings*, 619–624.

Bailey, M. G and Gillett, B. E, 1980. Parametric Integer Programming Analysis. Journal of the Operational Research Society, 31, 253–262.

Chaudhuri, A. and Bhattacharyya, M., 2008. 'A Combined QFD and Integer Programming Framework to Determine Attribute levels for Conjoint Study, Accepted, *International Journal of Production Research*.

Chen, Y., Fung, R. Y. K. and Yang, J., 2005. Fuzzy expected value modelling approach for determining target values of engineering characteristics in QFD. *International Journal of Production Research*, 43(17), 3583–3604.

Chen, L. H., Weng, M. C., 2006. An evaluation approach to engineering design in QFD processes using fuzzy goal programming models. *European Journal of Operational Research*, 172(1), 230–248.

Franceschini, F. and Rossetto, S., 1998. Quality Function Deployment: How to improve its use. *Total Quality Management*, 9(6), 491–500.

Fung, R. Y. K., Tang, J., Tu Y. and Wang, D., 2002. Product design resources optimisation using a non-linear fuzzy quality function deployment model. *International Journal of Production Research*, 40(3), 585–599.

Herrera, F and Verdegay, J. L (1995). Three models of fuzzy integer linear programming. *European Journal of Operational Research*, 83(1), 581–593.

Iranmanesh, S. H. and Salimi, M. H., 2003. An investigation of rank reversal when using fuzzy importance levels in QFD analysis. *International Journal of Reliability, Quality and Safety Engineering*, 10(2), 195–203.

Kahraman, C., Ertay, T. and Büyüközkan, G., 2006. A Fuzzy Optimization Model for QFD Planning Process Using Analytic Network Approach. *European Journal of Operational Research*, 171, 390–411.

Karsak, E.E., 2004. Fuzzy multiple objective programming framework to prioritise design requirements in quality function deployment. *Computers & Industrial Engineering*, 47, 149–163.

Khoo, L.P. and Ho, N.C., 1996. Framework of a fuzzy quality function deployment system. *International Journal of Production Research*, 34(2), 299–311.

Kim, K., Moskowitz, H., Dhingra, A. and Evans, G., 2000. Fuzzy multicriteria models for quality function deployment. *European Journal of Operational Research*, 121(3), 504–518.

Lai X., Tan, K.C., Xie, M. (2007), Optimizing product design using quantitative Quality Function Deployment: a case study, Quality and Reliability Engineering International, 23(1), 45–57.

Liu, S.T., 2005. Rating Design Requirements in Fuzzy Quality Function Deployment via a Mathematical Programming Approach. *International Journal of Production Research*, 43(3), 497–513.

Marsten, R.E., Morin, T.L. (1976). Parametric Integer Programming: The right-hand side case. Working Paper No. OR 050–76, Massachusetts Institute of Technology.

Pullman, M.E., Moore, W.L. and Wardell, D.G., 2002. A comparison of quality function deployment and conjoint analysis in new product design. *The Journal of Product Innovation and Management*, 19, 354–365.

Raharjo, H., Xie, M. and Brombacher, A.C., 2006. Prioritizing quality characteristics in dynamic Quality Function Deployment. *International Journal of Production Research*, 44(23), 5005–5018.

Vanegas, L.V. and Labib, A.W., 2001. Fuzzy Quality Function Deployment (FQFD) model for deriving optimum targets. *International Journal of Production Research*, 39(1), 99–120.

Van de Poel, I., 2007. Methodological problems in QFD and directions fo0072 future development. *Research in Engineering Design*, 18(1), 00321–36.

Yager, R.R., 1981. A procedure for ordering fuzzy subsets of the unit interval. *Information Science* 24, 143–161.

Chapter 14
Project Risk Modelling and Assessment in New Product Development

Jian-Bo Yang, Dawei Tang, Dong-Ling Xu, and Kwai-Sang Chin

Abstract New Product Development (NPD) is a crucial process to maintain the competitiveness of a company in an ever changing market. In the process of developing new products of a high level of innovation, there are various types of risks, which should be properly identified, systematically analyzed, modeled, evaluated and effectively controlled. In this paper, the Bayesian Network (BN) method will be investigated to assess risks involved in NPD processes. A systematic method is examined to generate probabilities in BNs. Both the probability generation method and the BN model are demonstrated by a case study about the design of electronic products for a multinational flashlight manufacturing company.

Keywords Probability generation · Bayesian network · New product development · Risk evaluation

14.1 Introduction

As market competition and product technology advancement become increasingly intense (Di Benedetto 1999; Nadkarni and Shenoy 2001), the process of New Product Development (NPD) has become more important for the success of companies than ever before (McCarthy et al. 2006). However, due to the constant changes in market environments, the originalities in NPD process as well as the uncertainties involved in NPD, NPD operates with relatively high risks (Kahraman et al. 2007). Therefore, evaluating NPD risks becomes very important for a company to decide which project should be selected for further investment in face of several alternatives. However, it is found that risk evaluation in NPD projects in many

J.-B. Yang (✉), D. Tang, and D.-L. Xu
Manchester Business School, The University of Manchester, Manchester, M15 6PB, UK
e-mail: jian-bo.yang@mbs.ac.uk

K.-S. Chin
Department of Manufacturing Engineering and Engineering Management, City University of Hong Kong, Hong Kong

N.R.S. Raghavan and J.A. Cafeo (eds.), *Product Research: The Art and Science Behind Successful Product Launches*, DOI 10.1007/978-90-481-2860-0_14,

organizations is often done by informal and non-systematic methods and based largely on management perceptions (Calantone et al. 1999; Cooper 2006). There is an increasing need to develop analytic methods to evaluate NPD project risks for helping designers to make decisions among design alternatives from project risk point of view.

A number of decision methods and tools have been developed for risk evaluation in NPD processes, such as Behavioral model (Monti and Carenini 2000), Failure Mode and Effects Analysis (FMEA) (Carbone and Tippett 2004), Analytical Hierarchy Process (AHP) (Chen et al. 2007), Analytical Network Process (ANP) (Ayag and Ozdemir 2007), etc. However, these tools, to some extent, suffer from certain underlying weaknesses when applied in complex NPD environments.

Another potential method for risk analysis in NPD is Bayesian Network (BN) (Pearl 1988; Cooper 2000), which can represent the dependant relationships among the factors of NPD in a network structure and quantitatively model uncertainties involved in risk evaluation in NPD. In most of the current research, however, the emphases were put on how to determine the structure of BN, how to perform inference in BN and how to analyze inference results. Little attention has been paid on how to generate conditional probabilities in BN, which is a precondition for BN inference.

For generation of conditional probabilities in BN, the most classic approach is the noisy OR model (Saaty 1980) and its generalizations (Diez 1993; Cozman 2004). However, such methods can only handle cases where the states of nodes are binary and the parents of nodes are assumed to be independent of each other. Monti & Carenin proposed another way to generate conditional probabilities using pair-wise comparisons (Mullins and Sutherland 1998). However, it only generates the conditional probabilities of a node with a single parent, while in BN a node can have multiple parents.

This paper is aimed to develop a systemic probability generation method and also attempts to propose a BN based evaluation method to evaluate NPD project risks.

14.2 The Proposed Approach to Generate the Probabilities in Bayesian Network

14.2.1 Generation of Probabilities of the Nodes without Parent

The probabilities of the node without parents are the prior probabilities of the node's states. Therefore, if node N has n states $S_1, S_2, \ldots, S_n$, the probability of each state S_i, i.e., $P(S_i)$ needs to be specified.

Traditionally, $P(S_i)$ is specified directly by experts. However, if a node has many states, direct estimation of probabilities of all states at a time may inevitably involve bias and inaccuracy. Therefore, an alternative way is to perform pair-wise comparisons between states to generate the probabilities. Since there are only two instead of n states to be considered at a time, the bias and inaccuracy of judgments can be

Table 14.1 Pair-wise comparison matrix to generate prior probabilities

	S_1	S_2	......	S_n	ω
S_1	a_{11}	a_{12}	......	a_{1n}	ω_1
S_2	a_{21}	a_{22}	......	a_{2n}	ω_2
......	......	......	......	......	......
S_n	a_{n1}	a_{n2}	......	a_{nn}	ω_n
λ_{max} =		CI =		CR =	

reduced. Specifically, the prior probability of each state at a node can be determined by the following pair-wise comparison matrix (Table 14.1):

In the above matrix, a_{ij} $(i = 1, 2, \ldots, n; j = 1, 2, \ldots, n)$ can be specified by questions like "comparing S_i with S_j, which state is more likely to appear and how much more likely?" and the value of a_{ij} represents the multiple of the likelihood of the presence of S_i over that of S_j. Note that from the meaning of a_{ij}, $a_{ji} = 1/a_{ij}$ and $a_{ii} = 1$, so there are $n\ (n-1)/2$ different comparisons in the above pair-wise comparison matrix. However, it is sufficient to provide $(n-1)$ interrelated comparisons rather than all the $n\ (n-1)/2$ different comparisons, although it is useful to have more comparisons for checking consistency.

Similar to AHP, the relative priorities of S_i can be generated from the maximum eigenvector $\omega = (\omega_1, \omega_2, \ldots, \omega_n)^T$ of matrix $(a_{ij})_{n \times n}$ and the consistency of the judgment can be checked by the consistency ratio CR (Saaty 1980).

Since $\sum_{i=1}^{n} \omega_i = 1$ and ω_i represents the relative importance of the state S_i among all the states, it is natural to interpret ω_i as the prior probability of the state S_i. In other words, we have

$$P(S_i) = \omega_i$$

14.2.2 Generation of Probabilities for Nodes with a Single Parent

The probabilities of a node with a single parent are the probabilities of its states conditional on its parent's states (Table 14.2). If a node N has a single parent node M and there are n and m states for the node N and the node M respectively, which can be represented by $S_{N1}, S_{N2}, \ldots, S_{Nn}$ and $S_{M1}, S_{M2}, \ldots, S_{Mm}$ respectively, the probability of each state of the node N conditional on each state of the node M, i.e., $P(S_{Ni}|S_{Mj})$ $(i = 1, 2, \ldots n, j = 1, 2, \ldots m)$ should be estimated.

Similar to the above estimation, the pair-wise comparison method is used here to estimate $P(S_{Ni}|S_{Mj})$. When the node M is in the state of S_{Mj}, the corresponding comparison matrix is given by:

In the above matrix, a_{pq} $(p = 1, 2, \ldots, n; q = 1, 2, \ldots, n)$ can be specified by questions like "if node M is in the state of S_{Mj}, comparing node N's state S_{Ni} with S_{Nj}, which one is more likely to appear and how much more likely?". After we get

Table 14.2 Pair-wise comparison matrix to generate conditional probabilities of nodes with a single parent

M is in the state of S_{Mj}	S_{N1}	S_{N2}	...	S_{Nn}	ω_N
S_{N1}	a_{11}	a_{12}	...	a_{1n}	ω_{1j}
S_{N2}	a_{21}	a_{22}	...	a_{2n}	ω_{2j}
...	...	...	...	...	...
S_{Nn}	a_{n1}	a_{n2}	...	a_{nn}	ω_{nj}
$\lambda_{max} =$		CI =		CR =	

ω_{ij} $(i = 1, \ldots, n)$, we can set:

$$P\left(N = S_{Ni} | M = S_{Mj}\right) = \omega_{ij}$$

Since the node M has m states, m matrices should be constructed to get all ω_{ij} for $i = 1, 2, \ldots, n; j = 1, 2, \ldots, m$.

14.2.3 Generation of Conditional Probabilities for Multi-Parent Nodes

If a node N with n states $S_{N1}, S_{N2}, \ldots, S_{Nn}$ in a BN has k parents, namely, $M_1, M_2, \ldots, M_k$, and the node M_j has the states of $S_{M_j1}, S_{M_j2}, \ldots, S_{M_jm_j}$ $(j = 1, \ldots, k)$, it will be very difficult for experts to directly estimate the probability of each state of N conditional on the combination of its parents' states, which is defined by

$$\begin{aligned}&P\left(N = S_{Ni} | M_1 = S_{M_1p_1}, M_2 = S_{M_2p_2}, \ldots, M_k = S_{M_kp_k}\right);\\&i = 1, 2, \ldots, n;\ p_j = 1, 2, \ldots, m_j;\ j = 1, \ldots, k\end{aligned}$$

In the above section, conditional probability of a node with a single parent is generated. So, a natural question arises as to how to generate the node's probability conditional on each of its parent first and combine those conditional probabilities to get the node's probability conditional on different combinations of its parents' states? This is investigated as follows.

In Kim and Pearl (1983), when a node A in a BN has two parents B and C, its probability conditional on B and C can be approximated by: $P(A|B,C) = \alpha P(A|B)\,P(A|C)$, where α is a normalization factor which is used to ensure $\sum_{a \in A} P(a|B,C) = 1$. The above result can be generalized as follows:

$$P(A|X_1, X_2 \ldots, X_n) = \alpha P(A|X_1)\,P(A|X_2) \ldots P(A|X_n) \tag{14.1}$$

In Eq. (14.1), α is a normalized factor to ensure $\sum_{a \in A} P(a|X_1, X_2, \ldots, X_n) = 1$.

From Eq. (14.1), we can see that the probability conditional on multi-parents can be given by the product of the probabilities conditional on each single parent. Then, in our model,

$$P\left(N = S_{Ni}|M_1 = S_{M_1 p_1}, M_2 = S_{M_2 p_2}, \ldots, M_k = S_{M_k p_k}\right);$$

can be calculated as follows:

$$P\left(N = S_{Ni}|M_1 = S_{M_1 p_1}, M_2 = S_{M_2 p_2}, \ldots, M_k = S_{M_k p_k}\right)$$
$$= \alpha \prod_{j=1}^{k} P\left(N = S_{Ni}|M_j = S_{M_j p_j}\right) \quad (i = 1, 2, \ldots, n;\ p_j = 1, 2, \ldots, m_j;$$
$$j = 1, \ldots, k)$$

where α is a normalized factor to ensure that

$$\sum_{i=1}^{n} P\left(N = S_{Ni}|M_1 = S_{M_1 p_1}, M_2 = S_{M_2 p_2}, \ldots, M_k = S_{M_k p_k}\right) = 1$$

and $P\left(N = S_{Ni}|M_j = S_{M_j p_j}\right)$ can be generated from the method proposed in the above section.

Since the estimation of $P\left(N = S_{Ni}|M_j = S_{M_j p_j}\right)$ is much easier than the direct estimation of $P\left(N = S_{Ni}|M_1 = S_{M_1 p_1}, M_2 = S_{M_2 p_2}, \ldots, M_k = S_{M_k p_k}\right)$, the method proposed in this section is more reliable.

14.3 Application of the Method in Risk Evaluation of NPD

14.3.1 Case Description

The developed approach to determine the probabilities of BN was validated in the context of NPD project risk evaluation for assessing two alternative product design concepts of a new 2AAA penlight in a multinational flashlight manufacturing company. The two design alternatives, Push Button Switch (PBS) design and Rotary Switch (RS) design, are more or less able to meet the customer needs. They may, however, differ in material used, product structure, manufacturing processes, optional functional and aesthetic features, expected quality, reliability performance, etc. The required information and knowledge were collected from the Product Development Director of the company.

The validation process consists of three steps: to build a BN, to specify conditional probabilities and to determine prior probabilities. Conditional probabilities in BN reflect general relations among risk factors during the company's NPD process, which remain the same for different alternatives, while prior probabilities in

BN reflect the specific features of different alternatives, which may be different for different alternatives. After the probabilities are determined, inference can be performed to find which alternative is of low risk and alternatives with lower risk are preferred.

14.3.2 Bayesian Network Construction

Based on the relationships among different risk factors in NPD and the experience of the expert, the BN to assess NPD project risk in terms of the company's technical considerations is built as follows:

The meanings of abbreviations of the nodes in Fig. 14.1 are summarized in the following table and each node in the BN has three states: High (H), Medium (M) and Low (L) (Table 14.3).

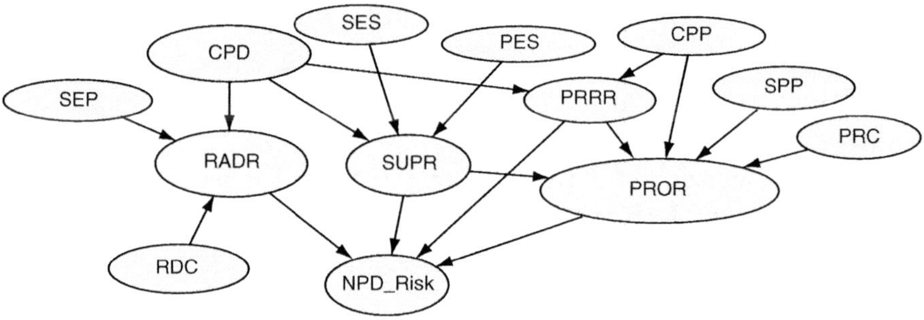

Fig. 14.1 BN for risk evaluation of NPD project

Table 14.3 Explanation of the Abbreviations in Fig. 14.1

Abbr.	Meaning
SEP	Similarity of the existing product
CPD	Complexity of the product design
RDC	R&D capability
RADR	Research and development risk
SES	Similarity of the existing supply
PES	Supplier performance
SUPR	Supply risk
PRRR	Product reliability risk
CPP	Complexity of the production process
SPP	Similarity of the production process
PRC	Production capability
PROR	Production risk

14.3.3 Generation of Conditional Probabilities in BN

The node 'PRRR' will be selected as an example when generating conditional probabilities in BN. 'PRRR' has two parent nodes, namely, 'CPD' and 'CPP'. According to the method proposed, $P\ (PRRR|CPD)$ and $P\ (PRRR|CPP)$ should be calculated first (Tables 14.4 through 14.7).

When generating $P\ (PRRR|CPD)$ on the condition that CPD is in the state of 'H', the following matrix is obtained from the experts' opinions:

The number in the above matrix can be specified by answering the questions like: "Without considering the influence of the other parent on 'PRRR', when 'CPD' is in the state of 'H', which state of 'PRRR' is more likely to occur, and how much more likely?" For instance, in the matrix, we can see that given that 'CPD' is at high

Table 14.4 Evaluation of the probabilities of PRRR conditional on CPD is in the state of H

CPD = H	H	M	L	ω
H	1	2*	3*	$\omega_H = 0.5396$
M	1/2#	1	2*	$\omega_M = 0.2970$
L	1/3#	1/2#	1	$\omega_L = 0.1634$
CR = 0.0079		CI = 0.0046		$\lambda_{\max} = 3.009$

#Expert's judgments, * reciprocal of the expert's judgments.

Table 14.5 Probabilities of PRRR conditional on CPD's different states

PRRR	CPD = H	CPD = M	CPD = L
H	0.5396	0.1947	0.0333
M	0.2970	0.4895	0.2502
L	0.1634	0.3158	0.7165

Table 14.6 Probabilities of PRRR conditional on CPP's different states

PRRR	CPP = H	CPP = M	CPP = L
H	0.4895	0.1947	0.0302
M	0.3158	0.4895	0.2263
L	0.1947	0.3158	0.7435

Table 14.7 Probabilities of PRRR conditional on different state combinations of CPD and CPP

CPD	H		
CPP	H	M	L
H	0.6777	0.3478	0.0795
M	0.2407	0.4814	0.3279
L	0.0816	0. 1708	0. 5926
CPD	M		
CPP	H	M	L
H	0. 3060	0. 1004	0. 0167
M	0.4965	0.6352	0.3152
L	0. 1975	0. 2644	0. 6681
CPD	L		
CPP	H	M	L
H	0. 0694	0. 0182	0. 0017
M	0.3365	0.3448	0.0959
L	0. 5941	0. 6370	0. 9024

(H) level, the probability of 'PRRR' being at the medium (M) level is 1/2 times the probability of 'PRRR' at the high (H) level, and the probability of 'PRRR' at the low (L) level is 1/3 times the probability of 'PRRR' at the high (H) level. This is reasonable since higher complexity in the product design may lead to the higher chance of suffering from performance problems in the product life and thus may lead to a higher level of risks in product reliability.

According to the discussion in the above section, we can get:

$$
\begin{aligned}
P\,(PRR = H|CPD = H) &= \omega_H = 0.5396\\
P\,(PRR = M|CPD = H) &= \omega_M = 0.2970\\
P\,(PRR = L|CPD = H) &= \omega_L = 0.1634
\end{aligned}
$$

Similarly, we can get the probability of states of node 'PRRR' on the condition that the state of 'CPD' is M and L, and the results are summarized as follows:

In the same way, the probabilities of the states of the node 'PRRR' on the condition of different states of the node 'CPP' are listed in the following table:

After the probabilities of all the states of the node 'PRRR' conditional on each state of its parent node have been generated, the probabilities conditional on the state combinations of both its parent nodes can be estimated.

For example, when both the state of 'CPD' and the state of 'CPP' are H, we will have:

$$
\begin{aligned}
P\,(PRRR = H|CPD = H, CPP = H) &= \alpha P\,(PRRR = H|CPD = H)\\
&\quad P\,(PRRR = H|CPP = H)\\
P\,(PRRR = M|CPD = H, CPP = H) &= \alpha P\,(PRRR = M|CPD = H)\\
&\quad P\,(PRRR = M|CPP = H)\\
P\,(PRRR = L|CPD = H, CPP = H) &= \alpha P\,(PRRR = L|CPD = H)\\
&\quad P\,(PRRR = L|CPP = H)
\end{aligned}
$$

with $\alpha = \frac{1}{K}$ where

$$
\begin{aligned}
K = \; & P\,(PRRR = H|CPD = H)\,P\,(PRRR = H|CPP = H)\\
& + P\,(PRRR = M|CPD = H)\,P\,(PRRR = M|CPP = H)\\
& + P\,(PRRR = L|CPD = H)\,P\,(PRRR = L|CPP = H)
\end{aligned}
$$

From the above equations, we can get the following results:

$$
\begin{aligned}
P\,(PRRR = H|CPD = H, CPP = H) &= 0.6777\\
P\,(PRRR = M|CPD = H, CPP = H) &= 0.2407\\
P\,(PRRR = L|CPD = H, CPP = H) &= 0.0816
\end{aligned}
$$

In a similar way, the probabilities of the state of the node 'PRRR' conditional on the other state combinations of its parent nodes (i.e., the conditional probability table of the node 'PRRR') can also be generated and the results are shown as follows.

The Conditional Probability Table of the other nodes with parents in the BN can be specified in a similar way.

14.3.4 Generation of Prior Probabilities in BN

The prior probabilities can be generated through the pair-wise comparison matrix based on the features of two alternatives. For example, for alternative 1, which is the penlight design with PBS, the prior probability of the node SEP can be specified by the expert in the following matrix:

From the matrix, we can get:

$$P\,(SEP = H) = \omega_H = 0.7334$$
$$P\,(SEP = M) = \omega_M = 0.1991$$
$$P\,(SEP = L) = \omega_L = 0.0675$$

Similarly, the prior probabilities of other nodes of alternative 1 and the prior probabilities of nodes of alternative 2 can be generated as follows (Tables 14.8 through 14.11):

Table 14.8 The evaluation of probabilities of SEP for alternative 1

SEP	H	M	L	ω
H	1	5*	8*	$\omega_H = 0.7334$
M	1/5#	1	4*	$\omega_M = 0.1991$
L	1/8#	1/4#	1	$\omega_L = 0.0675$
CR = 0.0812		CI = 0.0470		$\lambda_{\max} = 3.0940$

#Expert's judgments, * reciprocal of the expert's judgments.

Table 14.9 The prior probabilities of different node for alternative 1

Alternative 1	SEP	CPD	RDC	SES
H	0.7334	0.7120	0.2052	0.7120
M	0.1991	0.2498	0.5251	0.2498
L	0.0675	0.0382	0.2697	0.0382
Alternative 1	PES	SPP	CPP	PRC
H	0.3914	0.6519	0.6519	0.2697
M	0.4893	0.2862	0.2862	0.5251
L	0.1193	0.0619	0.0619	0.2052

Table 14.10 The prior probabilities of different node for alternative 2

Alternative 2	SEP	CPD	RDC	SES
H	0.0428	0.1778	0.2500	0.1685
M	0.4358	0.3985	0.5000	0.4766
L	0.5214	0.4237	0.2500	0.3549
Alternative 2	PES	SPP	CPP	PRC
H	0.2500	0.0428	0.1685	0.2697
M	0.5000	0.4358	0.4766	0.5251
L	0.2500	0.5214	0.3549	0.2052

Table 14.11 The risk analysis result for alternative 1 and alternative 2

NPD Risk Level	Alternative 1	Alternative 2
H	0.2652	0.5090
M	0.5666	0.4437
L	0.1682	0.0474

14.3.5 Result

After all the probabilities are specified, the BN inference can be performed to generate the risk levels of the two alternatives. The inference is performed by the software named Hugin® and the result is shown as follows:

If we assign the utility of the states as: $U(H) = 0, U(M) = 0.5, U(L) = 1$, we can get the utility of alternative 1 and 2 are 0.4515 and 0.2693 respectively. Therefore, alternative 1 should be selected.

14.4 Conclusion

A probability generation method in BN is proposed in this paper, which can reduce the bias in expert judgments when conditional and prior probabilities are estimated in BN. The method is applied in a real case study for risk evaluation in NPD.

The proposed method has two major shortcomings: it suffers from the problem of "curse of dimension" when BN is very complex and when there are many states at a node in BN; and it can only accommodate inputs represented by prior probabilities of nodes, which will limit the application scope of the methods since information provided by experts may have various inherent features (e.g., qualitative, quantitative, interval, fuzzy, etc.). For our future work, possible methods to overcome the above shortcomings will be sought. A framework which can accommodate various forms of input and can offer a flexible way for experts to express their judgments will be investigated

Acknowledgement The work is supported by EPSRC under Grant No.: no. EP/F024606/1 and City University of Hong Kong under SRG project no. 7001971. It is also supported by the Natural Science Foundation of China under the Grant No.: 60734026 and the Grant No.: 70631003.

References

Ayag, A., & Ozdemir, R. G. (2007). An analytic network process-based approach to concept evaluation in a new product development environment. Journal of Engineering Design, 18(3), 209–226.

Calantone, R. J., Di Benedetto, C. A., & Schmidt, J. B. (1999). Using the analytic hierarchy process in new product screening. Journal of Product Innovation Management, 16(1), 65–76.

Carbone, T. A., & Tippett, D. D. (2004). Project risk management using the project risk FMEA. Engineering Management Journal, 16(4), 28–35.

Chen, H. H., Lee A. H. I., & Tong, Y. (2007). Prioritization and operations NPD mix in a network with strategic partners under uncertainty. Expert Systems with Applications, 33(2), 337–346.

Cooper, L. G. (2000). Strategic marketing planning for radically new products. Journal of Marketing, 64(1), 1–16.

Cooper, R. G. (2006). Managing technology development project. Research Technology Management 49(6), 23–31.

Cozman, F. G. (2004). Axiomatizing noisy-OR. Proceedings of the European Conference on Artificial Intelligence, Valencia, 979–980.

Di Benedetto, C. A. (1999). Identifying the key success factors in new product launch. Journal of Product Innovation Management, 16(6), 530–544.

Diez, F. J. (1993). Parameter adjustment in Bayes networks: the generalized noisy OR-gate. Proceedings of the 9th Annual Conference on Uncertainty in Artificial Intelligence, San Francisco, 99–105.

Kahraman, C., Buyukozkan, G., & Ates, N. Y. (2007). A two phase multi-attribute decision-making approach for new product introduction. Information Sciences, 177(7), 1567–1582.

Kim, J. H. & Pearl, J. (1983). A computational model for combined causal and diagnostic reasoning in inference systems. Proceedings of the 8th International Joint Conference on Artificial Intelligence Karlsruhe, 380–385.

McCarthy, I. P., Tsinopoulos, C., Allen, P., & Rose-Anderssen, C. (2006). New product development as a complex adaptive system of decisions. Journal of Product Innovation Management, 23(5), 437–456.

Monti, S. & Carenini, G. (2000). Dealing with the expert inconsistency in probability elicitation. IEEE Transactions on Knowledge and Data Engineering, 12(4), 499–508.

Mullins, J. W., & Sutherland, D. J. (1998). New product development in rapidly changing markets: an exploratory study. Journal of Product Innovation Management, 15(3), 224–236.

Nadkarni, S. & Shenoy, P. P. (2001). A Bayesian network approach to making inferences in causal maps. European Journal of Operational Research, 128(3), 479–498.

Pearl, J. (1988). Probabilistic reasoning in intelligent systems: networks of plausible inference. Morgan Kaufmann, San Mateo.

Saaty. (1980). The analytic hierarchy process: planning, priority setting, resource allocation. McGraw-Hill, New York.

Chapter 15
Towards Prediction of Nonlinear and Nonstationary Evolution of Customer Preferences Using Local Markov Models

Satish T.S. Bukkapatnam, Hui Yang, and Foad Madhavi

Abstract A dynamic vehicle design process hinges on one's ability to forecast how the customer preferences and the exogenous variables evolve over time. Prior research mentions and it is understood in practice that the dynamics of evolution of these variables is highly nonstationary, and can be nonlinear. This paper aims to provide a mathematical model based on Markov models for predicting future states of such nonlinear nonstationary dynamic processes. The state space of a dynamic system is reconstructed from the historic data using delay embedding principles, and a Markov model is derived therefrom. The reconstructed state space is clustered into different neighborhoods to form nodes of a graph by using a set covering formulation. The probabilities of transition between every node-pair were obtained for this graph. Next, the system trajectory is segmented into multiple near-stationary time intervals based on pattern analysis of the transition matrix. Assuming stationary behavior of the system within each segment, the corresponding Markov transition matrix can be used for prediction inside that segment. Due to the proprietary nature of actual customer preference data, the present model was validated using swapped Rossler attractor data as well as a time-series data that mimics, in the sense of nonlinear and nonstationary patterns, the trends of willingness to pay (WTP) data acquired for a customer group for particular sets of vehicles in a portfolio. One and five step ahead prediction results indicates that the proposed model is 5–25% more accurate (in terms of R^2 statistic) than the models commonly used in practice.

Keywords Markov modeling · State space reconstruction · State space segmentation

15.1 Introduction

The auto industry has been working towards deriving tools and procedures to effectively incorporate the Voice of the Customer into the advance vehicle development process. Such a tool should capture the trends in customer preferences and external

S.T.S. Bukkapatnam (✉), H. Yang, and F. Madhavi
Professor of Industrial Engineering and Management, Oklahoma State University, Stillwater, OK 74078
e-mail: bsatish@okstate.edu

N.R.S. Raghavan and J.A. Cafeo (eds.), *Product Research: The Art and Science Behind Successful Product Launches*, DOI 10.1007/978-90-481-2860-0_15,

and internal determinants, and project out the most desirable features to impart into the various options within a portfolio of the vehicles to be developed for release after 3–4 years.

Current vehicle development practice involves capturing the customer preferences gathered at the present time juncture and using the mathematical quantifications of these for designing the vehicles for the future. This approach critically hinges on the availability of adequate data on the historic customer preferences for various design feature options, as well as the various endogenous and exogenous variables. The recent efforts at GM CDAVD has shown that various alternative data sources, including Internet Blogs, on-line surveys, post-sale surveys, are available to provide a fairly extensive collection of the relevant data.

Many decision theoretic models have been developed, some of which have already been applied to the vehicle design process. Although they capture the various interrelationships among the vehicle features and the exogenous variables, they are limited by their static view of the vehicle design process. A typical vehicle concept-to-release cycle takes anywhere between 3 and 4 years. One would expect the customer preferences to have evolved during this time. As a result the vehicle design features selected based on past customer preferences would perform sub-optimally in terms of market shares in the future. Hence there appears to be a clear need to capture the dynamics of the customer preferences. A few recent works mention the need to treat the design of complex systems, such as an automotive vehicle as a complex adaptive process, but the analytical and computational tools to facilitate this process remain elusive.

A dynamic design process hinges on one's ability to forecast how the preferences and the exogenous variables evolve over time. Prior research mentions and it is understood in practice that the dynamics of evolution of these variables is highly nonstationary, and can be nonlinear. However many of the prediction approaches do not capture the complex dynamics of how the customer preferences, feature offerings, exogenous variables and endogenous variables jointly evolve.

In the open literature, many approaches have been proposed for this problem, such as Volterra models, nonlinear auto-regressive moving average with exogenous variable (NARMAX) models, regression methods, exponential smoothing, Kalman Filtering, non-parametric methods and others (Marmarelis and Marmarelis 1978; Marmarelis 1981; Leontaritis and Billings 1985; Bunn and Farmer 1985; Moghram and Rahman 1989; Chen et al. 1990; Narendra and Mukhopadhyay 1997; Noriega and Wang 1998; Bunn and Vassilopoulos 1999; Taylor 2003; Taylor and Buizza 2003; Al-Hamadi and Soliman 2004; Asber et al. 2007). Also, artificial intelligent (AI) techniques such as Expert Systems (Ho 1990; Rahman and Hazim 1996), Fuzzy inference models (Huang and Shih 2002; Al-Kandari et al. 2004) and Artificial Neural Networks (ANNs) (Niebur 1995; Senjyu et al. 2004; Baczynski and Parol 2004; Saini and Soni 2002; Kumluca and Erkmen 2003; Desouky and Elkateb 2000; Kermanshashi and Iwamiya 2002) have received a great deal of research interests in this area. While these models do not need complex mathematical formulations or quantitative input–output correlations, they are limited by (1) the sparseness of the available data, (2) their ability to provide physical insights into the underlying relationships, as well as (3) the heuristic nature of the methods to optimize topologies and avert overfitting. Researchers have also used conventional stationary models for prediction applications in nonlinear nonstationary systems. The prediction accuracy of these models deteriorates significantly overtime because of the dynamic and nonstationary nature of these systems (McNames et al. 1999). Under nonstationary conditions, it is necessary to determine the time intervals and state space subsets under which the identified model and its parameter values are valid and powerful enough for state prediction. The recent nonlinear dynamic forecasting approaches applied to weather and financial market predictions can offer a way forward on this issue (Hon 2005; Van Wezel and Potharst 2007; Chen et al. 2007; Einstein et al. 1998; Hettrich and Rosenzweig 2003; Chartier et al. 2008). They generally address the nonstationary issues with either in a piecewise quasi-stationary fashion or by means of adaptive recursive algorithms that track the time-varying parameters of the system (Iatrou et al. 1999).

This paper presents a methodology to model and predict evolutions of nonlinear nonstationary systems based on a piecewise stationary model. Here, a time delay reconstruction is used for deriving the state space of a nonlinear nonstationary system (Marwan et al. 2007). The Markovian nature of the model of the state vector is used to segment the state space attractor to near stationary time intervals using this Markov property of reconstructed state space and then using the Markov model order one transition matrix for each segment for prediction within that segment. The organization of the paper is as follows. Section 15.2 presents the Markov modeling approach for prediction in nonlinear dynamic systems under highly nonstationary conditions. In Section 15.3, the present approach is compared with commonly used stationary modeling methods using different time series from nonlinear dynamic systems that mimic the complexity associated with the evolution of customer preferences. The results show considerable prediction accuracy improvements from the present approach.

15.2 Markov Modeling Approach

The local Markovian prediction approach (see Fig. 15.1) consists of three basic steps, namely, (1) nonlinear dynamic characterization, (2) pattern analysis and segmentation and (3) state prediction. Nonlinear characterization involves reconstructing the state space of the dynamic system underlying the measured time series (here, WTP) data using delay reconstruction methods (Takens 1981; Casdagli et al. 1991). A Markov transition matrix derived from the reconstructed state space is applied for system pattern analysis and to segment the trajectory into piecewise stationary segments. For each stationary segment, Markov model transition matrix is used to provide predictions inside that segment. For model validation, the prediction results of the present approach are compared with those from models commonly used in practice.

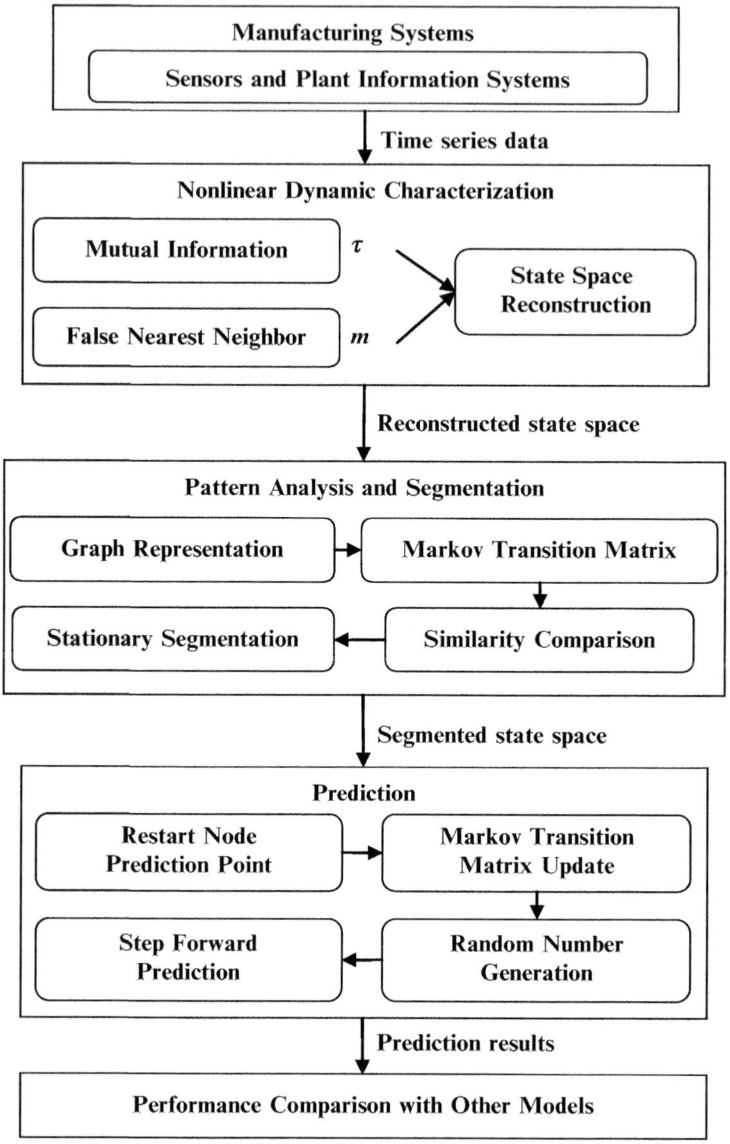

Fig. 15.1 General framework of the proposed methodology

15.2.1 Nonlinear Dynamic Characterization

Formally, a dynamical system is defined by a triad of a state space **X**, a continuous or discrete time set **T**, and a time evolution law **F**. The elements or "points" of the state space represent possible states $\vec{x}$ of the system. If the state of a system at a time t can be specified by d components

$$\vec{x} = (x_1(t), x_2(t), \ldots, x_d(t))^T \tag{15.1}$$

which form a vector in a d-dimensional state space of the system. The time-evolution law is a rule that allows determining the state of the system at each moment of time t from its states at all previous times. For time-continuous systems the time evolution is given by a set of differential equations

$$\dot{\vec{x}}(t) = \frac{d\vec{x}(t)}{dt} = \vec{F}(\vec{x}(t)), \quad F : R^d \rightarrow R^d \tag{15.2}$$

The vectors $\vec{x}(t)$ define a trajectory in state space (Marwan et al. 2007).

In most real-world settings, typically not all relevant components to construct the state vector are known or measured. Often we are confronted with a time-discrete measurement of only one or a few observables. In such a case, the state space has to be reconstructed. A frequently used method for the reconstruction is the time delay method (Marwan et al. 2007).

$$\hat{\vec{x}}_i = \sum_{j=1}^{m} u_{i+(j-1)\tau} \vec{e}_j \tag{15.3}$$

where τ is the time delay, m is the embedding dimension. u denotes the original time series data, and the vectors $\vec{e}_j$ are unit vectors that span an orthogonal coordinate system $(\vec{e}_i . \vec{e}_j = \delta_{i,j})$ (Marwan et al. 2007).

For the analysis of time series, embedding parameters, the delay τ and the dimension m have to be chosen appropriately. Different approaches for determining an appropriate time delay τ (e.g., the auto-correlation function, the mutual information function), as well as for estimating the smallest sufficient embedding dimension (e.g., the false nearest-neighbors algorithm) have been detailed in the literature (Marwan et al. 2007).

15.2.2 Pattern Analysis and Segmentation

If for all positive n, the nth-order PDFs of a random sequence do not depend on the shift parameter k, then the random sequence is said to be stationary (Stark and Woods 1994), i.e.,

$$\begin{aligned}
&\text{for all } n \geq 1 \\
&f(a_0, a_1, \ldots, a_{n-1}; k, k+1, \ldots, k+n-1) = f(a_0, a_1, \ldots, a_{n-1}; 0, 1, \ldots, n-1) \\
&\text{for all } k \text{ and } a_0 \text{ thru } a_{n-1}
\end{aligned} \tag{15.4}$$

The nth-order PDF's of a random sequence can be evaluated as

$$f(a_0, a_1, \ldots, a_{n-1}) = f(a_{n-1}) \times f(a_{n-2} \,|a_{n-1}) \\ \times f(a_{n-3} \,|a_{n-2}\,, a_{n-1}) \times \ldots \times f(a_0 \,|a_1, a_2, \ldots, a_{n-1}) \tag{15.5}$$

A random sequence is stationary if the transition matrices of all orders do not depend on the shift parameter k. The essential feature of stochastic Markov processes in dynamical systems is their strictly finite memory and the transition probabilities to the future states may only depend on the present state, not on the past realizations (Kantz and Schreiber 1997). So it can be assumed that reconstructed state space has Markovian property of order one in which S_{i+1} just depends on S_i. Considering this fact, for stationary definition, it is enough that the transition matrix of Markov model in reconstructed state space does not depend on the shift parameter k. Towards determining such stationary segments in the time-series from the reconstructed state space, we develop a Markov model, and use the patterns in the transition matrix of the Markov model for segmentation, as detailed in the following subsection.

15.2.2.1 Markov Model Derivation

To evaluate the transition matrix of a Markov model defined over a state space, state space needs to be represented as a graph comprising of state clusters as nodes and the transitions between the states as arcs. In this work we recognize that each point in state space denotes the state of the system at a specified time t. A suitable radius r of hyperspheres in the state space can be defined such that, two points in the state space separated by a distance of less than r can be treated to denote the same, quantized, state. Thus, the original state space is clustered into a simplified graph in which each node contains all state space points that are within an r-neighborhood. Towards this end, it is important to find the number and the locations of the centers of the r-balls to minimize the number of nodes in the graph, subject to the constraint that the graph nodes should cover all the state space points. State space is continuous, so this problem can be modeled as a continuous set covering problem (Drezner 1995). Figure 15.2 shows a feasible solution for graph representation of a two dimensional state space in which if the distance between two state space points is less than $2r$, they can be considered to have the same state.

In a continuous set covering problem, the objective is to find and place the minimum number of server nodes (simplified graph nodes in state space graph representation) that can cover all demand nodes (state space points in state space graph representation) that their number and place are fixed and determined. This model has two major constraints (Drezner 1995):

(1) A server node can cover a demand node when the distance between them is less than a determined serving distance (r)
(2) All demand nodes should be covered

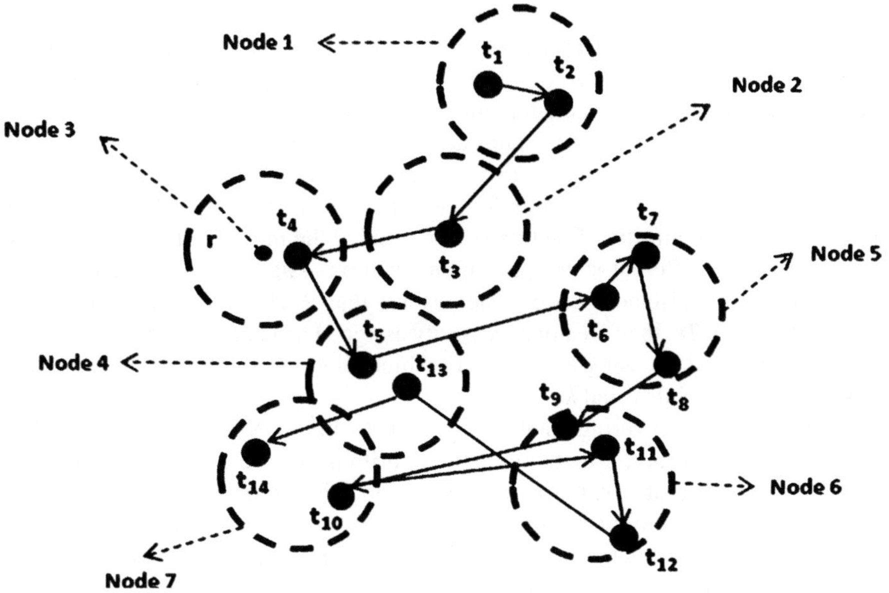

Fig. 15.2 A feasible solution for graph representation of a two dimensional state space in which if the distance between two state space points is less than $2r$, they can be considered to have the same state

The mathematical model for state space graph clustering problem is formulated as follows based on continuous set covering problem:

$$Min \sum_{j=1}^{n} y_j \tag{15.6}$$

$$subject\ to:$$

$$\sum_{j=1}^{n} x_{ij} = 1 \qquad i = 1, 2, \ldots, n \tag{15.7}$$

$$y_j \le \sum_{i=1}^{n} x_{ij} \le n y_j \qquad j = 1, 2, \ldots, n \tag{15.8}$$

$$x_{ij} + x_{kj} \le 2z_{ik} + 1 \qquad i, j, k = 1, 2, \ldots, n; i \neq k, \tag{15.9}$$

$$d(A_i, A_k) \le 2r \Rightarrow z_{ik} = 1, otherwise\ z_{ik} = 0$$

$$y_j = \{0, 1\} \qquad j = 1, 2, ..., n \tag{15.10}$$

$$x_{ij} = \{0, 1\} \qquad i, j = 1, 2, ..., n \tag{15.11}$$

The maximum number of graph nodes is the same as the number of state space points (n), So n virtual graph nodes are defined. The state space points are assigned to these virtual graph nodes by introducing a decision variable x_{ij} so that if state space point i be assigned to virtual graph node j, x_{ij} is one otherwise it's zero (Eq. 14.11). Each state space point should be assigned to just one virtual graph node but a virtual graph node can contain more than one state space points (Eqs. 15.7 and 15.8). If a virtual graph node contains at least one state space point, it is considered as a real graph node. To distinguish between real and virtual graph nodes, y_j is defined to be one for a real and zero for a virtual graph node (Eq. 15.10). Considering this notation, the objective function is to minimize the number of real graph nodes (Eq. 15.6). Two different state space points can be assigned to a particular virtual graph node if the distance between them is less than $2r$, but if the distance between them is less than $2r$, they are not necessarily assigned to a common graph node. This fact is formulated by using (Eq. 15.8) in which $d(A_i, A_k)$ is the distance between two state space points i and k.

Continuous set covering model is known to be an NP-hard problem (Owen et al. 1998). Therefore, a meta-heuristic algorithm may be effective for obtaining globally optimal solutions. In this study, a genetic algorithm (GA) is utilized. A GA algorithm seeks to minimize a fitness function given by Eq. 15.9 though an ensemble search where the solution space is encoded in the form of chromosomes. Each chromosome consists of m genes where genes represent the state space points. Each chromosome represents an order under which the original state space graph is clustered. To evaluate fitness function for a chromosome, Starting from the first gene of the chromosome and continuing according to the chromosome genes order, this gene and all other genes that their distances from each other is less that the determined threshold are placed in one cluster. Next cluster determination starts from the first gene in chromosome order which is not assigned to any cluster yet. This process will continue till all genes are assigned to clusters and the fitness function value is equal to the total number of clusters yielded by this clustering order.

For generating the population, randomized genes are assigned to all chromosomes. To decide between performing mutation or crossover, two variables are utilized: P_C *(probability of doing crossover)* and P_M *(probability of doing mutation)*. These variables are supplied by random number generator. Parents that take part in mutation and cross-over are randomly selected. To implement mutation, after selecting two genes of a parent randomly, they will exchange their places in chromosome order by each other. This gene exchanging will occur k times and k is a random number between zero and one tenth of chromosome length. To implement the crossover operation, two parents are selected randomly. Next, two children are generated from these two parents. To generate child 1, the first gene of this child comes from parent 1 and its second gene will be the first unassigned gene of parent 2 which is not already in child 1 genes. Again the third gene of this child will be the first unassigned gene of parent 1, and so on. This process will be continued till all child 1 genes are assigned. For generating child 2, the gene assignment process will be the same but it starts from parent 2.

15.2.2.2 Segmentation by Pattern Analysis of the Markov Transition Matrix

The transition matrix of a Markov model defined in the state space is used as an indicator of the system's stationary behavior. In time interval $T = (t_1, t_2)$, the system behavior can be considered to be stationary if Markov model order one transition matrix in system's state space does not undergo a statistically significant variation over this time interval.

A sliding window method is used to segment the state space to near-stationary time intervals. As shown in Fig. 15.3, three parameters are defined:

(1) Sliding window length, the time interval over which state space's transition matrix order one is evaluated
(2) Shifting forward length, the time through which the sliding window is shifted forward to update the transition matrix

Transition matrix similarity threshold of the difference between the transition matrix computed at two different times that determines if the patterns are statistically different. Here, the difference between two transition matrixes is given by

$$d(T_i, T_{i+1}) = \sum_{n=1}^{N} \sum_{m=1}^{N} (p_i[n,m] - p_{i+1}[n,m])^2 \tag{15.12}$$

where $d(T_i, T_{i+1})$ is the difference between two consecutive transition matrixes T_i and T_{i+1}. $p_i[n,m]$ is the element in row n and column m of T_i which is equal to the probability that system trajectory goes to node m after passing node n during the time interval that T_i has been calculated. Similarly, $p_{i+1}[n,m]$ is defined for transition matrix T_{i+1}.

After computing the transition matrix of the first segment, transition matrixes of the subsequent segments are calculated by shifting the sliding window forward

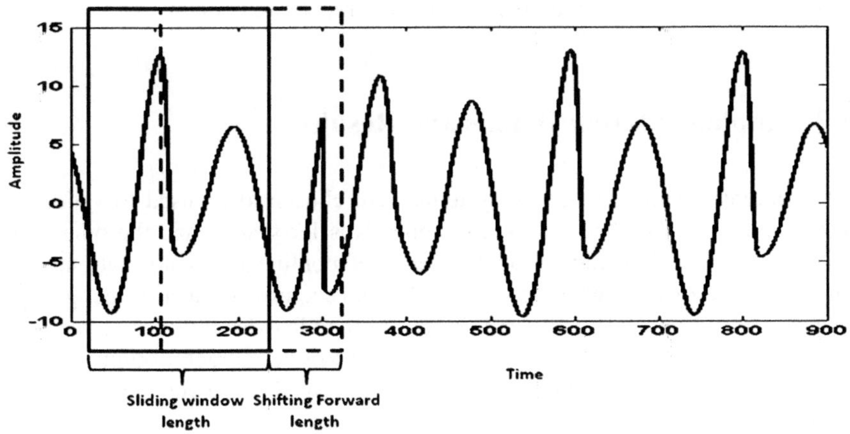

Fig. 15.3 Sliding window method for segmenting the state space to near-stationary time intervals

as determined by the shifting forward length parameter. Whenever the difference between a new transition matrix and the first exceeds transition matrix pattern similarity threshold, the system behavior in second transition matrix time interval has been changed significantly, and a segment boundary is marked at the beginning of the new segment. This process is continued by comparing the first transition matrix of this new segment with transition matrices of the subsequent segments till the whole state space trajectory is partitioned into multiple near-stationary time intervals.

15.2.3 State and Performance Prediction

To predict the system's behavior after time t, just the segment that contains system's state at time t is considered and the rest of segments are disregarded. This data selection makes sense by considering this fact that the behavior of the system is different in the other segments so using them for prediction will not result in good prediction results. Assuming stationary behavior of the system in each segment, Markov transition matrix order one in state space graph is calculated from the start point of considered segment to time t. This transition matrix can be used to simulate the system's behavior in this segment. By having the probability distribution of going from each node of state space to the others, the system trajectory after time t till the end of its segment is simulated and provided as the prediction of the state space in the rest of segment's time points.

In prediction using Markov transition matrix, there is one issue and that is sometimes the last node, after which the prediction is required, is a new node and this is its first appearance in system trajectory. So there is no history of the system behavior after passing this node which results in the model disability to predict the next nodes. This problem can be solved by considering a bigger state similarity threshold in graph representation stage because bigger state similarity threshold will increase the probability of recognizing the last node as a repetitive node.

15.3 Implementation Details and Results

Two different nonlinear dynamic systems were chosen to be used to validate the present approach. The first one is a swapped Rossler system and the other one is a time-series data that mimics, in the sense of nonlinear and nonstationary patterns, the trends of willingness to pay (WTP) data acquired for a customer group for particular sets of vehicles in a portfolio. A swapped Rossler time series is obtained by shifting the last 300 points of 900 points of a Rossler attractor output to its first part. Moving average type methods typically employed in the vehicle development practice do not capture such types of nonstationaries.

As shown in Fig. 15.4, time delay $\tau = 4$ and dimension $m = 3$ were found to be optimal for swapped Rossler. The reconstructed state space for the time series in a 3-dimensional space is shown in Fig. 15.5.

After solving the proposed state space graph clustering model for swapped Rossler time series, the stationary segmentation clusters are obtained and the Markov transition matrices derived based on the evolutions between the clusters. Figure 15.6 shows the segmented time series and state spaces for swapped Rossler time series. Here a 230 data points long window length and a threshold of 150 signal units were used. The resulting segmented time series and its corresponding state space representation is shown in Fig. 15.7. Figure 15.8 illustrates the prediction of the system's behavior after time $t = 700$ by using Markov transition matrix. Time point $t = 700$ belongs to the 4th segment of state space and the Markov transition matrix, calculated from the first point of this segment ($t = 430$) to $t = 700$, is used for system trajectory prediction in the rest of this segment points.

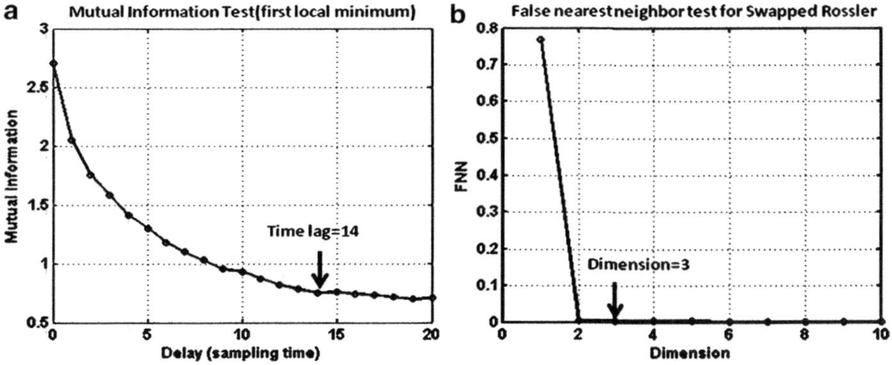

Fig. 15.4 (a) Time delay τ and (b) dimension m parameters for swapped Rossler time series

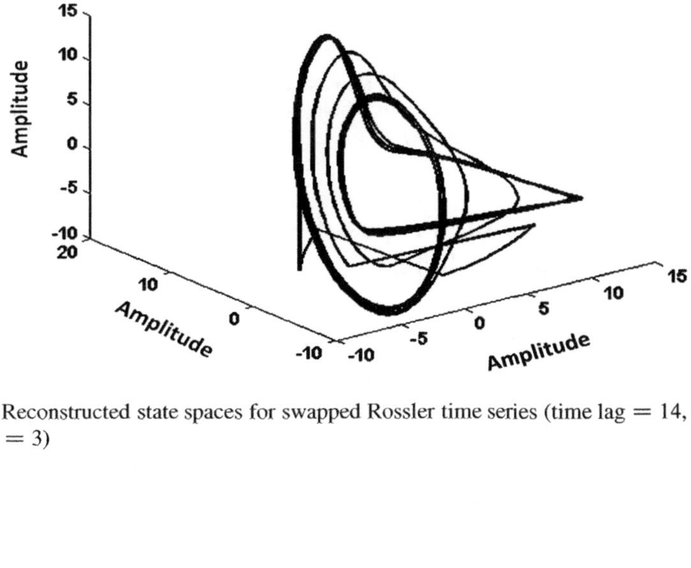

Fig. 15.5 Reconstructed state spaces for swapped Rossler time series (time lag = 14, embedding dimension = 3)

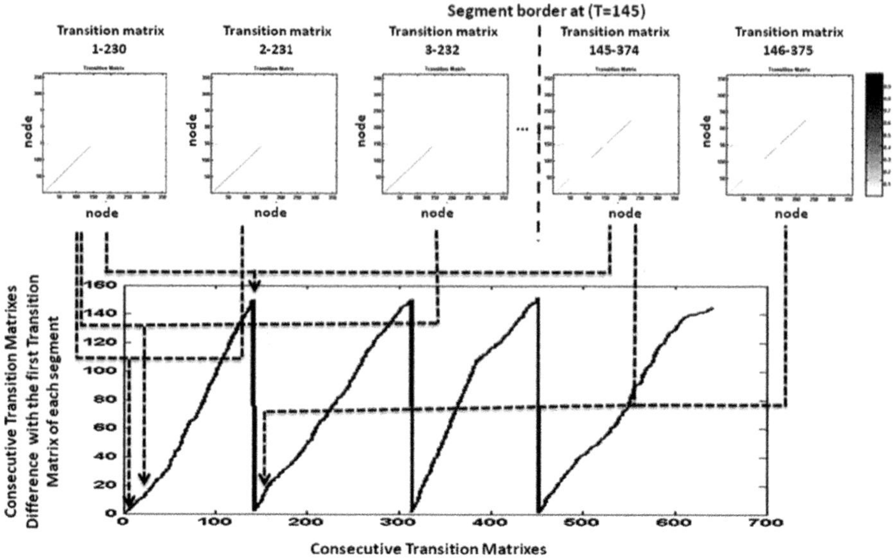

Fig. 15.6 Pattern analysis and segmentation process for swapped Rossler time series. (Window length = 230, Shift forward = 1, Pattern similarity threshold = 150)

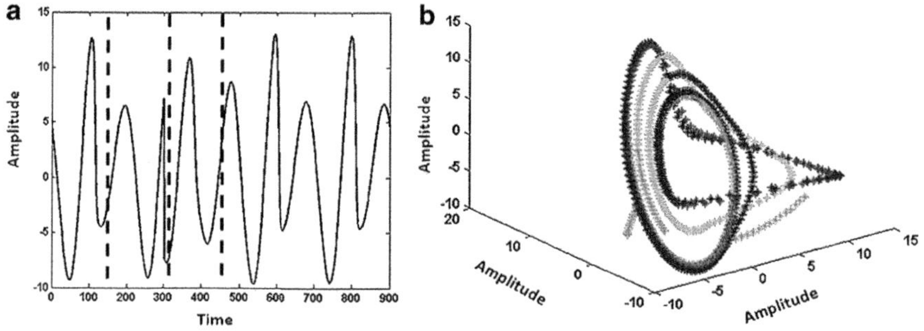

Fig. 15.7 (**a**) Segmented time series and (**b**) state space for swapped Rossler (Window length = 230, Shift forward = 1, Pattern similarity threshold = 150) (4 different colors in the state space portrait show 4 distinct system's behavior)

15.4 Comparison of the Proposed Model with Commonly Used Stationary Models

For illustration of the proposed model's ability to predict the future behavior of nonlinear dynamic systems, three different commonly used stationary models are chosen to be compared with the proposed model. The chosen models are ARMA (Autoregressive moving average model), grand average (denotes baseline model BM1) and a 20 point moving average (denoted BM2). One and five step ahead prediction results of these models for swapped Rossler and five WTP-mimicking data

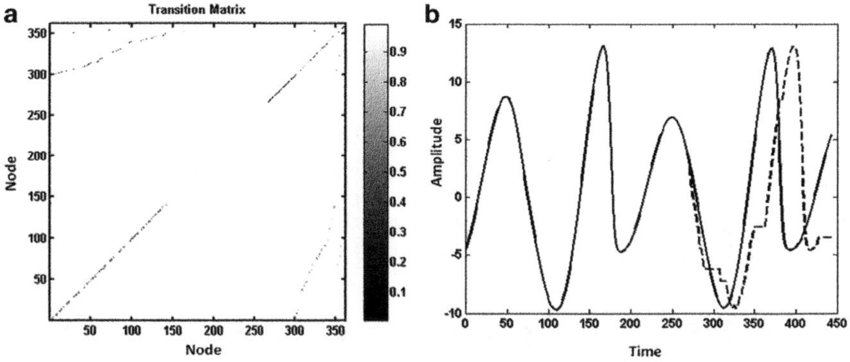

Fig. 15.8 (**a**) Markov transition matrix calculated from t = 430 to t = 700 for Swapped Rossler data, (**b**) Prediction of the system's behavior after time t = 700

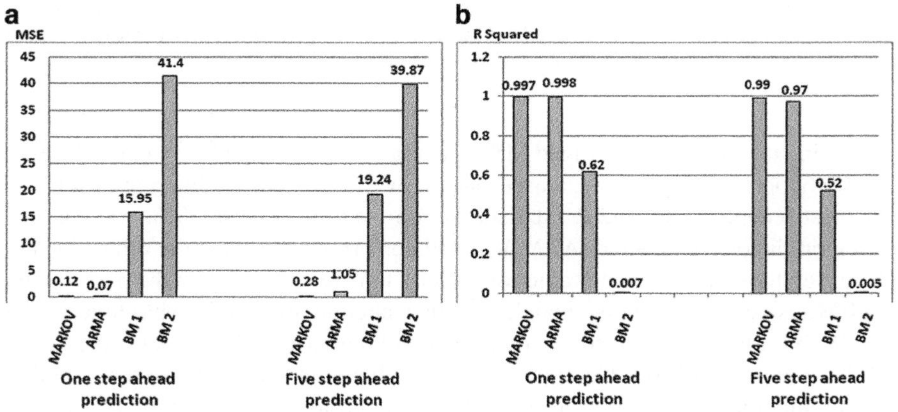

Fig. 15.9 Statistical comparison of commonly used stationary prediction models with proposed Markov model based on (**a**) MSE and (**b**) R-squared values for one and five step ahead prediction in Swapped Rossler data

sets at 25 different points are evaluated and compared with the prediction results of proposed model. Statistical comparison of these models prediction results measured in terms of MSE (Mean Squared Error) and R^2 values are summarized in Figs. 15.9 and 15.10 for the swapped Rossler and the WTP mimicking data.

For swapped Rossler data, ARMA as well as the Markov modeling approach outperform the two baseline models. In short term prediction using this time series, ARMA performs slightly better than our model. But the difference is not meaningful enough considering the high levels of noise associated with the data as well as the finite sampling limitations. The results for long term prediction using swapped Rossler data shows present approach is better than the other methods tested. The prediction accuracy (MSE) of Markov based prediction and ARMA models are meaningfully (at least 90%) better than two baseline models. For WTP-mimicking time series data, the present approach was found to be better than the others in terms

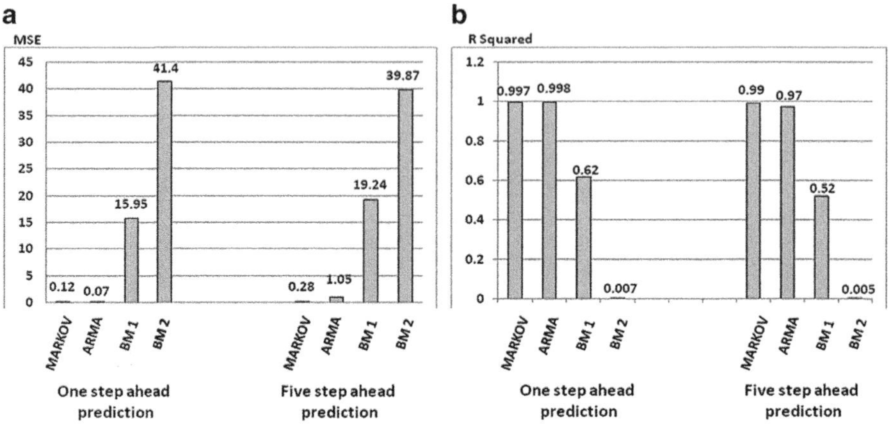

Fig. 15.10 Statistical comparison of commonly used stationary prediction models with proposed Markov model based on (**a**) MSE and (**b**) R-squared values for one and five step ahead prediction in average sense over five different WTP-mimicking data

of having the minimum averaged MSE and maximum averaged R^2 values. By using present approach MSE is decreased one an average by >29% in comparison with the best second model and R^2 statistic is improved by at least 7.9%. Among other three predictors considered, ARMA model performs better than the others in short (one step look-ahead) term prediction. Over a long term prediction the performance of BM2 is very close to this model. A statistical comparison of the baseline models for WTP-mimicking data shows that for both long term and short term prediction, BM2 outperforms BM1. Compared to the prediction results of these three models, the prediction accuracy improvement by our approach is found to be consistently higher, which shows the better suitability of the new method for WTP prediction compared to the methods commonly used in practice.

15.5 Conclusions

This paper provides a modeling and prediction approach for nonlinear and non-stationary systems by constructing a local Markov transition matrix to capture the recurrent nonlinear dynamic system behavior. The model is validated by implementing the model on two different nonlinear dynamic systems. For long term prediction, proposed methodology performs much better than the other stationary models with at least 10% improvement in MSE and 2% improvement in R^2 values. In short term prediction, the local Markov based model provides better predictions (lower MSE) in 83% of prediction runs. The second best prediction model is ARMA model which in average provides better predictions than Baseline model 1 and Baseline model 2 (100% of prediction runs). Considering the prediction results for swapped Rossler, it can be concluded that Baseline models are not efficient in prediction applications for time series data with high value of data variance. For comparison of two Baseline models, Baseline model 1 performs much better in swapped Rossler system but the performance of Baseline model 2 is better in WTP dynamic system.

Based on the present statistical analysis, it may be concluded that the present local Markov modeling approach can lead to meaningful improvements in the prediction accuracy (at least 29% decrease in MSE) compared to the commonly used stationary models like ARMA. It was also noted that the prediction results seem to be dependent on the value of the graph representation and the sliding window parameters. So the future research work will investigate a methodology by which the optimum value of these parameters for a particular time series data can be chosen.

Acknowledgement The authors acknowledge the National Science Foundation (Grant DMI-0428356, CMII 0729552) for the generous support of this research. The support from General Motors on a parallel effort is also greatly appreciated. The first author would like to thank Parameshwaran S. Iyer and N.R. Srinivasa Raghavan of GM R&D, India Science Lab, for the useful discussions.

References

H. M. Al-Hamadi and S. A. Soliman, "Short-term electric load forecasting based on Kalman filtering algorithm with moving window weather and load model," Electr Power System Research; Vol. 68, No. 1, pp. 47–59, 2004.

A. M. Al-Kandari, S. A. Soliman, and M. E. El-Hawary, "Fuzzy short-term electric forecasting," *Int J Electr Power Energ Syst*; Vol. 26, pp. 111–22, 2004.

A. J. Einstein, H.-S. Wu, and J. Gil, "Self-Affinity and Lacunarity of Chromatin Texture in Benign and Malignant Breast Epithelial Cell Nuclei," Physical Review Letters, Vol. 80, No. 2, pp. 397, 1998.

A. Hettrich and S. Rosenzweig, "Multivariate statistics as a tool for model-based prediction of floodplain vegetation and fauna," Ecological Modelling, Vol. 169, No. 1, pp. 73–87, 2003.

D. Asber, S. Lefebvre, J. Asber, M. Saad, and C. Desbiens, "Non-parametric short-term load forecasting," Int J Electr Power Energ Syst; Vol. 29, pp. 630–5, 2007.

D. Baczynski and M. Parol, "Influence of artificial neural network structure on quality of short-term electric energy consumption forecast," In: IEE proceedings – generation, transmission and distribution, Vol. 151; pp. 241–5, 2004.

D. W. Bunn and E. D. Farmer, "Comparative models for electrical load forecasting," New York: John Wiley; 1985.

D. W. Bunn and A. I. Vassilopoulos, "Comparison of seasonal estimation methods in multi-item short-term forecasting," *Int J Forecasting*; Vol. 15, pp. 431–43, 1999.

M. Casdagli, T. Sauer, and J. A. Yorke, "Embedology," *J. Stat. Phys.*, Vol. 65, no. 3/4, pp. 579–616, 1991.

S. Chartier, P. Renaud, and M. Boukadoum, "A nonlinear dynamic artificial neural network model of memory," New Ideas in Psychology, Vol. 26, pp. 252–277, 2008.

S. Chen, S. A. Billings, and P. M. Grant, "Nonlinear system identification using neural networks," Int. J. Contr., Vol. 6, pp. 1191–1214, 1990.

T.-L. Chen, C.-H. Cheng, and H. J. Teoh, "Fuzzy time-series based on Fibonacci sequence for stock price forecasting," Physica A: Statistical Mechanics and its Applications, Vol. 380, pp. 377–390, 2007.
A. A. Desouky and M. M. Elkateb, "Hybrid adaptive techniques for electric-load forecast using ANN and ARIMA," In: IEE proceedings – generation, transmission and distribution, Vol. 147, pp. 213–7, 2000.
Z. Drezner, "Facility Location: A Survey of Applications and Methods," Springer series in operations research, pp. 309–310, 1995.
K. L. Ho, "Short-term load forecasting of Taiwan power system using a knowledge based expert system," IEEE Trans Power Syst; Vol. 5, No. 4, pp. 1214–9, 1990.
K. K. B. Hon, "Performance and Evaluation of Manufacturing Systems," CIRP Annuals, Vol. 54, pp. 675, 2005.
S. Huang and K. Shih, "Application of a fuzzy model for short-term load forecast with group method of data handling enhancement," Int J Electr Power Energ Syst; Vol. 24, pp. 631–8, 2002.
M. Iatrou, T. W. Berger, and V. Z. Marmarelis, "Modeling of Nonlinear Nonstationary Dynamic Systems with a Novel Class of Artificial Neural Networks," IEEE transactions on neural networks, Vol. 10, No. 2, 1999.
H. Kantz and T. Schreiber, "Nonlinear time series analysis," Cambridge University Press, 1997.
B. Kermanshashi and H. Iwamiya, "Up to year 2020 load forecasting using neural nets," Int J Electr Power Energ Syst; Vol. 24, pp. 789–97, 2002.
A. Kumluca and I. Erkmen, "A hybrid learning for neural networks applied to short term load forecasting," Neurocomputing; Vol. 51, pp. 495–500, 2003.
J. Leontaritis and S. A. Billings, "Input–output parametric models for nonlinear systems, Part I: Deterministic nonlinear systems; Part II: Stochastic nonlinear systems," Int. J. Contr., Vol. 41, pp. 303–344, 1985.
V. Z. Marmarelis, "Practicable identification of nonstationary nonlinear systems," Proc. Inst. Elect. Eng., Vol. 5, pp. 211–214, 1981.
P. Z. Marmarelis and V. Z. Marmarelis, "Analysis of Physiological Systems: The White-Noise Approach," New York: Plenum, 1978. Russian translation: Mir Press, Moscow, 1981. Chinese translation: Academy of Sciences Press, Beijing, 1990.
N. Marwan, M. C. Romano, M. Thiel, and J. Kurths, "Recurrence plots for the analysis of complex systems," Physics Reports; Vol. 438, pp. 237–329, 2007.
J. McNames, J. A. K. Suykens, and Vandewalle, "Winning Entry of the K. U. Leuven, Time Series Prediction Competition," International Journal of Bifurcation and Chaos, Vol. 9, No. 8, pp. 1485–1500, 1999.
I. Moghram and S. Rahman, "Analysis and evaluation of five short-term load forecasting techniques," IEEE Transactions on Power Systems; Vol. 14, No. 4, pp. 1484–91, 1989.
J. R. Noriega and H. Wang, "A direct adaptive neural-network control for unknown nonlinear systems and its application," IEEE Trans. Neural Networks, Vol. 9, pp. 27–33, 1998.
K. S. Narendra and S. Mukhopadhyay, "Adaptive control using neural networks and approximate models," IEEE Trans. Neural Networks, Vol. 8, pp. 475–485, 1997.
D. Niebur, "Artificial Neural Networks in the power industry," survey and applications. Neural Networks World; Vol. 6, pp. 945–50, 1995.
S. H. Owen and M. S. Daskin, "Invited Review: Strategic Facility Location: A Review," European Journal of Operations Research, Vol. 111, pp. 423–447, 1998.
S. Rahman and O. Hazim, "Load forecasting for multiple sites: development of an expert system based technique," Electrical Power System Research, Vol. 39, No. 3, pp. 161–9, 1996.
L. M. Saini and M. K. Soni, "Artificial neural network based peak load forecasting using Levenberg–Marquardt and quasi-Newton methods," In: IEE proceedings – generation, transmission and distribution, Vol. 149; pp. 578–84, 2002.
T. Senjyu, P. Mandal, K. Uezato, and T. Funabashi, "Next day load curve forecasting using recurrent neural network structure," In: IEE proceedings – generation, transmission and distribution, Vol. 151; pp. 388–94, 2004.

H. Stark and J. W. Woods, "Probability, Random Processes, and Estimation Theory for Engineers," 2nd edition: pp. 334, 1994.

F. Takens, "Detecting strange attractors in turbulence," in Dynamical Systems and Turbulence, Warwich 1980, Lecture Notes in Mathematics, 898, D. Rand and L. S. Young, Eds. New York: Springer-Verlag, 1981, pp. 366–382.

J. W. Taylor, "Short-term electricity demand forecasting using double seasonal exponential smoothing," J Oper Res Soc; Vol. 54, pp. 799–804, 2003.

J. W. Taylor and R. Buizza, "Using weather ensemble predictions in electricity demand forecasting," Int J Forecasting; Vol. 19, pp. 57–70, 2003.

M. Van Wezel and R. Potharst, "Improved customer choice predictions using ensemble methods," European Journal of Operational Research, Vol. 181, No. 1, pp. 436–452, 2007.

Chapter 16
Two Period Product Choice Models for Commercial Vehicles

Atanu Chaudhuri and Kashi N. Singh

Abstract Choosing products to launch from a set of platform based variants and determining their prices are some of the critical decisions involved in any new product development (NPD) process. Commercial vehicles are products, whose sales are closely tied with the economic conditions. The manufacturers have to choose the variants of the commercial vehicles to launch and sequence the product launches in such a way that profitability is maximized. We develop a two period model to choose the platform based variants, their prices and launch sequences with the two periods, spanning two economic conditions, for example boom and recession. Our model helps in determining realistic prices of products in different economic conditions.

Keywords Choice based modeling · Two period model · Commercial vehicles

16.1 Introduction

Consumers attach importance to specific product attributes while consideringproduct choice decisions. Under many circumstances, consumers would like to weigh the option of buying a product in period 1 or postponing the purchase decision and buying the same or another product in period 2. When an automotive manufacturer has to plan its product line, consisting of platform based variants, understanding these consumer choice preferences become critical for choosing the products to launch, in determining their launch sequence and prices.

Commercial vehicles are products, whose demand is closely related to the fuel prices and the health of the economy. The choice of product line extensions to launch

A. Chaudhuri (✉)
Deloitte Support Services India Pvt. Ltd
e-mail: atanuch1@yahoo.com

K.N. Singh
Indian Institute of Management, Lucknow
e-mail: kns@iiml.ac.in

N.R.S. Raghavan and J.A. Cafeo (eds.), *Product Research: The Art and Science Behind Successful Product Launches*, DOI 10.1007/978-90-481-2860-0_16,

from the set of feasible product options becomes difficult because of possibilities of one product cannibalizing the sales of the other (Green and Krieger 1987). But when such product launch decisions have to be taken considering change in economic conditions from boom to recession or from recession to boom, the tasks of choosing the right products to launch become all the more complex. Even when the firm chooses the right variants to launch, failure to price them appropriately or sequence their launches can seriously undermine its profitability. In this paper, we develop a model, which would act as a decision support for practicing managers in choosing platform based variants to launch for commercial vehicles, their prices and launch sequences considering different economic conditions and duopolistic competition.

16.2 Literature Review

Zufryden (1982), Green and Krieger (1985), Mcbride and Zufryden (1988), Kohli and Sukumar (1990) considered different forms of the product line design problem using consumer utilities. Dobson and Kalish (1988, 1993) considered the problem of selecting the products to offer, from a set of potential products and determining their prices to maximize profit. None of the above studies considered launch sequence of products. Moore et al. (1999) described an application of conjoint analysis to design product platforms. Mussa and Rosen (1978) considered self-selection constraints while addressing the monopolist's product design problem. Moorthy (1984) also included self-selection constraints in a problem of segmentation through product design, with a single dimension for product quality. Kim and Chhajed (2002) recognized the limitation of the single attribute analysis and extended Mussa and Rosen (1978) and Moorthy's (1984) work by considering the product design problem with multiple quality-type attributes, for which more is always better. Urban et al. (1990) developed a consumer flow model, which monitors and projects key consumer transitions in response to marketing actions, provided in a test versus control consumer clinic. The model was used to forecast sales of a new car. But the approach suffers from application challenges in terms of creating expensive clinics and extensive data requirement.

There is another stream of literature, which deals with pricing and launch sequence determination and pricing and positioning issues. Moorthy and Png (1992) considered durable products and studied the impact of market segmentation and cannibalization on timing of product introduction. They showed that for sequential introduction, introducing a low end model first aggravates cannibalization as it also becomes attractive for the high end segment. Desai (2001) considered segmentation of markets based on 'quality' and generated useful insights on the impact of cannibalization on product line design in monopoly as well as duopoly.

The choice of product line and pricing literature dealt with choosing optimum levels of each attribute using integer programming models or developed specialized heuristics to solve them. But launch sequences were not considered in these models. Pricing and launch sequence literature considered only single attributes or 'quality' type attributes and advocated the launch of 'high-end' variants ahead of 'low-end' products to avoid cannibalization. None of these studies considered different economic conditions like 'boom' or 'recession' or willingness to wait to own customized products, which may affect product choices and pricing decisions.

16.3 Formulating Two Period Product Choice Models: Application in Commercial Vehicles

A commercial vehicle manufacturer, which already has a base product for each platform in the market, has to decide whether to launch an upgraded version or a derivative version when the launch periods can coincide with either boom or recessionary conditions. The feasible product options include the base products, which can be upgraded, for example, to meet new emission norms and derivative products, which can undergo performance improvements and/or styling changes. The products have multiple attributes like fuel economy, payload, and power-to-weight ratio, maximum cruising speed, gradability (ability to climb slopes) and price. All the attributes cannot be considered as "quality" type attribute, i.e., higher levels of the attributes for the products will not always generate higher utilities. For example, higher maximum cruising speed may not always generate higher utilities for all segments of customers. The company also has to determine the prices and launch sequences of the products.

16.3.1 Input to the Models

Data required for the model are the ratings for the different product profiles under different economic conditions, the average unit cost of the products and the risk factors for the customers and the firm. The ratings are obtained by generating 100 random numbers between 1 and 10 assuming normal distribution with mean of 5 and then rounding off the numbers. Such numbers are generated for the different product profiles of each platform under different economic conditions like boom, recession or stable. The other data are obtained through discussion with concerned engineers and managers in the departments of Engineering Design and New Product Introduction of the firm.

Part-worth utilities from the Conjoint Analysis are used as inputs to the model. Price is also included as attribute for the products. Refer to Green et al. (2001), Wittink (1989) and Page and Rosenbaum (1989) for comparison of different conjoint analysis methodologies. We use the customer rating data for different products under economic conditions and develop the two-period profit maximizing models to choose the product/s, determine their prices and the launch sequences. The periods are characterized by boom, recession or stable economic conditions. The different economic conditions are captured by varying volumes of the product platform and the different utilities, which the customers attach to the same product under different economic conditions. Overall demand for the individual product platforms were taken from the firm's own forecasting technique relying on past data and some growth projections.

We first determine the utilities for each of the 100 customers for each level of attribute from the randomly generated ratings. Then, for each customer we determine the importance, she attaches to each attribute. Importance attached to an attribute by a customer is calculated as her range of utility for the attribute divided by the range of utility for all the attributes. Then we cluster the respondents using hierarchical clustering using their importance attached to each attribute. We designate each cluster in terms of whether the customers in that cluster are likely to buy an upgraded base product 'j' or a derivative product 'k'. For example, a cluster with high importance for fuel economy, payload and price was designated as a cluster likely to buy an upgraded base product 'j' while one with high importance for power to weight ratio and maximum cruising speed was designated as a cluster likely to buy a derivative product 'k'. It is important to cluster the customers and use the cluster utilities for product choice decisions as the utilities for the cluster of customers, who are more likely to buy a particular product profile will be different from the utilities, obtained from all the customers.

Then we rerun the conjoint model for the respondents in the different clusters and obtain the utilities of the attributes for each cluster. These utilities are then scaled using the approach outlined by Jedidi and Zhang (2002) and used as inputs to the product choice model.

16.3.2 Modeling the Customers' Product Choice Decision

The customer buys a particular product when the consumer surplus she obtains for that product exceeds the surplus from the other product in the same period as well as in the next period. These are captured by constraints called the 'self-selection' constraints as the customers choose the product, which is best for them. Also the customers will buy the product only when the monetary value of the utility derived from the product exceeds its price. Such constraints are called 'participation' constraints as these are necessary conditions to allow the customers participate in the buying process. In the first period, the customer will buy a derivative product if the net utility of that product exceeds the net utility for the upgraded product in both the first and second periods. The utility functions of price are used to convert the sum of non-price utilities of a product, for the prospective cluster of customers, into monetary values using the approach outlined by Jedidi and Zhang (2002). But since each attribute has some threshold level, below which its utility is negligible, we deduct the threshold utilities for each attribute to arrive at the net sum of non-price utilities. These utilities are used to check for the satisfaction of the self-selection and participation constraints for the customer clusters. Based on the satisfaction of the constraints we can determine which product can be launched in which period.

We assume that a customer is expected to buy the product, for which she has the highest utility. Market shares are then calculated for each individual product profile in a platform, assuming that all the feasible products will be in the market, either launched by the firm or by its competitors. The entire market for a particular platform in a period (say boom) varies, depending on whether it is boom followed by recession or recession followed by boom conditions in the planning horizon. In this context we assume that the market share of individual products in a particular period (say boom) remains invariant of the sequence of economic conditions in the planning horizon. This assumption enables us to determine the volume of the products from the market share calculations.

We discount the utilities for the customers who buy the products in the second period as they have to wait for the products. Similarly we discount the profits of the firm obtained in the second period. Thus for each feasible upgrade and derivative product pair for each platform we can determine the profit. We repeat the entire procedure for all the feasible product pairs in the same platform. Finally we choose the profit maximizing product or products for each platform. Similar but separate models are used for each sequence of economic condition like boom followed by recession and vice-versa, boom followed by stable, etc.

16.4 Choice of Product Line Model for Commercial Vehicles Over Two Periods-Boom and Recession

We describe the model for commercial vehicles below:

Indices – 'k' – derivative product, 'j' – upgraded base product-boom period, r – recession period

Parameters

C_k – Cost of product 'k', C_j – cost of product 'j', n_{kb} – Volume of 'k' in boom, n_{jb} – Volume of 'j' in boom, n_{kr} – Volume of 'k' in recession, n_{jr} – volume of 'j' in recession

BR_k – buyer's risk corresponding to late introduction of model k.

BR_j – buyer's risk corresponding to late introduction of model j.

SR_k – Seller's risk from introducing model k late, SR_j – Seller's risk from introducing model j late

P_{kb} – Price of 'k' in boom, P_{kr} – Price of 'k' in recession

P_{jb} - Price of 'j' in boom, P_{jr} – price of 'j' in recession

i – attribute, q – levels of each attribute for product 'k', q' – levels of each attribute for product 'j'

n – number of attributes in a product including price

Sum of non-price utilities for the product 'k' in boom for potential customers of 'k'

$= \sum_{i=1}^{n-1} v_{kb}{}^{*}q_i$, which is the sum of the utilities, which the potential customer segment of product 'k' attaches to each attribute 'i' with levels 'q' in boom period 'b'

From conjoint analysis, we get $v_{kb}{}^{*}q_i$ for each attribute and sum all those utilities over (n – 1) attributes, where the nth attribute is price. Here we assume that the utility function of each attribute is piece-wise linear in nature. For attributes with three levels, the middle level is chosen as the base level and for attributes with two levels, the lower level is chosen as base (Jedidi and Zhang 2002).

For simplicity, we use only V_{kkb} to represent $= \sum_{i=1}^{n-1} v_{kb}{}^{*}q_i$, which is the total utility which a potential customer of product 'k' derives from product 'k' in boom period (b) and V_{jjb} to represent $= \sum_{i=1}^{n-1} v_{jb}{}^{*}q'_i$, which is the total utility which a potential customer of product 'j' derives from product 'j' in the boom period.

16.4.1 Customer Choice Constraints

[for cluster of customers buying 'k' in 1st period]

$V_{kjb} - P_{jb} \leq V_{kkb} - P_{kb} \ldots$ (i) [ensuring customers of 'k' prefers 'k' over 'j' in 1st period]

$BR_j^*(V_{kjr} - P_{jr}) \leq V_{kkb} - P_{kb}$ (ii) [ensuring customers of' 'k' prefers buying 'k' in 1st period to 'j' in 2nd period]

[for cluster of customers buying 'j' in 1st period]

$V_{jkb} - P_{kb} \leq V_{jjb} - P_{jb}$ (iii)

$BR_k^*(V_{jkr} - P_{kr}) \leq V_{jjb} - P_{jb}$ (iv)

[for cluster of customers buying 'k' in 2nd period]

$V_{kjr} - P_{jr} \leq V_{kkr} - P_{kr}$ (v) [ensuring customers of 'k' prefers buying 'k' over 'j' in 2nd period]

[for cluster of customers buying 'j' in 2nd period]

$V_{jkr} - P_{kr} \leq V_{jjr} - P_{jr}$ (vi)

$P_{kb}, P_{kr}, P_{jb}, P_{jr} \geq 0$; $P_{jb} \leq V_{jjb}$, $P_{jjr} \leq V_{jjr}$ $P_{kb} \leq V_{kkb}$, $P_{kr} \leq V_{kkr}$..constraint set vii and viii

Note constraints (i) to (vi) are the self-selection constraints and constraint set vii and viii are the participation constraints. Constraints (ii) and (iv) are the look-ahead constraints as they allow customers to consider buying now or later. There can be two additional constraints in 2nd period

$V_{kjr} - P_{jr} \leq BR_k{}^*(V_{kkr} - P_{kr})$ [for customers of 'k', if only 'j' is launched in the first period]

$V_{jkr} - P_{kr} \leq BR_j{}^*(V_{jjr} - P_{jr})$ [for customers of 'j', if only 'k' is launched in the first period]

Constraints (i) to (vi) are 'self-selection' constraints and constraints (vii) to (x) are participation constraints.

Cannibalization can occur if a product satisfies the participation constraints but fails to satisfy the self-selection constraints. If all self section and participation constraints are satisfied, then the product gets launched without being getting cannibalized. If cannibalization conditions are satisfied, cannibalized volumes are estimated by finding the change in market share when one of the products is not present and in the scenario when both are present.

Prices are obtained by making the customers indifferent to buying one product and another, i.e., by converting some of the inequality constraints to equality constraints.

We start from P_{jr} and price 'j' in recession at the reservation price for a potential customer 'j' in recession. Reservation price is the maximum price, which a customer is willing to buy for a product. Now using P_{jr}, we make the potential customer of 'k' indifferent to buy 'k' or 'j' in recession and obtain P_{kr} using constraint (v). If we price P_{kr} at its reservation price and derive P_{jr} from it, then P_{jr} might exceed its reservation price. Thus potential customers of 'j' will not buy 'j' and if P_{jr} also becomes lesser than P_{kr}, then customers of 'k' will shift to 'j'. To avoid this cannibalization we price 'j' in recession at its reservation price and derive P_{kr} from it. Similarly using constraints (ii) and (iv), we obtain the other prices. We choose constraints (ii) and (iv) to obtain the prices because if the prices obtained by making a set of customers indifferent between buying a product now and buying the other product late also helps in choosing his preferred product over the other in the same period, it will capture his true intention to buy his preferred product. It will also give us the optimum prices, which can be charged without allowing the customer to shift to another product. The prices for the products are obtained by converting the utilities to monetary value using the utility function of price as shown in Jedidi and Zhang (2002).

The prices are shown below

$$
\begin{aligned}
P_{jr} &= V_{jjr}, \\
P_{kr} &= \underset{[r]}{V_{kkr}} - \underset{[r]}{V_{kjr}} + \underset{[r]}{V_{jjr}} \\
P_{kb} &= \underset{[b]}{V_{kkb}} - BR_j{}^*(\underset{[b]}{V_{kjr}} - \underset{[r]}{V_{jjr}}) \\
P_{jb} &= \underset{[b]}{V_{jjb}} - BR_k^* \underset{[b]}{V_{jkr}} + BR_k{}^*(\underset{[r]}{V_{kkr}} - \underset{[r]}{V_{kjr}} + \underset{[r]}{V_{jjr}})
\end{aligned}
$$

Note – the cluster or homogeneous group of customers likely to buy 'j' or 'k' varies with economic conditions. So [b] and [r] denote the utilities are obtained from clusters in boom and recession period respectively.

Using the prices and utilities, we check for the satisfaction of the constraints and determine the products to be launched and their launch sequences. Then we calculate the profit given by the objective function as stated below and choose the profit maximizing product pair.

$$(P_{kb}-C_k)^* n_{kb}+SR_k{}^*(P_{kr}-C_k)^* n_{kr}+(P_{jb}-C_j)^* n_{jb}+SR_j{}^*(P_{jr}-C_j)^* n_{jr}$$

16.5 Managerial Implication of the Results

The proposed model helps us in choosing the products to launch in determining a range of prices for them, their reservation prices and their launch sequences, based on customer ratings. In the process, we gain insights on the effect of economic conditions on prices and on the possibility of cannibalization that can take place. Our results show that considering the price level of the products, used to describe product profiles, as the final prices will be sub-optimal. Instead, prices determined using our model by incorporating economic conditions and self-selection constraints in product line decision making gives more realistic estimates of prices. Using our model, we can determine the reservation price, which is the maximum price the customers will be willing to pay for the product and also the prices of the products under different sequences of economic conditions with some given level of risk factors. Thus the project managers can provide a range of prices with the reservation price as upper bound, which are more realistic. Comparison of Prices is shown below. Prices are in '000 of Rupees (1 USD = Rupees 50) (Figs. 16.1 and 16.2).

The above figures clearly show that the price level of the profit maximizing product, as described while creating product profiles in Conjoint Analysis need not be the target price. It will be beneficial to determine the reservation price and a price

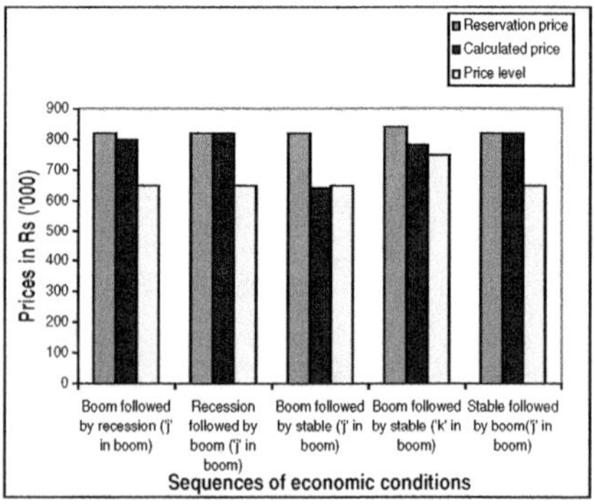

Fig. 16.1 Comparison of prices for multi-axle trucks

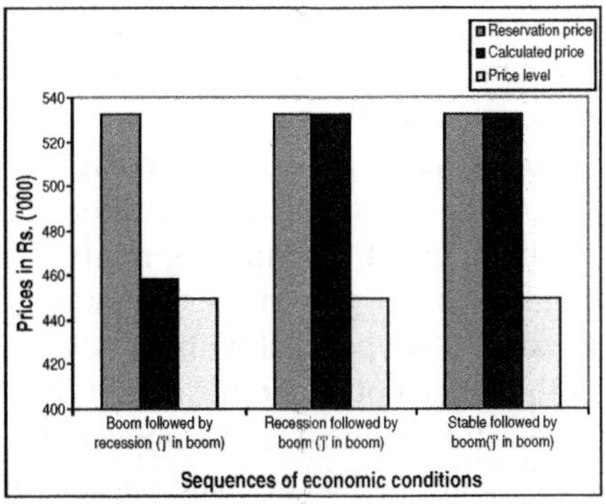

Fig. 16.2 Comparison of prices for 4 tonne light commercial vehicle

considering the sequence of economic conditions at the time of launch. This will provide the managers a range of prices, within which they should price the product.

Our model also helps in determining the impact of economic conditions on product choice and launch sequence decisions. Moorthy and Png (1992) showed that introducing a low end model first aggravates cannibalization as it also becomes attractive for the high end segment. But they didn't consider different economic conditions. Our results showed that an upgraded model can be launched ahead of a derivative product without being cannibalized. We show that such cannibalization can occur only when specific conditions in terms of utilities of the products are satisfied. Interested readers can refer to Chaudhuri and Singh (2005) for detailed proof of this proposition.

With different sequences of economic conditions, product choices, their prices and overall profits can either vary or remain invariant. For some product platforms like the 4 ton lcv, the product choice does not vary with the sequence of economic condition though the pricing changes. For such products, product planners can focus their efforts on the development of the particular product. For other products like the 15 ton bus, product choices vary with sequences of economic conditions. Thus it is important for the company to predict the likely sequence of economic conditions.

See Appendix for tables showing launch sequences and prices for the different platforms under different sequences of economic conditions.

16.6 Discussion

The product design literature, which deals with choosing optimum levels of attributes, does not seem to have considered the launch sequences of the products and 'self-selection' constraints over multiple periods. The pricing and launch sequence literature using 'self-selection constraints' considered only single attribute products or 'quality' type attributes, for which utilities increase with increasing levels. We add to the choice of product line literature by using a two period optimization model with self-selection constraints, to screen profit maximizing multi-attribute products for each platform. We do not assume the attributes to be of 'quality' type. We show how utilities obtained from Conjoint Analysis can be used to construct two period models for product choice decisions for commercial vehicles. Our two period model is characterized by different economic conditions and allow us to capture the buying preferences of the customers over the two periods.

Our results show that choosing the prices used to describe product profile as the final prices of the products will be sub-optimal. Our conjoint utilities based choice model helps in determining realistic estimates of prices under different economic conditions. We also show that an upgraded base product can be launched ahead of the derivative product without being cannibalized. We obtain such results because of the impact of economic conditions.

We did not consider the impact of interest rates, depreciation norms and revenues from future of sale of spare parts in our model. We have used full profile Conjoint Analysis for obtaining the utilities of customers. The accuracy of the model depends on the ability of the respondents to reliably provide the ratings of the products under hypothetical economic conditions. We have not validated the results in our current research and recognize that validation of our results using simulated data from multiple such experiments will add more credence to the findings. For products with more attributes and restrictions on feasible product profiles, we can use Analytic Hierarchy Process to obtain the utility functions (Scholl et al. 2005). Similarly our model allows pair-wise comparisons among different variants. Conducting all multiple pair-wise comparisons can be tedious if the number of possible product profiles, to be evaluated is large. We believe choice of appropriate attributes and their levels will result in more feasible product profiles and thus restricting the number of products. But if the number of possible product profiles still remains high our model can be converted to a non-linear mixed integer programming model with both the price and product choices as variables. We have also not considered the impact of competition on product choice decisions under different economic conditions which we do so in our later research.

Appendix

Results for 4ton lcv (with '10' as 'j')

Boom followed by recession	Recession followed by boom	Boom followed by stable	Stable followed by boom
Launch 'j' in boom, P_{jb} = Rs. 458.1 ('000s)	Launch 'j' in boom, P_{jb} = Rs. 532.6 ('000s)	None	Launch 'j' in boom, P_{jb} = Rs.532.6 ('000s)
Profit = Rs. 601.229 million	Profit = Rs. 817.988 million		Profit = Rs. 1433.426 million

'10' and '19' are the feasible upgraded base products and '5' and '25' are the feasible derivative products. But only the upgrades could get launched for 4 ton lcv. Out of the two feasible base products, '10' turned out to be more profitable.

Reservation price of 'j' in boom is Rs. 532.6 ('000s) while the price level for the product profile was Rs. 450 ('000s).

Costs

'j' – Rs. 400 ('000s)

Volumes

'j' in boom – 10340 (boom followed by recession), 'j' in boom – 8225 (recession followed by boom).

'j' in boom – 8608 (stable followed by boom)

Results for Tipper with '17' as 'j'

Boom followed by recession	Recession followed by boom	Boom followed by stable	Stable followed by boom
Launch 'j' in recession, P_{jr} = Rs. 956.5 ('000s)	Launch 'j' in boom, P_{jb} = Rs. 1016.7 ('000s)	Launch 'j' in stable, P_{js} = Rs. 850.9 ('000s)	Launch 'j' in boom, P_{jb} = Rs. 1016.7 ('000s)
Profit = Rs. 39.4676 million	Profit = Rs. 331.844 million	Profit = Rs. 47.522 million	Profit = Rs. 381.6212 million

Feasible product concepts for upgraded base are '9' and '17' as 'j' and that for derivative are '5', '6', '13' and '27'. But in all combinations only 'j' could be launched. Out of '9' and '17', '17' turned out to be more profitable.

Reservation price of 'j' in boom is Rs. 1016.7 ('000s), that in recession is Rs. 956.5 ('000s) and that of 'j' in stable is Rs. 850.9 ('000s).

Costs

'j' –Rs. 590 ('000s)
'k' –Rs. 625 ('000s)

Volumes

'j' in recession – 269 (boom followed by recession), 'j' in boom – 1111 (recession followed by boom), 'j' in stable – 365 (boom followed by stable), 'j' in boom – 1278 (stable followed by boom).

Results for 15t bus (with '9' as 'j' and '25' as 'k')

Boom followed by recession	Recession followed by boom	Boom followed by stable	Stable followed by boom
Launch 'k' in boom and 'j' in recession	None	Launch 'j' in stable,	None
P_{kb} = Rs. 731.9 ('000s)		P_{js} = Rs. 717.5 ('000s)	
Pjr = Rs. 707.4 ('000s)		Profit = Rs. 4142.71 million	
Profit = Rs. 3784.8 million			

For ‘15t’, the feasible upgraded base products are ‘9’, ‘17’, ‘27’ and the feasible derivative products are ‘5’, ‘25’. Reservation price for ‘k’ in boom is Rs. 962.39 (’000s), that of ‘j’ in recession is Rs. 707.4 (’000s) and that of ‘j’ in stable is Rs. 717.5 (’000s) Price levels of ‘9’ and ‘25’ are both Rs. 600 (’000s).

Though ‘17’ and ‘25’ also have equally high market shares individually, profit was lesser than that of ‘9’ and ‘25’ as only ‘j’ could be launched in that case. This happened as ‘17’ and ‘25’ were very close substitutes. This again showed that platform based variants have to be sufficiently separated to get launched together.

Costs

‘k’ – Rs. 650 (’000s)
‘j’ – Rs. 575 (’000s)

Volumes

‘k’ in boom – 6613, ‘j’ in recession – 912, ‘j’ in stable –2249
Note: 1 US Dollar = 50 Indian Rupees (Rs.)

References

Chaudhuri, Atanu and K.N. Singh (2005), ‘Incorporating Impact of Economic Conditions on Pricing and Launch Sequence Determination of Commercial Vehicles’, *Metamorphosis*, Vol. 4, No. 1, pp. 26–38.

Desai, Preyas S. (2001), ‘Quality Segmentation in Spatial Markets: When Does Cannibalization Affect Product Line Design?’, *Marketing Science*, Vol. 20, No. 3, pp. 265–283.

Dobson, G. and S. Kalish (1988), ‘Positioning and Pricing a Product Line’, *Marketing Science*, Vol. 7, No. 2, pp. 107–125.

Dobson, G. and S. Kalish (1993), ‘Heuristics for Pricing and Positioning a Product Line Using Conjoint and Cost Data’, *Management Science*, Vol. 39, No. 2, pp. 160–175.

Green, Paul E. and A.M. Krieger (1985), ‘Models and Heuristics for Product Line Selection’, *Marketing Science*, Vol. 4, No. 1, pp. 1–19.

Green, Paul E. and A.M. Krieger (1987), ‘A Consumer based approach to designing product line extensions’, *Journal of Product Innovation Management*, Vol. 4, pp. 21–32.

Green, Paul E, A.M. Krieger and Yoram Wind (2001), ‘Thirty Years of Conjoint Analysis: Reflections and Prospects’, *Interfaces*, Vol. 31, No. 3 Part 2, S56–S73.

Jedidi, Kamel and J.Z. Zhang. (2002), ‘Augmenting Conjoint Analysis to Estimate Consumer Reservation Price’, *Management Science*, Vol. 48, No. 10, pp. 1350–1368.

Kim, Kilsun and Dilip Chhajed (2002), ‘Product Design with Multiple Quality-Type Attributes’, *Management Science*, Vol. 48, No. 11, pp. 1502–1511.

Kohli, R. and R. Sukumar (1990), ‘Heuristics for Product-Line Design Using Conjoint Analysis’, *Management Science*, Vol. 36, No. 12, pp. 1464–1478.

McBride, R.D., and F.S Zufryden (1988), ‘An Integer Programming Approach to the Optimal Product Line Selection Problem’, *Marketing Science*, Vol. 7, No. 2, pp. 126–140.

Moore, William L, Jordan J. Louviere and Rohit Verma (1999), ‘Using Conjoint Analysis to Help design Product Platforms’, *Journal of Product Innovation Management*, 16, pp. 27–39.

Moorthy, Sridhar K. (1984), ‘Market Segmentation, self-selection and product line design’, *Marketing Science*, Vol. 3, No. 4, pp. 288–307.

Moorthy, Sridhar K and I.P.L Png (1992), 'Market Segmentation, Cannibalization and the Timing of Product Introductions', *Management Science*, 38(3) 345–359.
Mussa, M and S. Rosen (1978), 'Monopoly and Product Quality', *Journal of Economic theory*, Vol. 18, No. 2, pp. 301–317.
Page, Albert L and Harold F Rosenbaum (1989), 'Redesigning product lines with conjoint analysis: A reply to Wittink', *Journal of Product Innovation Management*, Vol. 6, pp. 293–296.
Scholl, Armin, Manthey, Laura, Helm, Roland, Steiner, Michael (2005). Solving multi-attribute design problems with analytic hierarchy process and conjoint analysis: An empirical comparison, *European Journal of Operational Research*, Vol. 163, No. 4, 760–777.
Urban, Glen L., Hauser, John R and Roberts, John H (1990), 'Prelaunch forecasting of new automobiles', Management Science, Vol. 36, No. 4, pp. 401–421.
Wittink, Dick R (1989), 'Redesigning Product Lines using with Conjoint Analysis: A Comment', Journal of Product Innovation Management, Vol. 6, pp. 289–292.
Zufryden, F.S. (1982), 'Product Line Optimization by Integer Programming', Proceedings of Annual Meeting of ORSA/TIMS, San Diego, CA.

Index

V

W

GPSR Compliance

The European Union's (EU) General Product Safety Regulation (GPSR) is a set of rules that requires consumer products to be safe and our obligations to ensure this.

If you have any concerns about our products, you can contact us on ProductSafety@springernature.com

In case Publisher is established outside the EU, the EU authorized representative is:

Springer Nature Customer Service Center GmbH
Europaplatz 3
69115 Heidelberg, Germany

Batch number: 10755284

Printed by Printforce, the Netherlands